Everyday Evolution: Unveiling Life's Hidden Patterns

Barbora Trubenová • Kristína Hudáková •
Himani Sachdeva
Editors

Everyday Evolution: Unveiling Life's Hidden Patterns

Springer

Editors
Barbora Trubenová
D-USYS
ETH Zürich
Zürich, Switzerland

Kristína Hudáková
eduB
Bratislava, Slovakia

Himani Sachdeva
Faculty of Life Sciences and
Faculty of Mathematics
University of Vienna
Vienna, Austria

ISBN 978-3-032-07049-4 ISBN 978-3-032-07050-0 (eBook)
https://doi.org/10.1007/978-3-032-07050-0

This Springer imprint is published by the registered company Springer Nature Switzerland AG
The registered company address is: Gewerbestrasse 11, 6330 Cham, Switzerland

To Melinda, in memory of her commitment to conservation and science communication.

Foreword

Life evolved soon after the Earth became habitable, and over the next four billion years, it elaborated the intricate biochemistry that is shared by all present-day organisms. Organisms have diversified to find an extraordinary variety of ways to make a living. The Sun's light is harvested by bacteria, algae and plants, and microbes use almost every feasible source of chemical energy. Life is found in extreme environments, from Antarctic rocks to volcanic vents, while intricate ecological interactions have accumulated, ranging from the parasites that live within our genomes, through to the dependence of plants and animals on microbes and fungi supply nutrients.

These extraordinary adaptations appear as though carefully designed, and this appearance of design was long taken as evidence of an intelligent creator. We now know that evolution rests on the process of natural selection, which simply picks out those organisms better able to survive and reproduce. Understanding the evolutionary process is crucial if we are to see our proper place in the world, but it also has practical relevance for a wide range of issues: slowing the spread of antibiotic resistance, assessing the risks and benefits of genetically manipulated organisms, understanding sex, gender and ageing, valuing biodiversity, and so on. Yet, despite the central importance of evolution, it is often taught inadequately, and in some countries, there is even resistance to teaching it at all.

This book is therefore especially welcome. It introduces the evolutionary process and its implications in an accessible and engaging way, and should be valuable for a wide audience: high school students, undergraduates studying biology and beyond, and interested laypeople in general. It will also be a resource for those teaching evolution, whether to students or to the wider public, and should provide nourishing food for thought to all readers.

Institute of Science and Technology Austria — Nick Barton

Preface

The theory of evolution is considered one of the most important unifying concepts in biology. Only by understanding universal evolutionary principles underlying various phenomena can we recognise patterns and connections between them, and appreciate the diversity and complexity of the world that surrounds us. Yet, around the world, the teaching and understanding of evolution remain inadequate and, in many cases, completely lacking.

Therefore, in this book, we embark on a journey through the most important concepts of evolutionary biology. We explore how evolutionary mechanisms shape the world around us, discovering along the way the ingenious solutions to the challenges of survival and reproduction that have emerged across time and space.

In the popular imagination, evolution is typically viewed as an exceedingly slow process that can only be discerned in the fossil record, and even standard textbooks often focus disproportionately on fossils as the main evidence for evolution. We believe that this largely misrepresents our modern understanding of evolution. While one can rarely do direct experiments to answer the big "why" questions (Why sex? Why distinct species?), even this is changing somewhat with the emergence of "experimental evolution" as a distinct subfield. In fact, experiments evolving microorganisms in the lab over tens of thousands of generations (or model organisms such as fruit flies over hundreds of generations) now offer an unprecedented window into the genetic basis of short- and medium-term evolutionary change. More broadly, evolution as a field of scientific enquiry increasingly relies on multiple lines of evidence: experimental evolution, morphological and behavioural observations from field studies of populations over space and time, phylogenetic comparisons, as well as molecular and genomic data. Moreover, evolution is rather unique in the role played by "theory" in its development—starting from the entirely verbal, detailed arguments of Darwin to the elegant mathematical theory that underlies population and quantitative genetics (which even led to the new field of statistics) to the heavily computational "Big Data" methods used to analyse genomic data today. By delving into specific topics that are active areas of ongoing research, the book also gives readers a sense of the diversity of approaches used to study evolutionary questions in the modern era.

The history of evolutionary thought—from Darwin's groundbreaking theories (and the ideas that foreshadowed them) to the more mechanistic understanding emerging from modern molecular genetics—is a fascinating one. However, here we avoid a historical account, instead highlighting how an understanding of evolution can give us a fresh perspective on our most everyday experiences: the food we eat, the diseases that affect us and the cures we attempt, our social behaviours, and the natural world we are part of. Another goal is to present evolution as a thriving, vibrant field of scientific enquiry and to convey something of the excitement of modern evolutionary research.

In keeping with this, the book is written by early-career scientists and educators, who convey their excitement in contemporary language, avoiding jargon and technical terms, in a way that is accessible to lay audiences. Many of the authors use examples from their own research to illustrate key concepts, thus also conveying to the reader a sense of their own unfolding scientific journey. Therefore, we firmly believe that this book will resonate with a wide readership, expanding their comprehension of the natural world in an engaging and accessible manner.

Who This Book Is For

This book has been written with the following groups of readers in mind:

Our primary audience is the advanced biology student—typically in high school or the early years of university—with a keen interest in the subject. The book aims to provide a deeper understanding of the principles governing the living world, offering a valuable resource for those preparing for final exams, university applications, or science competitions.

Second, the book is designed as a resource for high school teachers. It can serve as supplementary material to enrich classroom teaching, challenge motivated students with more advanced topics, or provide projects and questions that help connect various biological subjects into a coherent whole.

Third, this book offers substantial value for university students training to become biology teachers, and for their instructors. Future teachers can use it to deepen their own understanding of evolutionary concepts, while their university educators can draw on it to demonstrate effective approaches for teaching complex biological topics.

Ultimately, this book is intended for any member of the general public interested in the natural world. For readers who may already be familiar with many biological facts but seek a framework to understand how they are all interconnected, these chapters offer an opportunity to discover the underlying logic of the mechanisms of evolution.

How This Book Is Organised

The book consists of 20 chapters written by authors from various European universities and research institutions, working in different fields of evolutionary biology, in close collaboration with high school educators. The chapters are organised into four broad thematic areas: Basics of evolution; Evolution and humans; Broad-scale patterns and processes; and Genetics: a molecular perspective. Within each thematic area, chapters are organised by increasing difficulty. The first chapters start at a level accessible to high school students and assume no prior knowledge of evolutionary principles. Later chapters build upon the ideas introduced in earlier ones.

The chapters on the **Basics of evolution** introduce readers to the fundamental ingredients of evolution: natural selection, heredity, and the generation of novel variation through mutation. These chapters primarily deal with *micro-evolution*, i.e., changes in the genetic and phenotypic composition occurring *within* populations over relatively short time scales. The chapters also explore the nature of genetic variation in some detail, building up from the more familiar examples of Mendelian traits to the more ubiquitous but less understood continuous (polygenic) traits. Indeed, reconciling discrete Mendelian genetics with the observation of continuous variation is at the heart of the "modern synthesis", which remains one of the most influential developments in modern genetics, underlying much of animal and plant breeding and modern agricultural advances. These chapters thus provide readers with the necessary conceptual background to appreciate evolution in everyday, observable contexts, which also forms the focus of the next set of chapters.

The second thematic group **Evolution and humans** aims to understand the history and the present (challenges) of humanity through an evolutionary lens. It includes chapters on some of the most pressing challenges facing humans, such as declining biodiversity, infectious diseases, and drug resistance. These chapters go beyond proximal causes to address medium and long-term causes and consequences, which can only be understood from an evolutionary perspective. Such a long-term perspective is also invaluable for our understanding of our own origins, which is the subject of two other chapters in this section—one dealing with our hominin ancestors and the other with the migration of *Homo sapiens* out of Africa. Indeed, a longer view of human history can provide a much-needed counterweight to heated debates about contemporary issues.

The third set of chapters on **Broad-scale patterns and processes** zoom out even further, focusing on the longer timescales over which genome architecture, modes of reproduction, and social organisation can themselves evolve. They explore how evolutionary processes occurring over billions and millions of years have led to the amazing diversity of life forms—prokaryotes vs. eukaryotes, asexually vs. sexually reproducing organisms, hermaphrodites vs. species with separate sexes. These chapters also give a flavour of how over these very long timescales, evolutionary change is governed not only by systematic, directional processes such as natural selection (that may favour ecological adaptation or social behaviours such as altruism), but also by extremely unlikely, unpredictable events—such as those that led to the emergence of eukaryotes.

The fourth group of chapters on **Genetics: a molecular perspective** are somewhat different in content and style, focusing heavily on molecular mechanisms and techniques. They illustrate how the field of genetics has been revolutionised in the last two decades by the availability of genome sequences, as well as advances in genetic engineering, allowing us to examine and modify genetic material with unprecedented precision. This raises complex ethical concerns, e.g., related to genetically modified crops or the management of disease and ageing, which the chapters also explore.

Our classification of the chapters into these four broad themes is partly a matter of convenience. Many, if not most, of the contributions in this book span multiple themes. For example, the evolution of antibiotic resistance, while being a major biomedical and societal challenge, also presents a paradigmatic example of micro-evolution that can be observed over short timescales. Similarly, the chapter on sex differences, while tracing the evolution of separate sexes across diverse forms of life, also explores the complexities inherent in studying human sex differences and the implications of conflating biological sex and gender. Indeed, we believe that one of the strengths of the book lies in its attempt to link different themes, e.g., by situating human experiences within a broader biological and evolutionary context.

Within each chapter, an attempt is made to streamline the text by using "Boxes" to clarify more complex ideas and explain connections between concepts. Important terminology and concepts are highlighted throughout the text and also summarised in "Key Definitions" at the end of each chapter. To make the text more accessible and interesting, we have used a variety of illustrations, including schematic diagrams that illustrate abstract concepts, graphs that visualise data from experiments or field observations, and photographs. Finally, cross-referencing between chapters and references for further reading provides broader contextualisation for each chapter.

A unique feature of the book is the extensive educational material included in each chapter. This material consists of question sets, hands-on activities, and simple simulations collectively referred to as "projects". The projects have already been used extensively by high school students in Slovakia, making them a potentially valuable resource for educators in a broader setting. These projects are designed to help readers actively explore important evolutionary principles and ideas, investigate limitations, and discover interesting dependencies.

For example, the chapter on the evolution of sex differences is complemented by a simulation that uses different-sized pieces of paper representing eggs and sperm. Readers can design hypotheses for various scenarios, conduct experiments, analyse results, and draw conclusions in a guided exercise. This not only deepens their understanding of the evolutionary forces that favour isogamy or the evolution of separate sexes but also gives insight into the scientific method, fostering analytical thinking. Furthermore, it provides a fresh perspective on the concept of sex and the differences between sexes. As another example, the chapter on human viruses is supplemented with a project simulating the effects of vaccination, prompting readers to think about the importance of vaccination for their own health and the protection of vulnerable individuals in the population. During the COVID-19 pandemic, this activity was adapted to create an educational board game for children, called Virus Alarm in Stayhompton (DE: 'Virusalarm in Bleibhausen', SK:'Poplach! Vírus v Nižných Domasedoch!'), which was widely distributed across schools in Germany, Austria, and Slovakia.

Beyond This Book

Evolution is a vast subject relevant to almost all aspects of the living world, and a book of this sort can only cover some of these aspects, inevitably leaving gaps. While some of these gaps are specific to our book, others reflect critical gaps within the broader field. One such gap concerns the links between microevolution and macroevolution, which remain largely obscure. A big open question is: to what extent do population-level processes constrain evolution at macroevolutionary scales, e.g., the rates of speciation and extinction across different domains of life? Or is long-term evolution largely governed by unlikely events that alter its course in profound but fundamentally unpredictable ways?

Perhaps somewhat orthogonal to this is the long-standing debate about the extent to which biophysical constraints limit the "creativity" of evolution—leading, for example, to highly conserved developmental pathways across different species, and, more broadly, limiting the set of evolutionary strategies available to organisms. These questions are an active area of research in developmental biology, Evo-Devo and allied fields, but are beyond the scope of this book. More generally, we exist today in a scientific landscape characterised by intense activity in many different fields—ecology, evolution, development, and cell biology—that are inextricably intertwined, but operate, in practice, as rather separate fields, with their own specialised languages and literatures. This makes it difficult to distil a synthetic understanding of natural phenomena by combining insights from different fields. This kind of synthesis thus remains an exciting challenge for books, such as ours, that aim to bridge the gap between basic concepts and cutting-edge research.

From Our History

Evolutionary thinking, despite its fundamental significance, is often underrepresented or completely excluded from high school curricula globally. Even when taught, it's frequently presented in isolation rather than as an overarching principle connecting diverse biological topics. As scientists, educators, and science communication enthusiasts, we recognised this gap and decided to take action.

Inspired by the success of correspondence seminars in mathematics, physics, and computer science that supplement formal education, we established a similar initiative focused on evolutionary biology in Slovakia (where sisters Kristína and Barbora are from). However, due to the lack of suitable educational material, we had to create our own.

Fortunately, the Institute of Science and Technology Austria (ISTA) provided a collaborative environment, and many young scientists were willing to contribute and write articles for our educational programme. Kristína's experience in teaching and developing educational materials proved crucial in making these articles accessible to a high school audience and in designing hands-on activities for students.

The first five series engaged almost a hundred students from Slovakia. Thanks to funding from the European Society for Evolutionary Biology and ISTA's support, we were able to organise a meeting for the most successful students at ISTA, where they could observe research being conducted at a leading research institute.

This extracurricular activity has consistently received very positive responses from both students and teachers over the years, encouraging us to continue. Since its inception in 2017, the programme has benefited hundreds of students and has gained increasing popularity among high school students and their biology teachers in Slovakia.

Over time, some students who completed the programme returned as organisers, contributing to its growth. More scientists volunteered their expertise to write articles, and the competition briefly expanded into Austria. This sustained enthusiasm, coupled with continued support from ISTA for our annual meetings, has enabled this programme to continue. Today, many former participants are pursuing PhDs in biology or medicine, embarking on careers as scientists and doctors. We are proud to have fostered their interest in biology, nature, and science.

Due to the many positive reactions we have received over the years, we firmly believe that the material we have collected—rigorously shaped by the experiences and insights of competition participants and organisers—deserves a wider audience. By publishing it as a book, we aim to make this valuable resource more accessible to a wider audience, enabling them to benefit from the dedicated efforts and knowledge gained through years of running the program. We are driven by the conviction that sharing this knowledge and fostering a deeper understanding of evolutionary biology will inspire and empower more individuals to appreciate the complexities of the living world and engage in meaningful scientific exploration.

As of this writing, our competition is still ongoing, now entering its ninth year. While some years have presented more challenges than others, we are determined to continue our work on this project. We intend to continue expanding our coverage of exciting biological topics and developing additional engaging educational materials. This is a living project, and who knows—there may very well be a second volume of this book in the future.

Zürich, Switzerland — Barbora Trubenová
Bratislava, Slovakia — Kristína Hudáková
Vienna, Austria — Himani Sachdeva

Acknowledgements This book is the result of the effort of many people and has received support from even more. We are very grateful to all of them.

We would like to extend special thanks to Professor Nick Barton for his continuous support and motivation, and for fostering an environment in his group where this type of work was valued.

We also appreciate the team from ISTA, particularly Yvonne Kemper, who spent countless evenings polishing our English, as well as Magdalena Steinruck and Georg Heilig, who helped us organise events at ISTA over the years. Furthermore, we are thankful to Nathalie Gontier for her encouragement and suggestions in the early stages of this project.

The questions included in this book were developed with the invaluable assistance of various volunteers who contributed to the smooth running of the competition: Fedor Župník, Tomáš Jurčík, Katarína Karáseková, Katarína Majerová, Samuel Pauliny, Tamara Schenner, Veronika Smolková, Emma Čačková, Júlia Orlovská, Lucia Deáková, Michaela Matušovová, Linda Valenčíková, Lara Mia Holéczyová, and Veronika Plachá.

Kristína and Barbora would like to thank their parents, Jaroslava and Róbert, for their support and for instilling in them the value of education, hard work, and responsibility. They would also like to thank Peter for his help in creating online material used in this book and Martin, Ondrej, and Hana for their support and patience.

Himani would like to thank her family for their support over the years and Aditya for all his encouragement and help.

Daria Shipilina would like to dedicate her chapter to the loving memory of two of her collaborators Vjacheslav Fedorov and Valerii Alekseev. She would also like to express her gratitude to her scientific advisors Irina Marova and Vladimir Ivanitskii.

Hildegard Uecker would like to thank Edward C. Schrom for providing invaluable answers to all our questions on the human immune system, Roland Regös for helpful discussions, Lindi Wahl for many constructive comments on a previous version of the text of Chap. 7, Sally Otto for valuable comments on the section about the COVID-19 pandemic, Kristína Hudáková and Barbora Trubenová for making many of the figures and Christine and Manfred Rohde for the kind permission to reproduce Fig. 17.2.

The editors are also grateful to all the participants of the STEB competition, whose feedback has greatly improved the material and motivated us to continue our work.

Finally, the editors would like to acknowledge the various funding bodies that helped us organise the STEB competition, thereby contributing to this book. The competition initially received support from the Institute of Science and Technology Austria (ISTA) and the European Society for Evolutionary Biology (ESEB). As the competition gained momentum and recognition, it also secured funding from the Society for the Study of Evolution (SSE) and generous contributions from private donors.

Competing Interests The editors declare no competing interests.

Declaration of AI Use

Generative AI (Gemini and Grammarly) was used to proofread the manuscript and to improve the clarity of writing.

Contents

Editors and Contributors

About the Editors

Barbora Trubenová is a theoretical evolutionary biologist specialising in population and quantitative genetics who, as an assistant professor, leads a research group at ETH Zürich, funded by a Swiss National Science Foundation Starting Grant. Her group's primary focus is modelling the evolution of drug resistance in pathogens by integrating theoretical frameworks with empirical data. After completing her PhD on the evolution of social behaviours at the University of Manchester, she held postdoctoral positions at the Institute of Science and Technology Austria (ISTA), working on evolutionary computation, and at ETH Zürich, where she began her research on drug resistance. A strong advocate for interdisciplinary science, she consistently collaborates with empirical biologists and is incorporating machine learning into her research. Beyond academia, she is actively involved in science education and co-founded the civil association eduB to engage high school students with contemporary biology.

Kristína Hudáková is a biology educator who leads the development of educational activities for eduB, the civil association she co-founded to transform biology education in Slovakia. A former scientist with a PhD in Zoology, her work centres on creating innovative, experience-based natural science education. She connects high school students with contemporary science and scientists to help them understand biology as a dynamic science with practical significance. She actively collaborates with scientific institutions to bridge the gap between cutting-edge research and classroom learning.

Himani Sachdeva is a theoretical geneticist at the University of Vienna. She uses mathematical models to study long-term evolutionary outcomes such as adaptation, speciation, and extinction in natural populations. After a doctorate in physics from TIFR Mumbai, she switched to theoretical evolutionary genetics for her postdoctoral work at IST Austria. She currently works in the Mathematics and Biosciences Group at the University of Vienna. She has a keen interest in understanding how collaboration and communication can work between disciplines, as well as between science and society.

Contributors

Barbora Trubenová D-USYS, ETH Zürich, Zürich, Switzerland
Eawag, Dübendorf, Switzerland
eduB, Bratislava, Slovakia

Daria Shipilina Department of Botany and Biodiversity Research, University of Vienna, Vienna, Austria

F. Gözde Çilingir Eidgenössische Forschungsanstalt für Wald, Schnee und Landschaft (WSL), Birmensdorf, Switzerland
Department of Evolutionary Biology and Environmental Studies, University of Zurich, Zürich, Switzerland

Gemma Puixeu Sala Institute of Science and Technology Austria, Klosterneuburg, Austria

Hildegard Uecker Institute of Integrative Biology, ETH Zürich, Zürich, Switzerland
Department of Theoretical Biology, Max Planck Institute for Evolutionary Biology, Plön, Germany

Himani Sachdeva Faculty of Life Sciences and Faculty of Mathematics, University of Vienna, Vienna, Austria

Jarmila Šimková eduB, Bratislava, Slovakia
KBVE FEE Technical University in Zvolen, Zvolen, Slovakia

Katarína Bod̆ová Department of Mathematical Analysis and Numerics, Faculty of Mathematics, Physics, and Informatics, Comenius University in Bratislava, Mlynska Dolina, Bratislava, Slovakia

Katarína Juríková Department of Genetics, Faculty of Natural Sciences, Comenius University in Bratislava, Mlynská dolina, Bratislava, Slovakia

Kristína Hudáková eduB, Bratislava, Slovakia

Melinda Pickup Greening Australia, Native Seed Centre for Restoration Western Sydney University, Richmond, NSW, Australia

Michal Hledík Institute of Science and Technology Austria, Klosterneuburg, Austria

Réka Kelemen Institute of Science and Technology Austria, Klosterneuburg, Austria

List of Figures

List of Tables

Part I

Basics of Evolution

1 Evolution in Action

Himani Sachdeva, Barbora Trubenová, and Kristína Hudáková

▶ Most of us experience a sense of wonder when we think about the amazing diversity of life on our planet. Scientists today agree that all living organisms evolved from a single common ancestor that lived around 3.8 billion years ago—a unicellular creature nicknamed LUCA (Last Universal Common Ancestor). But how did complex organisms such as humans, which are made up of trillions of cells working together, *evolve* from this simple, single-celled organism? In this chapter, we will explore the basic principles of evolution by natural selection—without which we would not be here.

Let's start with a simple question: Is it possible to *directly* observe evolution in a human lifetime? Surprisingly, the answer is yes! In this chapter, we will dive into three different arenas where evolution can be observed *in action*. Along the way, we will learn what evolution is and how we might track evolutionary change over a few generations of a single population.

H. Sachdeva (✉)
Faculty of Life Sciences and Faculty of Mathematics, University of Vienna, Vienna, Austria
e-mail: himani.sachdeva@univie.ac.at

B. Trubenová
D-USYS, ETH Zürich, Zürich, Switzerland

Eawag, Dübendorf, Switzerland

eduB, Bratislava, Slovakia
e-mail: barbora.trubenova@env.ethz.ch

K. Hudáková
eduB, Bratislava, Slovakia
e-mail: kristina.hudakova@edub.sk

1.1 Evolution and the Food We Eat

Most of the plants and animals we consume evolved to their present form relatively recently (between 5000 and 13000 years ago), when ancient humans (perhaps accidentally) became evolutionary engineers. To understand how this occurred, let us rewind 9,000 years and visit Mexico during a dry spell. Imagine the following scenario: The fruits of the forest and the small game animals of the grassland are both scarce. Humans turn to the teosinte—a wild grass with thin "ears" or "cobs" covered with small, hard fruit cases, each of which contains an edible seed. Some people notice a few plants with slightly larger cobs and take them back to the cave to eat. After a while, they find more teosinte plants growing where they discarded the leftover seeds. Some of these plants also have large cobs.

Excited, they take the largest of these cobs and plant its seeds. After a few years of repeatedly selecting the largest cob and replanting its seeds, the cave dwellers have teosinte plants with cobs that

B. Trubenová et al. (eds.), *Everyday Evolution: Unveiling Life's Hidden Patterns*,
https://doi.org/10.1007/978-3-032-07050-0_1

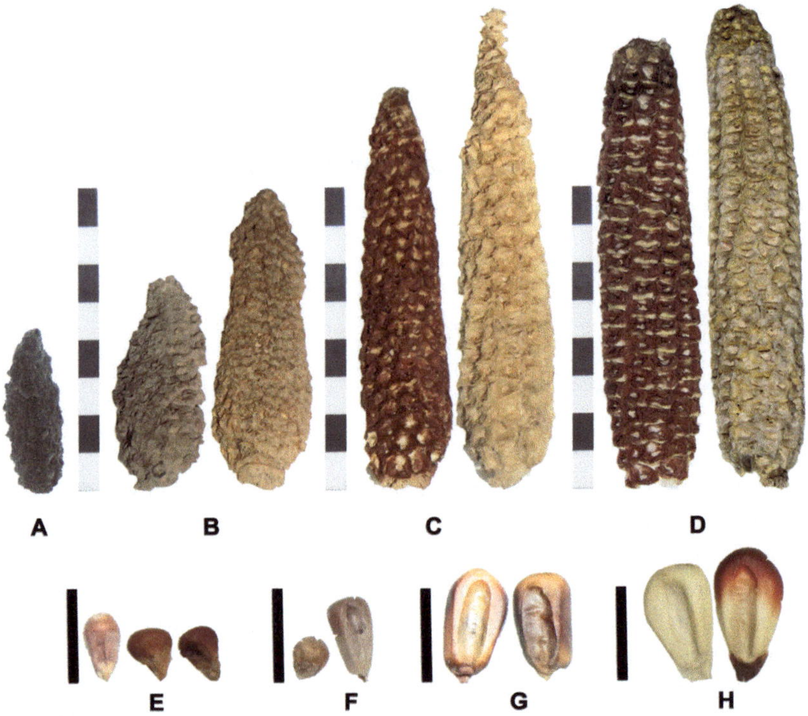

Fig. 1.1 Cobs and kernels from archaeological and modern populations. Cobs (A–C) and kernels (E–G) from archaeological sites compared to modern cobs (D) and kernels (H) suggest a progressive increase in size for both cobs and kernels. (Picture taken from: Vidal Elgueta A, Hinojosa LF, Pérez MF, Peralta G, Rodríguez MU (2019). Genetic and phenotypic diversity in 2000 years old maize (Zea mays L.) samples from the Tarapacá region, Atacama Desert, Chile. Under CC BY 4.0 license creativecommons.org/licenses/by/4.0/)

are bigger than the largest cobs they found in the wild—they have witnessed and, in fact, caused evolutionary change within their lifetimes!

Generations later, their descendants notice that some cobs have bigger, sweeter-tasting seeds and even find a rare teosinte plant whose seeds are not enclosed by hard outer shells—this cob is delicious and easy to eat! They plant its seeds and the seeds of its offspring until, eventually, most of the teosinte plants have large cobs with seeds that are sweet-tasting and lack outer shells. Thousands of years later, teosinte has evolved into a plant that would perhaps be unrecognisable to the ancient accidental farmers—maize, a primary food source cultivated on every continent (except Antarctica) (see Fig. 1.1).

Scientists have learned a lot about the **domestication** of teosinte by comparing the DNA sequences of modern-day maize, teosinte (which still grows in Central America) and 5000-year-old fossils of plants that are intermediate between teosinte and maize (and share features with both). Such comparisons enable scientists to identify the genes responsible for the differences between teosinte and maize, determine what these genes do and make educated guesses about what traits the ancient farmers were selecting for.

Not only maize but most of the plants and animals bred in farms today were domesticated from their wild ancestors. Archaeological and genetic evidence suggests that agriculture started independently in at least eleven regions of the world. Two of the earliest and most intensively cultivated sites were the Tigris-Euphrates Valley in the Middle East and the Nile Valley in Africa. It was here that many modern-day lentils, beans and cereals, including wheat, originated. As in the case of maize, ancient farmers selected (often accidentally) for traits that made cultivation and storage easier. For example, while wild wheat shatters spontaneously and scatters seeds onto the ground, mutant wheat retains its seeds, allowing farmers to harvest these seeds by hand. Thus, more and more mutant wheat was cultivated until it had largely replaced its wild ancestor near human settlements.

Ancient farmers were not the only ones who domesticated a wide variety of plants and animals for their use: Modern farmers and animal breeders are also evolutionary engineers! Even today, they continue to rely on the same basic evolutionary principles to improve crop yields and create new breeds.

Darwin's Theory of Evolution

The story of maize suggests that ancient farmers may have had an intuitive understanding of evolutionary change. Millennia later, the British naturalist Charles Darwin would draw on his observations of the wildlife of the Galápagos Islands, as well as those of more modern farmers (actually pigeon breeders), to formulate his theory of evolution. The key features of this theory already lie hidden in our maize-teosinte story.

First, evolution requires natural **variation**—individual teosinte plants must differ from each other in important characteristics such as height, cob size or number of seeds. Such observable differences arise due to minor variations in DNA sequence between individuals. Variation between individual plants is what allowed ancient farmers to select some plants over others. As we will see in the next section, selection often occurs naturally—for instance, if taller plants are more likely to survive in a particular environment or if their flowers are visited more often by bees. Thus, the second key ingredient of evolutionary change is **selection**—either natural or by humans. In this case, human selection favoured plants with larger cobs and larger seeds, making it more likely that these plants would survive and reproduce.

The third crucial ingredient of evolution is **heredity**: Many characteristics, such as shape, size or even susceptibility to certain diseases, are passed on from parent to offspring via DNA. With teosinte, we saw that long-term change occurred because plants with large cobs produced offspring that, on average, also had large cobs. Thus, even if the evolutionary change from one generation to the next is small (and sometimes it is not), these changes can add up over several generations to produce very dramatic effects.

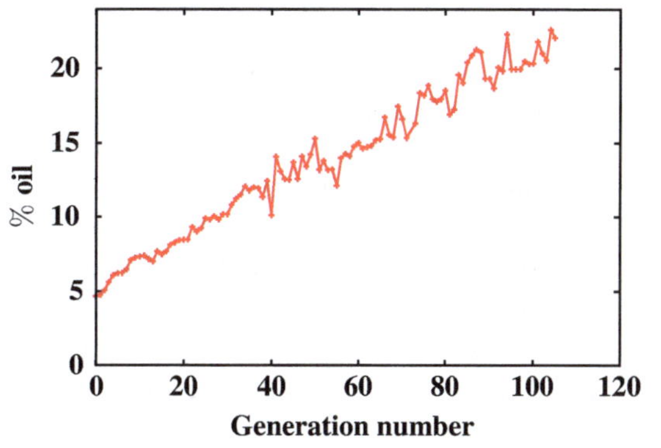

Fig. 1.2 Illinois long-term selection experiment in maize. The average oil concentration in corn is increasing even after more than 100 generations of selection. (University of Illinois at Urbana-Champaign: the Illinois long-term selection experiment for oil and protein in corn. Data taken from: http://mooselab.cropsci.illinois.edu/longterm.html)

Of course, one might wonder if a given characteristic of a plant can keep changing indefinitely with repeated rounds of selection—it is hard to imagine maize cobs the size of elephants! Nevertheless, it can take a surprisingly long time to reach the limits of evolutionary change. This has been demonstrated by the Illinois Long-term Selection Experiment, which has been running since 1896. In this experiment, scientists have been trying to increase the oil (or protein) content of maize by choosing plants with the most oil-rich (or protein-rich) seeds and crossing them with each other to produce the next generation. The oil and protein content of maize kernels has increased at a nearly constant rate for over 110 generations of selection by humans (see Fig. 1.2)!

Cauliflower and Its Cousins

The wild cabbage, which grows all over Europe, also has an interesting story to tell. A few thousand years ago, farmers in Germany started collecting wild cabbage plants with the widest stems and resowing their seeds, while Italian farmers sowed plants with the smallest buds and started selecting for these. After many generations of hu-

(continued)

man selection, the descendants of the wild cabbage looked quite different in the two regions—German fields were covered in kohlrabi, while Italian farmers were enjoying broccoli. In fact, ancient farmers produced a variety of vegetables—cabbage, broccoli, Brussels sprouts, kohlrabi, kale, cauliflower—by selecting for and amplifying different traits of the same plant: the wild cabbage.

Selection by humans is responsible not only for the nutritious, easy-to-harvest crops that feed our large population but also for the domestic pets and livestock that form the majority of large animals today. For instance, deliberate or unconscious selection by humans for traits such as docile behaviour, smaller tooth size, smaller brain size, etc. led to the evolution of dogs (from a wolf-like ancestor), pigs (from boars), modern-day cattle (from the now-extinct aurochs of Turkey and India) and chicken (from the Asian jungle fowl).

1.2 Rapid Evolution in the Wild

From the point of view of a plant or an animal, there is no difference between humans who allow certain individuals in the population (but not others) to reproduce and the hail or frost which kills off some (but not other) individuals.[1] Both impose an environmental challenge on the plant or animal population—if this challenge is extreme enough, and if the population has some members who perform better than others in the new environment, then what follows is rapid and sometimes dramatic evolutionary change.

The environmental challenges confronting wild populations can take very bizarre forms, as illustrated by the curious case of field crickets in Hawaii. Till the early 1990s, male crickets in Hawaii attracted female crickets for mating by flapping their wings, which produced a distinctive chirping sound. Unfortunately for the cricket, this chirping also started attracting females of another species, *Ormia ochracea*, a parasitic fly that invaded the Hawaiian islands sometime before 1990. The fly would lay thousands of eggs on the male cricket, and then the growing larvae would proceed to eat the cricket alive. This caused a severe decline in the cricket population until a rare mutant cricket with flattened wings appeared on the scene. This mutant produced no chirping sound and could thus escape the parasitic fly and survive, even though it was a bit more difficult for him to find a mate and reproduce.

Its descendants, who were also silent, gradually took over the population due to the strong natural selection favouring them. Over time, the cricket population revived, but now consists mostly of mutated, silent males! By capturing both silent and chirping males in the population and comparing their DNA sequences, scientists could identify the genetic change responsible for the adaptation—a mutation they named "flatwing". The whole story played out over approximately 20 cricket generations (about 5–7 years)—the time it took for the flatwing mutation to appear and then spread through the population.

The case of the cricket is not the only example of rapid evolutionary change in the wild. When drought struck the Galápagos Islands, it severely reduced the supply of the small seeds eaten by the medium ground finch *Geospiza fortis*—a bird species living on the island. The birds had to turn to large seeds, which could only be cracked open by individuals with bigger beaks. This gave a strong advantage to large-beaked birds, who could survive the drought and pass on their genes to the next generation. As a result, the average beak size increased in just two generations!

Of course, adaptation to an environmental challenge requires that the population already has a mutant individual that can thrive in the new

[1]These two fundamentally similar processes are referred to respectively as artificial selection (when humans select for or against particular traits of organisms) and natural selection (when organisms with certain traits survive or reproduce better in, and are thus selected for by, their natural environment).

environment or that such a mutant arises before the population goes extinct. This is not always the case, especially when the environment changes too much or too quickly or when populations are too small and lack genetic variation. The race between adaptation and extinction is going on at this very moment for millions of species. One example is that of the Tasmanian devil, whose population has shrunk by 80% in just 20 years due to the devil facial tumour disease—a highly contagious cancer that spreads via contact between individuals. In recent years, however, mutations that increase immunity to this cancer have also spread through the population. Will the endangered Tasmanian devil evolve to become resistant to this cancer and thus escape extinction? We will have to wait and see.

1.3 Evolution in the Lab

A typical human observer can only track evolutionary change over a few dozen generations in the wild, since most large organisms take at least a few months and often several years to mature and reproduce. However, microbes such as bacteria and yeast can reproduce every few hours (or even minutes, when conditions are favourable) by duplicating their DNA and dividing into two! Moreover, millions of microbes can be grown in a single test tube in the lab and observed under a microscope. The environment can also be easily controlled by changing the contents and temperature of the test tube. Bacteria and other microorganisms, such as yeast, have thus emerged as model organisms for the study of evolution.

The longest-running bacterial evolution experiment involves the bacterium *Escherichia coli* (most commonly found in the gut of humans and other warm-blooded animals) and has been running for over 80,000 generations (approximately 37 years). During this period, numerous mutations emerged in the different experimental populations, including both beneficial and harmful ones. The mutants that could reproduce faster managed to outcompete and replace their non-mutated siblings. Some even evolved the ability to utilise and grow on a different class of chemicals called citrates (found in fruits and included in the test tubes), which normally cannot be utilised by *E. coli*—a step toward becoming a different species of bacteria.

An interesting feature of this experiment is that scientists began with twelve clonal (initially identical) populations and allowed them to evolve under identical conditions. Having copies or replicates of the same experiment allows researchers to ask: Do the different populations evolve in exactly the same way, or does chance play an important role in evolution, causing populations to become distinct over time despite identical environments? The answers are complex and still only partially understood. Some adaptations, such as an increase in cell size, were observed in all 12 replicates, but other changes, such as the ability to metabolise citrate, evolved in only one of the 12.

Interesting Fact

Because of their short generation times, we can select for microorganisms that perform many useful functions for us—from strains of yeast that produce the best beer to bacteria that can break down toxic chemicals in our wastewater to microorganisms that are extremely efficient at producing biofuels. There is even a strain of bacteria that can extract gold from seawater!

Studying bacterial evolution in the lab has also given scientists vital clues about how bacteria can accumulate genetic mutations that make them resistant to antibiotics. Understanding and combating antibiotic resistance is one of the most pressing scientific and medical challenges in the world today. We will dive into the fascinating world of bacteria and their ability to outsmart modern medicine in Chap. 6.

What have we learned by directly observing evolution across a variety of organisms and environments? We have seen that evolutionary change across diverse evolutionary arenas has several common features: First, it requires genetic variation within a population. Second, these variations must be heritable—passed on from generation to generation. Finally, evolution involves the action of natural (or human) selection on these variations.

In later chapters, we will consider timescales much longer than the human lifespan and explore how evolutionary processes have led to the amazing biodiversity around us (as well as how biodiversity is threatened today). We will also explore important differences between evolution in organisms that reproduce primarily without sex (e.g. bacteria) and those that reproduce sexually (e.g. humans, where each individual inherits genetic material from two different parents). Our journey will take us to human origins, and we will see how evolutionary thinking can help us better understand some of the challenges faced by modern societies. So welcome aboard and bon voyage!

DEFINITIONS

Biological evolution: heritable change in the characteristics of groups of organisms over the course of generations due to changes in genetic material.

Selection: a process by which organisms that are better suited to their environment survive and reproduce more successfully.

Adaptation: an evolutionary process by which populations become (gradually) better suited to their environment through natural selection acting on heritable variation.

Mutation: a genetic change (error while copying DNA sequence) that results in variation among individuals; these genetic changes are heritable, i.e. passed on from parent to offspring.

Fitness: the expected number of surviving offspring of an individual.

Domestication: a process by which an organism (typically humans) gradually selects for specific traits in wild populations of plants and animals, thus increasing their suitability for association with humans and/or as food sources.

SUGGESTED READING

- Hake S, Ross-Ibarra J. Genetic, evolutionary and plant breeding insights from the domestication of maize. Elife. 2015 Mar 25;4: e05861.
- Zuk M, Rotenberry JT, Tinghitella RM. Silent night: adaptive disappearance of a sexual signal in a parasitized population of field crickets. Biol Lett. 2006 Dec 22;2(4):521–4.
- Boag PT, Grant PR. Intense Natural Selection in a Population of Darwin's Finches (Geospizinae) in the Galapagos. Science. 1981 Oct 2;214(4516):82–5.
- Lenski RE. Experimental evolution and the dynamics of adaptation and genome evolution in microbial populations. ISME J. 2017 Oct;11(10):2181–2194.
- Vidal Elgueta A et al. Genetic and phenotypic diversity in 2000 years old maize (Zea mays L.) samples from the Tarapacá region, Atacama Desert, Chile. PLoS One. 2019 Jan 30;14(1):e0210369.

1.4 Questions

1. **Organisms inhabiting the Earth today...**
   (A) ...are the result of evolutionary processes that have occurred over millions and billions of years.
   (B) ...have essentially the same form they have always had.
   (C) ...all came into existence at about the same time.
   (D) ...are all descendants of a single common ancestor.
   (E) ...can be arranged in a linear order from "least evolved" to "most evolved".
2. **Evolutionary adaptations...**
   (A) ...increase the ability of a species to survive and reproduce in a specific environment.
   (B) ...increase the ability of a species to survive and reproduce under any circumstances.
   (C) ...always result in individuals becoming bigger, stronger or faster.
   (D) ...do not follow any particular direction.
   (E) ...continually increase intelligence in all animals.
3. **Which of the following changes is an example of evolution?**
   (A) The leaves of deciduous trees change colour from green to yellow, red and brown and fall down each year.
   (B) A volcano erupts on a small island. Only a few lizards survive. Over the following generations, the descendants of these lizards gradually become darker coloured.
   (C) A caterpillar hatches from an egg. After a period of rapid growth, it becomes a pupa, from which an adult butterfly emerges a few weeks later.
   (D) A mammalian zygote divides and develops into a fetus. The fetus then grows inside the female's uterus until the baby mammal is born.
   (E) After a glacier retreats, the exposed rock is first colonised by lichens, algae and fungi, which slowly make the area habitable for vascular plants.
4. **Cheetahs are the fastest land mammals. They evolved their ability because:**
   (A) In each generation, it was necessary for cheetahs to run faster in order to catch prey, so every new generation was born with improved running abilities.
   (B) Only those individuals that were the fastest were able to find mating partners and reproduce.
   (C) In each generation, individuals that happened to be faster runners due to chance mutations caught more prey and left more offspring on average than slower runners.
   (D) In each generation, the slowest half of the population of the individuals died out, as they could not catch any prey.
   (E) Cheetahs spend all their lives chasing prey, so they learn to run faster. When they reproduce, they pass this skill to their offspring. Each generation then runs slightly faster than the previous one.
5. **Figure 1.3 shows the distribution of plant height in three different wheat populations grown on three different fields. Each individual plant within a population experiences exactly the same environmental conditions (temperature, soil composition, etc.); that is, variation in height is due entirely to genetic differences between individual plants. Imagine that a farmer decides to increase the average height of her crop by breeding the tallest 20% plants**

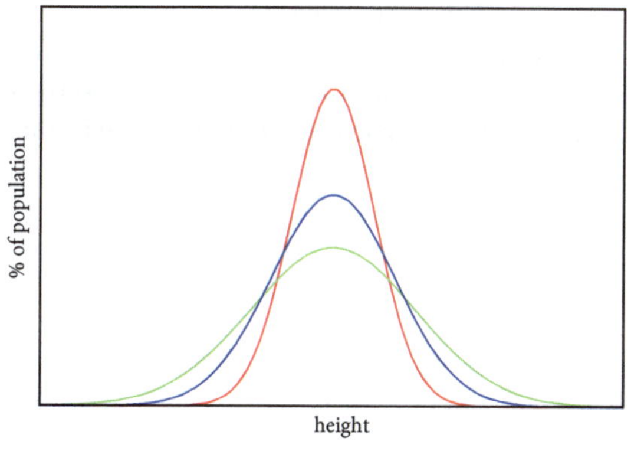

Fig. 1.3 Distribution of plant height in three different wheat populations. The x-axis shows height values, and the y-axis shows the percentage of individuals with a particular height. The red, blue and green curves show the height distribution in populations 1, 2 and 3, respectively

in every generation in each of the three populations. Which population(s) will show the most rapid increase in height over the next few generations?

(A) Population 1.
(B) Population 2.
(C) Population 3.
(D) All three populations will show exactly the same increase in height, as they originally have the same average height.
(E) The average height will *not* change in any of the three populations.

6. **Now imagine that all individuals within any given population are genetically identical (clones of each other). However, they are grown in soil patches of different quality, causing some individuals to be taller and others to be shorter. Suppose that the farmer again decides to increase the average height of her crop by breeding the tallest 20% of plants in every generation in each of the three populations. Which population(s) will show the most rapid increase in height over the next few generations?**

(A) Population 1.
(B) Population 2.
(C) Population 3.
(D) All three populations will show exactly the same increase in height, as they originally have the same average height.
(E) The average height will *not* change in any of the three populations.

7. **Two populations A and B of the same plant species inhabit two different islands. Population A consists of 10000 individuals, while population B consists of 500 individuals. The islands are widely separated, so that seeds or pollen cannot reach from one to the other. The two islands originally have the same environmental conditions. After a severe earthquake, the soil salinity increases (by the same amount) on both islands. Initially, both populations A and B decrease. However, population A recovers over the next few years and once again reaches 10000 individuals, while population B is completely wiped out. Which of the following factor(s) might explain the different fates of the two populations?**

(A) Mutation rates per individual are typically lower in smaller populations.
(B) Large populations are more likely to have at least one individual who already carries mutation(s) that might help it survive and reproduce in an altered environment.
(C) An adaptive mutation is more likely to occur in at least one member of a large population before the population goes extinct.
(D) Members of a small population have to compete more fiercely with each other, reducing their ability to survive and adapt to a new environment.
(E) Individuals in a small population are likely to be related to each other and hence are genetically more similar.

1.5 Project: Is Bigger Better?

In Fig. 1.4, you can see seeds of two tree species common in towns and cities in Europe. Look at them and think about the obvious feature they share. What is the function of this feature?

Many trees evolved wings to disperse their seeds over a broader area. Seed dispersal over a wide area has several benefits for the plant species: First, seedlings have a greater chance of survival if they sprout far away from the parent plant. Second, landing far away decreases the intensity of competition with the parent plant as well as with other seedlings of the same species. Third, seed dispersal provides a tree's offspring with a chance to colonise a new, potentially better habitat.

In this project, you will go to nature, observe and analyse variation among individuals in a population and perform a small experiment.

Part 1: Getting the Data Sample

Go for a walk in your village, town or city and look for trees producing winged seeds.

1. Take a picture of the trees and the location in which they grow. Describe the trees and their location in one sentence.
2. Collect 10 to 20 seeds[2] produced by one tree and another 10 to 20 seeds produced by another tree of the same species.[3] Take a picture of each of the two seed collections.

Part 2: Analysing the Data

At first glance, the seeds may look very similar to one another, but when you look closer, you will observe some differences. To measure the individual variability of the seeds, you will work on just one trait—the wing length.

1. Measure the wing length for each seed and record the data in Table 1.1.
2. Look at the seeds and their measurements: Which seed do you expect to be able to fly the best? Which would probably fly the worst? Indicate it in Table 1.1 and explain why you think so.
3. For each seed collection, calculate the arithmetic mean of the wing length and record it in Table 1.1.
4. For each seed collection, indicate the range of wing length values.
5. Looking at these basic statistics, would you say that the two seed collections are rather similar or rather different? Explain.
6. For each seed in each sample, calculate how different each wing length is from the mean wing length of its sample, and record the difference in Table 1.1.
7. Would you characterise the seeds produced by one tree as rather similar or rather different? Explain.

Part 3: Experiment—Flying the Seeds

In nature, when the seeds are shed by the tree, they spend various amounts of time in the air. The longer the time they can spend in the air, the higher their chance of being brought further away from the mother tree and from the competing siblings.

First, test how your seeds fly:

1. Take the seeds that you expected to fly the best and the worst in each seed collection.
2. Let each seed fall from a balcony, upper-floor window or other high place, and record the time it spends in the air.
3. For each seed, measure the time it spends flying three times and record the data in Table 1.2.
4. Summarise what you observed: Did the seeds that you expected to fly the best really do better than the ones you expected to fly the worst?

Second, imagine that a mutation happens, affecting the size of the wing.

[2] The more seeds you collect, the better results you get.

[3] If you are not entirely sure of the species, but both of them are of the same genus (e.g. maple), it would do for this exercise.

Fig. 1.4 Winged seeds of linden and maple

Table 1.1 Characteristics of the two seed collections

Tree 1:			**Tree 2:**		
Seed	Wing length (mm)	Difference from the mean	Seed	Wing length (mm)	Difference from the mean
#1			#1		
#2			#2		
#3			#3		
#4			#4		
#5			#5		
#6			#6		
#7			#7		
#8			#8		
#9			#9		
#10			#10		
...			...		
mean			**mean**		
range			**range**		

Table 1.2 Flying times of the four chosen seeds

Tree 1			
Seeds expected to be	Flight 1 (seconds)	Flight 2 (seconds)	Flight 3 (seconds)
best flying			
worst flying			
Tree 2			
Seeds expected to be	Flight 1 (seconds)	Flight 2 (seconds)	Flight 3 (seconds)
best flying			
worst flying			

Table 1.3 Flying times of the two modified seeds

	Flight 1 (seconds)	Flight 2 (seconds)	Flight 3 (seconds)
First modified seed			
Second modified seed			

1. From the four seeds that have spent time in the air you have just measured, choose two that will represent seeds changed by a mutation.
2. Model the effect of the mutation by either cutting from the wing or by adding to the wing using any suitable material.
3. Take pictures of both modified seeds.
4. For each modified seed, propose a hypothesis describing how the modification of the wing will affect its flying time.
5. Let each modified seed fly three times. Measure the time it spends flying and record the data in Table 1.3.

Analysis

1. How did the modifications affect the flying times of the seeds? Was your hypothesis supported? Explain.
2. What other seed traits, apart from the ability to fly, would provide a seed with an advantage, increasing its chances for survival? Give at least two examples.
3. What environmental factors would affect the survival of the seed after it lands? Give at least three examples.

This project was developed by Kristína Hudáková.

DNA, Mutations, and Evolution

2

Barbora Trubenová, Kristína Hudáková, Hildegard Uecker, and Himani Sachdeva

▶ DNA is the molecule that governs all life on Earth. Almost every cell of every living organism, from bacteria to humans, contains DNA, which encodes the information that directs an organism's development, reproduction, and even programmed cell death. What characteristics make the DNA molecule so unique and so essential?

DNA has three key properties that make the existence of a vast variety of living organisms possible. First, it contains very detailed and complex *instructions on how to build proteins*, which, in turn, compose the organism.[1] Second, it can *replicate* (copy) itself. DNA copies are passed on to daughter cells. In this way, genetic information is transmitted to the subsequent generations, and successful instructions for building proteins and organisms are conserved for millions of years. Finally, it can be *modified*. During DNA replication, errors (mutations) can occur that give rise to new variants and, ultimately, to new species. In this chapter, we will look in detail at this last property: the creation of new variant—mutations — and their role in evolution.

2.1 Creating Diversity

Before a cell divides, it replicates its DNA.[2] During this process, the two strands that make up a DNA molecule separate. Each strand serves as a template for the synthesis of a *new* strand of DNA, which is synthesised from the free nucleotides floating around (with extensive help from proteins). At the end of this process, two identical DNA molecules are made, each com-

B. Trubenová (✉)
D-USYS, ETH Zürich, Zürich, Switzerland

Eawag, Dübendorf, Switzerland

eduB, Bratislava, Slovakia
e-mail: barbora.trubenova@env.ethz.ch

K. Hudáková
eduB, Bratislava, Slovakia
e-mail: kristina.hudakova@edub.sk

H. Uecker
Institute of Integrative Biology, ETH Zürich, Zürich, Switzerland

Department of Theoretical Biology, Max Planck Institute for Evolutionary Biology, Plön, Germany
e-mail: uecker@evolbio.mpg.de

H. Sachdeva
Faculty of Life Sciences and Faculty of Mathematics, University of Vienna, Vienna, Austria
e-mail: himani.sachdeva@univie.ac.at

[1] This is true for all organisms except for some viruses, which contain only RNA as their genetic material.

[2] In this chapter, we will always talk about mitotic cell division. Cells of the germline undergo meiosis, which is a rather different process. We discuss meiosis in several other chapters, for instance, in Chap. 12.

B. Trubenová et al. (eds.), *Everyday Evolution: Unveiling Life's Hidden Patterns*,
https://doi.org/10.1007/978-3-032-07050-0_2

prising one new and one original strand. These two DNA molecules are then passed on to two daughter cells, which are identical (if everything goes right). In this way, genetic information is passed from generation to generation. This is a very simplified version of the complex process of DNA replication.

DNA is made up of four different nucleotides: adenine (represented by the letter A), thymine (T), guanine (G), and cytosine (C). An organism's **genome** is the collection of *all* of its DNA, most of which is organised in chromosomes. The human genome contains about three billion (3×10^9) pairs of nucleotides. All of these nucleotides must be copied precisely to ensure that when a cell divides mitotically, the two daughter cells are identical. When a human baby is born, it has about two trillion (2×10^{12}) cells—all produced by repeated cell divisions from a single cell, the fertilised egg. Thus, in total, approximately two trillion cell divisions must occur to make a human baby![3]

Numerous control mechanisms in the cell ensure that the long sequence of nucleotides is reproduced as precisely as possible. Despite these mechanisms, some errors are bound to occur when three billion nucleotides are copied two trillion times! For instance, a nucleotide might be lost, added, or replaced by another one, resulting in a few differences—**mutations**—between the parent and daughter cells.

2.2 Harmless Mistakes

Fortunately for us, most mutations have no or only very minor effects. Many mutations occur in fully specialised **somatic cells**. Somatic cells include all cells that make up the body other than reproductive cells (eggs or sperm), also known as gametes. Often, a particular mutated gene is not important for the function of that specialised cell. For example, a mutation within a gene involved in chlorophyll production might render a leaf cell non-functional. However, the same mutation in a root cell would have no effect, as the root cells of most plants do not produce chlorophyll. Scientists estimate that an average human accumulates hundreds to thousands of **somatic mutations** in a lifetime—and, fortunately, most of these don't really cause problems. An important exception is mutations in genes that regulate cell division or cell death. Somatic mutations that disrupt the cell cycle may cause cells to replicate rapidly in an uncontrolled manner, resulting in tumours and cancer.

However, if a mutation occurs in the sperm or eggs of the parent or during early embryonic development, all or most of the cells in the baby's body will carry that mutation. More importantly, these mutations will also be present in the reproductive cells of the new individual and can thus be transmitted to its offspring. These mutations are known as **germline mutations** (in contrast to somatic ones). As we will see later, germline mutations are the main drivers of evolutionary change as they have the potential to affect biological function over multiple generations. Nevertheless, many germline mutations are also harmless or *neutral*. Why is this the case?

First, exchanged nucleotides may still lead to the production of the *same* protein. Each amino acid is always encoded by a sequence of three nucleotides—also known as a codon or triplet. Several codons can code for the same amino acid: for instance, both codons CAT and CAC code for the amino acid histidine. Another amino acid, leucine, is encoded by six different codons! Therefore, when DNA is transcribed to mRNA and then translated into a sequence of amino acids, any of the codons that code for the same protein will give the same result. Thus, mutations between codons that encode for the same amino acids have no effect on the structure and function of proteins.

Interesting Fact

The set of rules specifying how a sequence of nucleotides in DNA determines the sequence

[3] If we take a single DNA molecule from any of those two trillion cells, this molecule is about a 40th copy of the original DNA molecule present in the zygote because there were about 40 generations of dividing cells since conception.

of amino acids in a protein is called the **genetic code**. In fact, the genetic code is *universal* across the vast majority of living species: for example, in almost every organism on Earth, the codons CAT and CAC encode histidine.[4] This is one of the strongest pieces of evidence for the common origin of life on Earth.

Second, mutations may occur in parts of DNA that are not directly translated into proteins. **Genes** (segments of DNA that code for a protein) do not appear one after the other on the DNA sequence, like words in a sentence. Between the genes are sequences of non-coding DNA, i.e. DNA that is not translated into proteins. Coding regions actually constitute only a tiny fraction of the DNA in many organisms. For example, a whopping 98% of the human genome is non-coding! A significant fraction of non-coding DNA has no known biological function—thus, mutations within such segments have little or no effect.

2.3 Mutations That Matter

Sometimes, however, just a single error in the wrong place can have dire consequences. Even some mutations in non-coding regions of DNA can matter. Let's look at one example where a mutation in a non-coding region makes a huge difference.

Lactase Persistence, the Opposite of Lactose Intolerance

In some parts of the world, such as Northern Europe, it might seem like lactose intolerance is something unusual, but from a biological point of view, it is actually the opposite. It is not lactose-intolerant people who have some weird mutation. Rather, adults who can still digest milk— typically a food for babies—are the ones who carry an unusual mutation. But how do we know what the mutation is and what the ancestral (original) state is? We can answer this question by looking at all other mammals. All baby mammals can digest lactose—the sugar found in milk—at first but lose this ability after weaning. Humans are the exception—some of us can digest lactose even as adults.

Non-coding and "junk" DNA

The structure of a single gene may itself be quite complicated. Genes can span long sections of DNA and may include many long, "useless" sequences—**introns**—within a single gene that need to be cut out during the process of protein synthesis. The remaining (coding) segments—**exons**—are then joined together and translated into the sequence of amino acids that forms the protein. Mutations in introns mostly do not affect the organism. In addition, genes are separated by large chunks of non-coding sequences that are not transcribed to RNA. Initially, scientists believed that most non-coding DNA was "junk" DNA—ancient sequences that had lost all biological function due to the accumulation of harmful mutations. However, more recent research suggests that non-coding DNA plays a critical role in *gene regulation*—by increasing or decreasing the production of proteins to match the needs of the cell. Calling it "junk" DNA is a regrettable choice of a name—many parts of this DNA are very important!

Interestingly, some 8% of human DNA is actually the remnant of ancient viral DNA! Some viruses incorporate their genomes into the genome of the host cell. In rare cases, the viral genome is passed on from generation to generation until it is integrated into the genome of the entire host population (see Chap. 7). Scientists are still trying to figure out if human DNA segments of viral origin may be linked to an increased risk of certain diseases. Understanding how much non-coding DNA is functional versus junk is an active area of research today.

[4]There are very, very few exceptions, e.g. ciliates.

Van Gogh and Mutant Sunflowers

The Dutch painter Vincent van Gogh painted an entire series of pictures showing a bouquet of sunflowers in a vase. The "head" of a sunflower consists of many small flowers—the disk and the ray flowers. A typical sunflower has a big, dark disk in the middle, where the seeds develop, surrounded by bright yellow ray flowers. However, in some of van Gogh's pictures, the flowers do not look like typical sunflowers. They only have small bright disks and are otherwise all yellow.

Just a few years ago, researchers identified a mutation in the gene *HaCYC2c* that leads to these all-yellow sunflowers. Without this mutation, the gene is turned *off* in the inner part of the sunflower, which therefore develops into small, dark disk flowers, resulting in a typical sunflower. The mutated gene contains an extra section of DNA that messes up the on-off switch of the gene. It, therefore, remains turned *on* everywhere, causing the inner florets to also be ray like. In nature, this is a harmful mutation since the yellow ray flowers do not have reproductive parts, which are only found in the dark disk flowers. The role of the yellow flowers is to attract pollinators. In the same study, the researchers found another mutation that turns all florets into disk-like florets (though the outermost ones are large and yellow).

Wild-type and mutated sunflowers. The upper row shows a wild-type (normal) sunflower and two mutants, and the lower row shows the individual florets. Van Gogh painted mutant sunflowers in a vase. (Picture taken from Chapman MA et al. (2012) Genetic Analysis of Floral Symmetry in Van Gogh's Sunflowers Reveals Independent Recruitment of CYCLOIDEA Genes in the Asteraceae under CC BY 4.0 license creativecommons.org/licenses/by/4.0/)

Lactose is a disaccharide that is broken down into monosaccharides by an enzyme called *lactase*, which is produced by cells lining the small intestine. Lactase is coded for by the gene *LCT*. Typically, as growing mammals switch to food sources other than milk, the production of lactase decreases, and eventually, the gene is switched off completely.

When an individual who no longer produces lactase consumes milk, the lactose remains intact and moves down the digestive system into the large intestine. There, it is fermented by bacteria, producing an increased amount of gas. Furthermore, it draws in water through osmosis. This, plus the gas, leads to pain, flatulence, and diarrhoea—the symptoms described as lactose intolerance. This can be avoided simply by taking the enzyme as a pill before consuming dairy products.

However, in some individuals, the *LCT* gene is not switched off, causing these individuals to produce lactase throughout adulthood. But why?

The *LCT* gene is actually the same in all individuals, whether or not they are able to digest lactose. However, scientists have discovered that most Europeans who are lactose tolerant have a particular mutation: a single cytosine replaced by thymine in the intron of a *different* gene near the *LCT* gene sequence. This single mutation attracts proteins that assist in the transcription of the *LCT* gene, thus increasing its production. It causes the transcription of the lactase gene to continue into adulthood, leading to the condition known as lactase persistence. This example illustrates an interesting point: many mutations affect biological function not by altering the structure of a protein but by changing when and how much of it is produced.

Changing the Script

When a mutation occurs in the coding part of the gene (exon), even a simple nucleotide exchange can result in a different amino acid being used in a protein, which, in turn, can change the properties of the protein. For instance, sickle cell anaemia—a blood disorder common mainly in some parts of Africa—is caused by a single error in a person's DNA: adenine replaced by thymine. Instead of the codon GAG coding for glutamic acid, the mutated codon GTG results in valine being used in the amino acid chain, leading to clumped haemoglobin and defective red blood cells.

However, not all mutations involve exchanged nucleotides. If a nucleotide is lost or added, this can have an even more powerful effect: it can shift the way that DNA is read and deciphered, giving a completely wrong sequence of amino acids! This is the cause of cystic fibrosis, a rare but severe disease.

In addition, sometimes, a larger segment of DNA can be deleted, inserted, duplicated, or inverted. Such mutations usually have severe consequences. For instance, some types of cancer are the result of duplications of a section of DNA. However, these larger rearrangements also underlie some of the biggest evolutionary innovations!

Not all mutations that lead to malfunctioning proteins are harmful. Sometimes, they create exciting variants that change an individual's appearance without affecting function. For instance, the main[5] gene for human eye colour is called OCA2. The original variant of this gene makes a protein necessary for the development of brown pigment. In general, a variant of a gene is called an **allele**. So, individuals with the original variant have the allele for brown eyes. Mutated versions of the gene result in a protein that is not fully functional, so not enough brown pigment is produced. This leads to blue eyes—thus, these mutated gene variants are considered alleles for blue eyes.

Let us consider in detail how a mutation can completely change the look and taste of a garden pea, one of the earliest model organisms in the study of genetics and evolution.

Now examine the dried pea seeds found in most kitchens. Are they wrinkled or nicely rounded? Seeds that wrinkle when dried contain higher levels of sucrose and more water and taste sweeter when fresh. On the other hand, seeds that retain their round shape even after drying have less water and sucrose but contain higher levels of starch. Thus, they contain more energy and are preferred for cooking or feeding animals.

But why do some seeds have more sucrose while others have more starch? The sugars that are produced during photosynthesis undergo a series of reactions catalysed by *enzymes*. The purpose of these reactions is to store energy in energy-rich molecules, such as starch, that can be used by seeds during germination. In the case of the pea, the last in this series of reactions is the branching of the starch molecule, which is catalysed by a starch-branching enzyme. If this enzyme is present, glucose and fructose produced during photosynthesis will be used up almost entirely to produce starch, most of it in a branched form. However, if this enzyme is *absent*, only a few unbranched starch molecules are made. Instead, the excess glucose and fructose molecules are combined to form sweet-tasting sucrose.

The starch-branching enzyme is coded for by the *SBE1* gene. This gene comes in two[6] alleles: *R*, which codes for the functioning enzyme, and *r*, which codes for a mutated, non-functioning enzyme. The *R*-allele contains about 3550 nucleotides that determine the sequence of amino acids in the protein. This amino acid strand will fold into a functioning starch-branching enzyme. The mutated *r*-allele contains about 800 extra nucleotides that are located almost in the middle of the gene. These additional nucleotides lead to additional amino acids, causing the chain to fold

[5] So far, scientists have discovered eight genes that influence eye colour. However, the remaining seven genes only play a minor role.

[6] Recently, more alleles have been described, but those were not available in Europe until recently.

differently and the protein to lose its ability to catalyse the branching reaction.

Sometimes, mutations can even drastically improve or save the organism. For instance, a bacterial gene *ompF* encodes a protein necessary for transporting some types of molecules through the cell membrane. A mutation in this gene damages the protein. It may sound like a typical harmful mutation; however, it may also prevent antibiotics from entering and killing the bacterial cell! A mutation in the *ompF* gene actually causes antibiotic resistance in bacteria!

Things Are Not That Simple

From the sections above, one might imagine that DNA completely determines all traits. But this is definitely not the case! While some traits, for instance, eye colour or the ability to digest lactose, are mostly determined by the alleles carried by an individual (alleles which may have been inherited from their ancestors or gained through a new mutation), most traits are strongly influenced by the environment. For instance, an individual's height depends on the alleles they inherit from their parents, as well as their nutrition and potentially other environmental factors. A lack of vitamins and minerals may lead to reduced bone growth and shorter height. The relative contribution of the genotype and the environment can vary drastically between different traits. To complicate matters, most traits, especially quantitative traits such as height, are not encoded by a single gene but by many different genes working together. We will talk about this in detail in Chap. 4. But for now, let's briefly discuss how a single mutation that typically arises in a single individual can spread through the population.

2.4 DNA and Evolution

DNA is passed on from generation to generation, conserving the information about how to build an organism. **Inheritance** is the first crucial ingredient of evolution. The second is **variation**. In this chapter, we discussed how small changes in DNA can result in different traits, such as an unusual ability to digest lactose. The final key ingredient for evolution is **natural selection**—the survival of the fittest.

What does fitness mean in the biological sense? Essentially, it means how likely an individual is to survive, reproduce, and have viable offspring. Organisms that have phenotypes most suitable to a given environment are most likely to survive and reproduce in that environment. The others may starve, be eaten, or fail to reproduce.

Scientists estimate that the human genome accumulates about 50–100 germline (heritable) mutations per generation. As we explained above, most of them are, luckily, neutral—either they have no effect at all or neither harm nor benefit the individual. Very harmful mutations are quickly lost—the organism is either unable to produce offspring or produces fewer of them.

On the other hand, a beneficial mutation helps its carrier by increasing the carrier's fitness, for instance, by helping them live longer or enabling them to survive in harsher environments. Such an individual is more likely to leave more offspring, which may inherit this new, mutated version of the gene. In the generations that follow, this mutated version will become more and more common.

Think again about lactase persistence. The ability to digest milk as an adult may seem obviously beneficial, but it is not that simple. Lactase persistence is only an advantage in societies where humans have domesticated dairy animals. The ability to digest lactose evolved together with cattle domestication. Scientists estimate that the mutation for lactase persistence is between 7000 and 9000 years old. Archaeological records and genetic data suggest that early Middle Eastern and North African human populations domesticated cattle between 7500 and 7000 years ago and later migrated to Europe. In these populations, the ability to digest milk and use it as a food source, instead of killing the animals for meat, could have provided the necessary advantage to outcompete local hunters.

DNA, RNA and the Origin of Life

In this chapter, we have talked about DNA as the (extremely important!) molecule in which genetic information is stored and transferred from generation to generation, but most biologists agree that before our world of DNA-based life, a different one must have existed. As scientists began to understand how the genetic information encoded by DNA is used by the cell, an important question arose: DNA stores the information necessary for making proteins, but the molecules required to replicate DNA and to use this information to synthesise all proteins are themselves protein based. Which one came first?

Already in the 1960s, Francis Crick speculated that RNA might be the answer to this problem. His ideas found support in the 1980s when it was discovered that certain RNA molecules are indeed able to catalyse chemical reactions—a function previously attributed only to proteins. That means that RNA would be able to store genetic information (as DNA does) as well as regulate the use of this information by the cell (as protein enzymes do). These findings support the idea that the oldest forms of life relied on RNA to perform both these processes—a hypothesis known as **the RNA world**.

Later, DNA took over RNA's role of storing genetic information, providing living organisms with more stable genetic material (DNA is less reactive than RNA, thanks to one less oxygen in the sugar). Protein enzymes replaced RNA as a catalyst, enabling organisms to catalyse a greater variety of chemical reactions (proteins are built out of twenty amino acids, in contrast with RNA—which is built from just four nucleotides).

Despite these changes, RNA still plays many important roles in present-day organisms. It is essential to the processes of DNA replication, protein production, and gene regulation.

This allowed them to spread their culture, as well as the lactase persistence gene, further into Central and North-western Europe.

Interestingly, humans on different continents seem to have domesticated dairy animals and evolved lactase persistence independently. The mutation(s) responsible for lactase persistence are different, but the selective force and the effect are the same. The different frequencies of the lactase persistence allele in various parts of the world (see Fig. 2.1) reflect the different histories of cattle domestication and migration.

Without mutation, there would be no variation. There would be no heritable differences between organisms, and, well, there would be no organisms to begin with. Mutation is the only process that creates new variants of life, and it has enabled the rise of the extraordinary diversity of living organisms around us.

DEFINITIONS

DNA (deoxyribonucleic acid): a macromolecule that contains all the information required to build an organism. An individual passes on its DNA to its offspring during reproduction.

Chromosome: a long, linear molecule of DNA coiled around proteins; DNA is packaged into one or more chromosomes in the cell.

Gene: segment of DNA that codes for a protein and thus influences trait(s).

Genome: complete genetic material of an organism, including all its genes.

Protein: macromolecule made up of amino acids, which may serve various functions in an organism (structural, regulatory, catalytic, etc.).

Allele: one variant of a gene; individuals that carry different alleles of a particular gene may

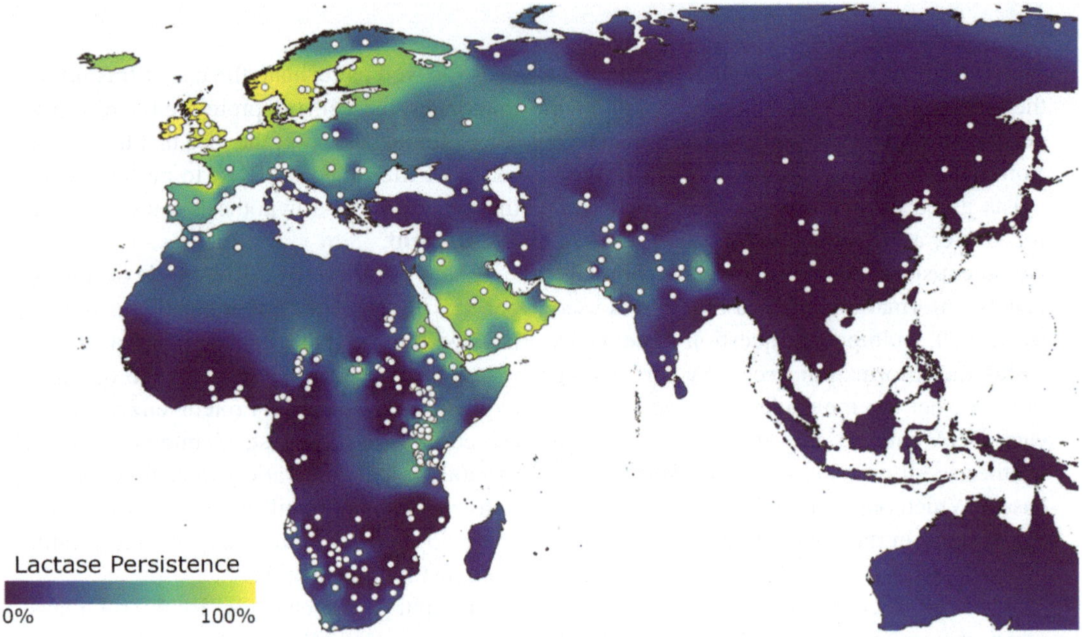

Fig. 2.1 Map of lactase persistence throughout Eurasia and Africa. Percentage of the adult population that is able to digest milk. (Picture taken from Stock JT, Wells JCK. Dairying and the evolution and consequences of lactase persistence in humans. Anim Front. 2023 Jun 14;13(3):7–13., under CC BY 4.0 license creativecommons.org/licenses/by/4.0/)

have different characteristics (e.g. blue versus brown eyes).

Codon: a sequence of three nucleotides which codes for a single amino acid (also called a triplet).

Exon: a segment of DNA or RNA that ultimately contributes to the mature RNA product of a gene, meaning it is the portion that is expressed.

Intron: a non-coding segment of DNA or RNA that is removed during RNA splicing and not included in the final mature RNA.

SUGGESTED READING

- Chapman MA et al. Genetic analysis of floral symmetry in Van Gogh's sunflowers reveals independent recruitment of CYCLOIDEA genes in the Asteraceae. PLoS Genet. 2012;8(3):e1002628.
- Moore L et al. The mutational landscape of human somatic and germline cells. Nature. 2021 Sep;597(7876):381–386
- Stock JT, Wells JCK. Dairying and the evolution and consequences of lactase persistence in humans. Anim Front. 2023 Jun 14;13(3):7–13

2.5 Questions

1. **Which of the following statements are true?**
   (A) Frameshift mutations (e.g. insertions or deletions) that change how a DNA sequence is read are always beneficial.
   (B) The same mutation can be beneficial in one environment but neutral or deleterious in another.
   (C) Mutations in somatic cells are passed on to the offspring of an organism.
   (D) Mutations that occur in non-coding DNA never have any effect on fitness, as they are not translated into proteins.
   (E) A single nucleotide substitution can have a drastic effect.
2. **The HBB gene codes for a protein called β-globin, which is a component of haemoglobin. Which of the following modifications of the DNA are likely to change the amino acid sequence of the haemoglobin protein?**
   (A) Deletion of one nucleotide from an exon in the HBB gene
   (B) Insertion of ten nucleotides into the second intron of the HBB gene
   (C) Changing a TCT triplet into a TCC triplet in an exon of the HBB gene
   (D) Insertion of ten nucleotides in the regulatory region of the HBB gene
   (E) A mutation in a skin cell that deletes ten nucleotides from the exon of the HBB gene
3. **The genetic code is universal across all living organisms. Which of the following statements are consistent with this fact?**
   (A) DNA was the first genetic material to arise.
   (B) All living organisms produce the same proteins.
   (C) The same codons translate into different amino acids in different organisms.
   (D) One amino acid is coded for by one specific codon across all organisms.
   (E) In all animals, AUG codes for the amino acid methionine.
4. **Polar bears share a common ancestor with brown bears. It has been hypothesised that a small population of bears became isolated from other members due to glacial movement. The white colour of the fur is caused by a mutation that was not present in the ancestral population. What could explain the fact that all polar bears today have white fur?**
   (A) Some individual(s) in the isolated population acquired the mutations for white fur, conferring higher fitness to their offspring.
   (B) At the time of the population split, there were few individuals carrying the mutation for white fur. Later, their offspring were able to better camouflage in the snow and had more surviving offspring.
   (C) The brown colour was disadvantageous for bears in the snowy environment, so a mutation causing white fur appeared in response.
   (D) The brown bears needed to blend in with their new environment, so their colour became lighter and lighter over the generations, ending in a bright white that is perfectly suited to the snowy region.
   (E) Newborn polar bears have brown fur that changes to white as they get older in response to environmental conditions.
5. **Which of the following statements is/are true?**
   (A) Comparing molecules of the same length, DNA contains twice as much genetic information as RNA.
   (B) Mutations are caused only by external factors.
   (C) Each amino acid is coded only by one specific codon (sequence of three nitrogenous bases).
   (D) Each codon codes for just one amino acid.
   (E) Inclusion of one nucleotide into the exon part of m-RNA can alter the resulting protein functionality.
6. **Some scientists suggest that RNA played an important role in the early evolution of life, because:**

(A) It can catalyse some chemical reactions.
(B) It can store genetic information.
(C) It contains ribose, instead of deoxyribose.
(D) It is single stranded, so there is twice as much of it as DNA, which is double stranded.
(E) Viral genes are found in human DNA.

2.6 Project: Obtaining DNA from Local Fruits or Vegetables

Since the discovery of DNA's structure and its crucial role in living organisms, molecular biology has advanced rapidly. Researchers have uncovered vast amounts of knowledge—not only about DNA, RNA, and proteins but also about the processes that drive life and evolution. At the heart of each of these discoveries was a compelling question and a biological sample from which scientists first needed to extract DNA for analysis.

In this activity, you will carry out the procedure of DNA extraction from biological material and consider the factors that influence both the amount of DNA extracted and the extraction process itself.

Part 1: Preparation

Questions

1. Consider the fruits and vegetables you can find in your local grocery store. Which of them are your favourites? Can you imagine being interested in any of these plants as a researcher? Would you be interested in studying their shape, size, colour, or taste?

Most eukaryotes are diploid, meaning they have two copies of each chromosome. However, several commonly grown crops are exceptions to this rule.

2. Choose three fruits or vegetables that differ in colour, shape, taste, and tissue texture of the edible part. Conduct a brief survey to determine the following for each of the studied fruits, and summarise your findings in a clear table.

- The number of chromosomes in one chromosome set—the haploid chromosome number (n)
- The ploidy—the number of sets of chromosomes in the nucleus of a single cell

3. How do these findings relate to the amount of DNA present in the cells of your chosen fruits or vegetables?

The quantity of DNA available from a sample is an important factor in molecular biology, since it determines whether further analysis can be performed successfully. In this activity, using only basic materials, you can begin to explore these differences first-hand.

4. Besides chromosome number and ploidy, what other factors could influence the amount of DNA you are able to extract from your samples?
5. Formulate a hypothesis predicting the relative amounts of DNA you expect to extract from equal-sized pieces of your three selected fruits or vegetables.
   - From which fruit do you expect to obtain the largest amount of DNA?
   - From which fruit do you expect to obtain the smallest amount of DNA?
   - Why do you think so?

6. Explain the requirement to use "fruit samples of the same size", and describe how you will ensure your samples are equal in size to allow a fair comparison.

Part 2: Work in the Lab (or Kitchen)

Test your hypothesis experimentally:
Isolate DNA from samples of the same size. For each fruit or vegetable, work with at least three different samples.

Materials and Equipment

- Resealable plastic bag
- Biological sample—DNA source
- Filtration equipment:
  - Cheesecloth (use in several layers) or coffee filter
  - Funnel
  - Test tube
- Ice-cold ethanol (chilled in the freezer for at least 1 hour before use)

- Beaker or a glass cup
- Thin wooden stirring rod
- DNA extraction buffer:
  - 4 ml (1 full teaspoon) of dishwashing detergent (e.g. dish soap)
  - 1.5 g kitchen salt (NaCl)
  - 5 g baking soda (NaHCO3)
  - 120 ml water (preferably distilled)

Procedure

1. Prepare the DNA extraction buffer in the beaker using the listed ingredients. Stir slowly and carefully to avoid creating too many bubbles.
2. Crush the sample by squeezing it in a plastic bag. Mash each piece of fruit or vegetable for at least 2 minutes.
3. Add 10 ml of the DNA extraction buffer to each piece of fruit or vegetable, and mix well for at least 1 minute.
4. Remove cellular debris by filtering each mixture through cheesecloth or coffee filter into a funnel, which leads into the test tube (Fig. 2.2).
5. Try to obtain 2–3 cm of filtered liquid in the test tube. Discard any extra mashed specimen pulp with the cheesecloth.
6. Slowly add cold ethanol along the wall of the test tube, until the ethanol layer is approximately 2 cm high. The ethanol should form a separate layer on top of the filtered extract.
7. Dip the stirring rod into the test tube where the ethanol and extract layers meet (Fig. 2.2). Gently rotate and move the rod. Hold the test tube at eye level to observe what is happening. Watch the DNA as it begins to precipitate (clump together).

Part 3: Results and Analysis

1. Take pictures of the test tubes with the results and reflect on what you observe.
   (a) Did DNA precipitate in each sample?
   (b) Did you observe qualitative or quantitative differences in the visible amount of DNA among your samples?
   (c) What differences did you observe among the three samples of the same fruit?
   (d) What differences did you observe among samples of different types of fruit?
   (e) What methods could a research laboratory use to quantify and compare DNA yields more accurately?
   (f) Do your results support your hypothesis? Explain.
2. Reflect on the individual steps of the procedure:
   (a) Explain the importance of each of the following steps in the DNA extraction procedure:

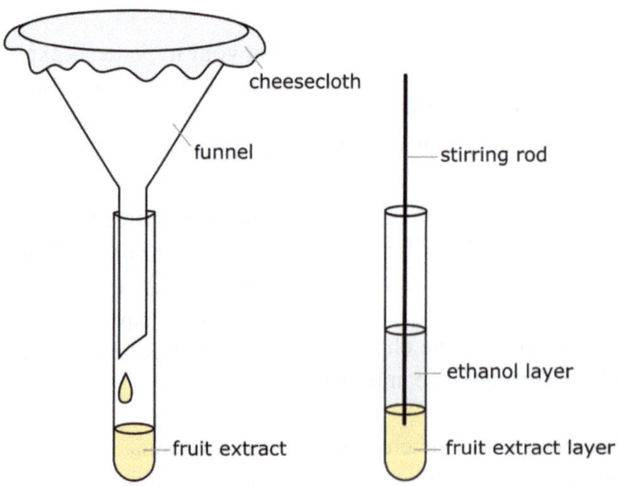

Fig. 2.2 Equipment needed

  - Mechanical crushing of the fruit
  - Mixing the crushed fruit with the DNA extraction buffer
  - Filtering the mixture through cheesecloth
  - Adding ethanol to the filtered liquid

(b) Did you follow the steps of the extraction procedure, or did you modify the procedure? If you modified it, explain how and why.

(c) How might skipping a step or changing the order of the steps affect the DNA extraction process?

3. Summarise what factors—including chromosome number, ploidy, cell type, and tissue composition— influence the amount of DNA in a single cell and in the whole sample.
4. Formulate a research question related to the plants you have been working with that could be addressed through DNA analysis.

This project was developed by Kristína Hudáková. The DNA extraction procedure and drawing of the equipment are adaptations of common methods widely used for educational purposes.

3 Mendelian Inheritance and Beyond

Michal Hledík, Kristína Hudáková, and Barbora Trubenová

▶ Individuals in any population differ in many respects—their physical characteristics, behaviours, or susceptibility to diseases. How can we identify the mutations that cause these differences? We have already learned about the connection between the genetic material we carry and our physical traits, the phenotype. We focused on the emergence of new traits through mutations. In this chapter, we will cover another essential ingredient of evolution: inheritance.

That some traits are heritable is no surprise to anyone. We resemble our parents and siblings, and twins can look so similar that we cannot tell one from the other. Inheritance is present in all animals and plants and is necessary for natural selection to have an effect: parents who possess characteristics that make them fitter than others must be able to pass these characteristics on to their offspring. Similarly, it is essential to agriculture—humans select the seeds from the most nutritious vegetables, hoping their offspring will also possess those qualities.

We already know that long molecules of DNA are the medium of inheritance; these DNA molecules vary slightly between individuals as a result of random mutations. DNA contains genes—instructions for the production of proteins, which are crucial for the functioning of cells and organisms. However, as we will see in this chapter, this is not the whole story. In fact, it just scratches the surface of the large and exciting field of genetics.

Modern technologies allow us to look at the DNA sequence of any cell of any organism in remarkable detail. But this was not always possible. The pioneers of genetics did not even know DNA existed, so observing the DNA sequences of individual animals or plants was out of the question. Instead, it was their ingenious experiments that uncovered the basic principles of genetics and hinted at properties that today we know are associated with DNA. Let's start with the fundamental concepts that emerged from the experiments of Gregor Mendel, a monk and scientist from Brno, in the 1850s.

M. Hledík
Institute of Science and Technology Austria, Klosterneuburg, Austria

K. Hudáková
eduB, Bratislava, Slovakia
e-mail: kristina.hudakova@edub.sk

B. Trubenová (✉)
D-USYS, ETH Zürich, Zürich, Switzerland

Eawag, Dübendorf, Switzerland

eduB, Bratislava, Slovakia
e-mail: barbora.trubenova@env.ethz.ch

B. Trubenová et al. (eds.), *Everyday Evolution: Unveiling Life's Hidden Patterns*,
https://doi.org/10.1007/978-3-032-07050-0_3

3.1 Mendel's Experiments

Mendel's experiments were laborious but simple: he repeatedly bred plants and animals, then recorded and counted the phenotypes. His most famous experiments involved garden peas' flower colour and seed shape. Mendel started with two different (natural) populations of pea plants: one with round and one with wrinkled seeds. Mendel crossed the two populations: he pollinated the flowers of the plants grown from round seeds with pollen from the plants grown from wrinkled seeds and vice versa. All the offspring in the first generation (called $F1$ **hybrids**) had round seeds. However, when he crossed these $F1$ hybrids with each other, their offspring (called $F2$ **hybrids**) had mostly round but sometimes wrinkled seeds. Even more interestingly, the ratio of the round-to-wrinkled seed was nearly always very close to 3:1!

Mendel and others observed this 3 : 1 ratio in $F2$ hybrids for other characteristics in peas and other plants and animals. But other ratios could also be consistently obtained in other crosses. For example, $F1$ hybrids crossed with the wrinkled parental population produce a 1 : 1 ratio of round to wrinkled offspring.

Discrete Alleles, Diploidy and Dominance

Mendel's observations hinted at the basic rules of inheritance. First, the properties of the offspring were not the average of their parents—all seeds were either round or wrinkled. This suggests that whatever substance mediates the inheritance of pea shape only exists in two distinct forms or variants. Today, we call these variants the **alleles** of one **gene**. Different alleles arise due to discrete differences in the DNA sequence.

Second, even though all $F1$ hybrids had round seeds when crossed with each other, they could still produce offspring with wrinkled seeds. This suggests that $F1$ hybrids contained an allele for shape from each of their parents (both round and wrinkled) and that they could pass those alleles on to their offspring—the $F2$ hybrids. Scientists later discovered that most sexually reproducing organisms have two copies of all genes (one from each parent), and we call such organisms **diploid**.

Inheritance Before Mendel

While people had been aware of inheritance long before Mendel, they did not know how it worked. One popular theory from the nineteenth century—the theory of blending inheritance—suggested that any characteristic of the offspring is intermediate between the characteristics of the two parents. Parents were assumed to pass on some heritable substance or "essence" to the offspring through their gametes. The essence would blend when the two parents' gametes combined to form the embryo. However, this theory has some obvious problems. For example, it cannot explain discrete characteristics like blood type or sex. Moreover, blending inheritance would cause any population to become more and more uniform over time, and thus cannot explain why populations maintain so much variety.

Third, not all alleles have equal influence. As explained above, $F1$ hybrids carry one allele for wrinkled and one for round shape, but they still produce round seeds. A wrinkled allele means that the gene that produces the starch branching enzyme in peas is dysfunctional. However, as long as individuals carry at least one allele for round shape, they can still produce enough quantity of the enzyme to convert sugars into starch, which makes their seeds round. Because the round allele masks the effect of the wrinkled allele, we call it **dominant** and denote it with a capital R. The wrinkled allele is called **recessive** and is denoted with a small r. Recessive alleles show an effect only if they are present on both chromosomes of a diploid individual. In that case, no starch branching enzyme is produced, and seeds become wrinkled.

This brings us to another important concept. Individuals who carry two copies of the same

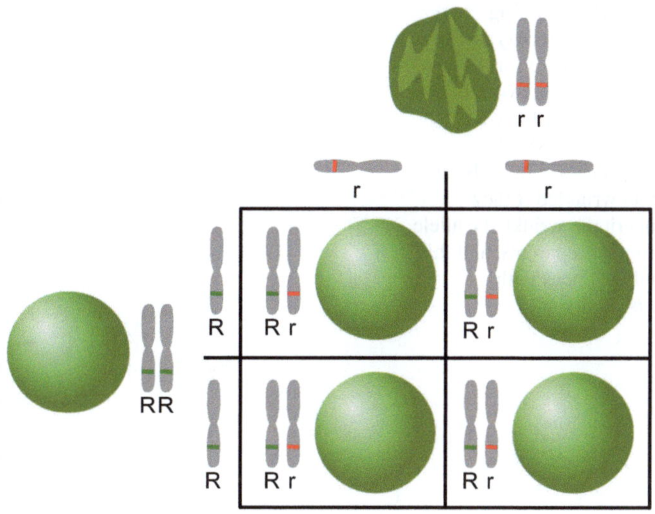

Fig. 3.1 Crossing between homozygotes. A cross between the two parental plants: round and wrinkled. Each parent can pass on either of the two alleles it carries; this is represented by the two rows and two columns of the table. Since the parents are homozygous, RR and rr, the offspring ($F1$ hybrids) are heterozygous, Rr, in all cases. This type of diagram is called a Punnett square

allele of a gene are called **homozygous** (at that gene); if the maternal and paternal alleles differ, they are **heterozygous**. The plants with round or wrinkled seeds in the parental populations were all homozygous for round or wrinkled alleles, respectively. That is why their cross, the $F1$ hybrids, were all heterozygous, inheriting one round allele from one parent and one wrinkled allele from the other, which resulted in round seeds (Fig. 3.1). Individuals in the second generation, the $F2$ hybrids, can inherit either a round or wrinkled allele from either parent, so some of them will be homozygous for recessive alleles, some for dominant alleles, while others are heterozygous. The 3 : 1 ratio of round to wrinkled offspring in the second generation of hybrids hints at the importance of probability: there are four possibilities for allele combinations, all equally likely, and only one results in a wrinkled seed—an $F2$ hybrid with two wrinkled alleles, one inherited from each of the $F1$ parents (Fig. 3.2).

Mendel's Luck, or It's Not that Simple!

Mendel was actually very lucky[1] to work with peas. First, parental populations in Mendel's pea experiments were homozygous for the traits he observed. Second, not all genes have a dominant and recessive allele; sometimes, heterozygous individuals have an intermediate trait value. For instance, the red pigment in snapdragon flowers is an example of what's known as incomplete dominance. When red (homozygous) snapdragons are crossed with white (also homozygous), their heterozygous hybrid offspring are all pink. Note that colours are not blended—there is just one copy of the allele for the red colour, resulting in less red pigment. The *incompletely* dominant trait is simply expressed less strongly.

Furthermore, the pea traits that Mendel observed are encoded by a single gene with only two alleles,[2] in a simple dominant-recessive relationship. For many traits, more than two alleles of a gene exist, and the relationships between them are often more complex. Other traits are encoded by multiple genes that can interact with each other. Moreover, some plants are not diploid but polyploid (e.g. tetraploid or hexaploid wheat), meaning that they have a larger number (e.g. four or six) of sets of chromosomes.

[1]Luck is one of the major ingredients in many scientific discoveries.

[2]Actually, more alleles have been recently found, but these were not available in Europe in Mendel's time.

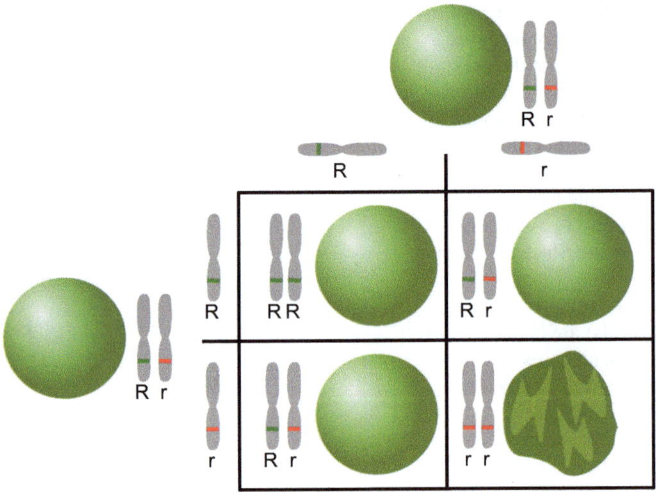

Fig. 3.2 Crossing between heterozygotes. A Punnett square for the cross between two $F1$ hybrids (which are Rr heterozygotes). The offspring ($F2$ hybrids) are only wrinkled if they inherit the wrinkled r allele from both parents, and this only happens with probability $\frac{1}{4}$

3.2 Linkage

A second important discovery was made by William Bateson and his colleagues while studying flower colour and pollen shape in Mendel's peas. They already knew that there were dominant alleles for purple flowers (P) and long pollen grains (L), while red colour (p) and round pollen grains (l) were recessive. When scientists looked only at pollen shape or only at flower colour, everything worked as expected, given Mendel's observations. All $F1$ hybrids were heterozygous (had both types of alleles) and had purple flowers or long pollen grains. The $F2$ hybrids also nicely followed the 3 : 1 ratios of purple to red flowers or long-to-round pollen grains.

But what happened when both traits were studied *together*? Bateson crossed two parent populations: one with purple flowers and long pollen ($PP\,LL$) and the other with red flowers and round pollen grains ($pp\,ll$). Thus, each population was doubly homozygous, i.e. homozygous for *both* traits. Now, suppose that alleles for flower colour and pollen grain are inherited *independently*. Then it is possible to show that, in the F2 generation, there should be a ratio of 9 : 3 : 3 : 1 for plants with purple flowers and long pollen, purple flowers and round pollen, red flowers and long pollen and red flowers and round pollen combinations (phenotypes) (see Box: The Punnett Square for details on how to create them).

However, that is not what Bateson observed. Instead, the ratio was 1528 : 106 : 117 : 381. Two parental combinations, the purple flowers and long pollen grains and red flowers and round pollen phenotypes, were much more common than expected. For some reason, the alleles that started out together tended to get inherited together—the alleles for flower colour and pollen shape appeared to be *linked* in some way.

Later, Thomas H. Morgan and his laboratory performed this experiment for various pairs of traits of the fruit fly (*Drosophila melanogaster*). The ratio 9:3:3:1 was found for some pairs of traits, but other pairs showed various degrees of *linkage*. Moreover, they were able to order all genes on a set of lines, such that unlinked pairs of genes were on different lines and closely linked pairs were near each other on the same line.

The lines, as was eventually discovered, correspond to chromosomes—long molecules of DNA that can be seen under a microscope, but only when they coil into compact shapes before the cell divides. In 1913, Morgan's student Alfred H. Sturtevant was the first to map the locations of different genes on the *Drosophila* chromosomes (see Fig. 3.3). When individuals produce gametes (sperm and egg), their maternal and paternal chro-

The Punnett Square

Punnett squares are an essential tool for predicting possible genotypes and phenotypes, as well as their frequencies. Fortunately, using them is not too difficult. Make a 2 × 2 table if looking at one trait or 4 × 4 if looking at two traits. Start by listing all alleles (or possible combinations of alleles, if you are looking at multiple traits at the same time) that offspring can inherit from their mother above the columns of the table. If the mother is homozygous, make sure to include both copies of that allele in the table. Then do the same for all paternal alleles (or combinations), and write them to the left of the rows of the table. Each cell in the table then represents one possible genotype—a combination of the maternal allele from above and paternal allele from the left. Note that some of them might be the same!

For two traits, it works very similarly—just make sure to list all possible combinations of alleles (assuming there is no linkage). Look at the example below. A heterozygous pea plant with yellow round seeds (YyRr, the yellow colour is encoded by a dominant allele Y) crossed with a homozygous plant with wrinkled green seeds (yyrr). The heterozygous plant can create four different combinations (gametes): YR, yR, Yr and yr. In the case of the homozygous plant, all four combinations are the same: yr.

RrYy

rryy

	RY	Ry	rY	ry
ry	RrYy	Rryy	rrYy	rryy
ry	RrYy	Rryy	rrYy	rryy
ry	RrYy	Rryy	rrYy	rryy
ry	RrYy	Rryy	rrYy	rryy

Punnett square with two traits. There are four different genotypes (marked in different colours). They all appear in the table four times, so the genotypic ratio will be 1:1:1:1. There are also four phenotypes present in this table: Yellow round pea (blue genotypes), yellow wrinkled (green genotype), green round (purple genotype) and green wrinkled (red genotype), and the phenotypic ratio is also 1:1:1:1 in this case

mosomes recombine to create new, reshuffled chromosomes that contain some segments from both parental chromosomes (see Box: Recombination, Linkage and Chromosome Distances). If two genes are close to each other on a chromosome, then they are less likely to be separated and will be more likely passed on together to the gamete on one chromosome.

3.3 Mendelian Diseases

Let's see how the basic principles we have learnt so far can be applied to study heritable human traits and diseases in particular. Understanding the genetic basis of a disease is useful for two reasons. First, it enables prediction. If we know that a specific mutation or mutations within a specific gene cause the disease, then genetic tests can reveal the danger before symptoms appear. Second, knowledge of the genetic cause of a disease can help in developing treatment.

Let's take, for instance, Huntington's disease—a rare disease that develops in an individual's thirties and runs in families. It affects the brain and causes psychological problems, involuntary movements and other symptoms. How can we understand its genetic causes? Unlike experiments with peas and flies, breeding

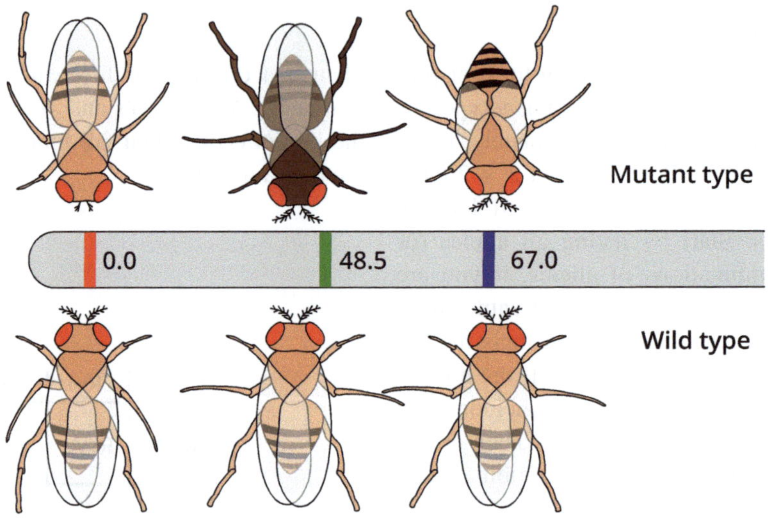

Fig. 3.3 Chromosomal map. Genes for different traits in the fruit fly can be mapped on the chromosome based on linkage. The positions are in centimorgans; see box Recombination, linkage and chromosome distances. (Figure based on work of Twaanders17, CC BY-SA 4.0 https://creativecommons.org/licenses/by-sa/4.0, via Wikimedia Commons and Madboy74, CC0 1.0 https://creativecommons.org/publicdomain/zero/1.0/deed.en via Wikimedia Commons)

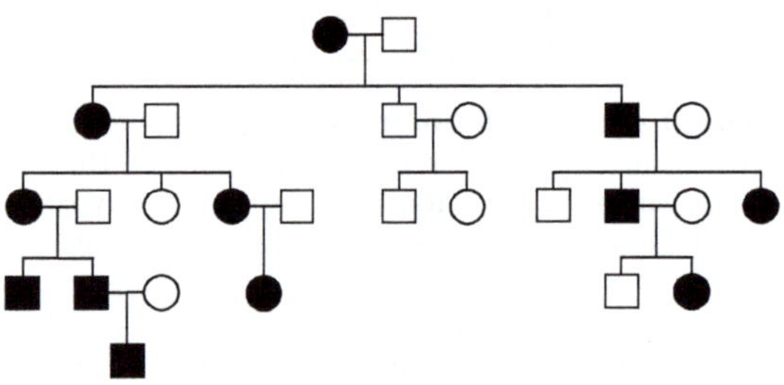

Fig. 3.4 Pedigree of a family with Huntington's disease. Circles represent females; squares represent males. A black symbol indicates that an individual is affected by the disease

experiments with humans are not ethical—instead, we look for patterns in the population.

The first step is to analyse the **pedigrees** of families affected by the disease (Fig. 3.4). For Huntington's disease, every affected individual has at least one affected parent—it does not skip generations. In families with one affected parent, about half of the children have the disease. This pattern exactly matches what we would expect for a dominant trait, as described by Mendel.

Second, we try to locate the responsible dominant mutation in the genome. Because the disease has a dominant inheritance pattern, we expect each patient to carry one disease-causing allele (rarely two). So, in so-called **linkage studies**, we look for polymorphisms in the DNA (see Box: Reading DNA sequences), where one of the alleles is carried by *all the affected* but *none of the healthy* individuals.

If we do this, we will most likely find several such polymorphisms next to each other on one of the chromosomes, forming a small block. This is because of linkage: relatives within each family share long blocks of chromosomes—the more closely they are related, the longer the blocks. The people with the disease will likely have inherited not just the allele that causes the disease but also a block of linked DNA around it. A linkage study can only identify this block—it may still be difficult to tell which one of the polymorphisms within it is responsible for the disease. Perhaps the block contains a small number of protein-coding genes, and researchers can focus their further research on those genes and proteins. If the block is too large, we can narrow it down by conducting studies with larger numbers of patients and healthy individuals.

The cause of Huntington's disease was eventually found. A gene called HTT contains repeats

Recombination, Linkage and Chromosome Distances

Gametes contain only one set of chromosomes—they are haploid. They are formed from diploid cells, and during their formation, recombination can occur. The two chromosomes (maternal and paternal) from each pair align, cross over and exchange DNA material. The result is two recombined chromosomes, each of which contains segments from both parental chromosomes. Each gamete then receives only one set of chromosomes with genetic material shuffled from both parents.

Suppose we pick two genes (or simply two locations) on one chromosome. If they are close to each other, crossovers between them will be rare, and as a result, the maternal (or paternal) alleles of those genes will usually be passed on to the gametes together. This is what Bateson and Morgan observed as linkage. If the genes are far apart, the maternal (and paternal) alleles at those genes may or may not stay together, with almost the same probability.

The average number of crossovers between two genes is a good way to measure the distance between them. Two genes are 1 centimorgan (cM) apart if there is, on average, 1 crossover between them every 100 generation. Human chromosomes are between about 50 cM and 280 cM long. Typically, in humans, 1 cM corresponds to about one million nucleotides, but this varies along the genome and between organisms.

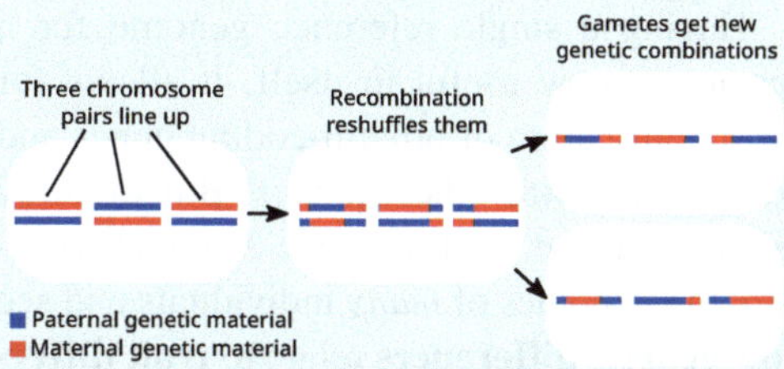

The principle of recombination. Maternally and paternally inherited chromosomes (shown in red and blue) recombine. The two resulting chromosomes contain a mosaic of parental genes. One chromosome from each pair ends up at random in the gamete

of the triplet *CAG*. A person with the disease has too many such repeats, which leads to additional amino acids being incorporated into the amino acid chain. This changes the properties of the resulting protein—it becomes toxic. The toxic allele is dominant since one such allele is enough to produce the toxic protein. Thus, individuals with even one mutated allele develop Huntington's disease.

Huntington's disease is one of the few examples of a dominant genetic disease. Most severe heritable diseases, such as cystic fibrosis, haemophilia and sickle cell anaemia, are recessive. Cystic fibrosis is a lung disease caused by the lack of a functional CFTR gene, which is needed to move salts in and out of cells. Sickle-cell anaemia is caused by a dysfunctional HBB gene, which is necessary for the production of healthy red blood cells.

In all organisms, in fact, many of the harmful mutations are recessive. It is more likely that a mutation makes an important protein-coding gene dysfunctional rather than toxic. Such a mutation doesn't have to be dangerous as long as we have another functional copy. A single non-mutated allele is sufficient to make enough functional protein, and the body will work essentially normally. Only when both our maternal and paternal alleles are dysfunctional do we get into trouble. The dysfunctional allele is, therefore, recessive.

Scientists estimate that on average, a human being carries disease-causing recessive alleles in five or more genes. If a person has only one such allele, he or she does not develop the disease—most people don't even know they carry it. We call such people **carriers**. Two carriers have about a 25% chance (recall the Punnett squares) of having a child with the disease—about one

Reading DNA Sequences

DNA sequencing—determining the sequence of A, T, G and C bases in DNA—is no easy task. Although the structure of DNA was described in the 1950s, sequencing technologies took decades to develop. The first complete human DNA sequence was finished in 2003—the project took 14 years, almost three billion USD, and the collective effort of many institutions. Bacteria with much shorter genomes had been sequenced much earlier. The full sequencing of wheat was only finished in 2018.

Having a single reference genome for a species is very useful in itself. It allows for the identification of protein-coding genes and comparison with other species. But to study the inheritance of traits, we need to compare the genomes of *many* individuals and see how **genetic differences** relate to **trait differences**! These genetic differences help us, for instance, to study family memberships or solve crimes.

This brings us to the *1000 Genomes Project* in which the genomes of over 2500 people from all around the world were sequenced. Scientists identified over 88 million sites on the genome that differed among individuals. These are called **polymorphisms**. Most of these, over 84 million, are **single nucleotide polymorphisms (SNPs)**; these correspond to a single nucleotide change, e.g. some people have a C where others have a T. To characterise another person's genome, we don't need to sequence their complete genome: it is enough to look at the polymorphisms because all the other parts are the same for all humans.

Moreover, because of linkage, nearby polymorphisms are generally inherited together. A subset of one million SNPs evenly spaced across the genome is enough to characterise the full genome of a person to a sufficient level of accuracy, and it is much faster and cheaper than looking at three billion base pairs. The 1000 Genomes Project has created a foundation for further large-scale genomics studies. Other projects, like the "Bird 10,000 Genomes (B10K) Project", are also working to create large-scale genomic data sets for non-human species. DNA sequencing is continuously becoming cheaper, faster and more accessible.

in four of the offspring inherits recessive alleles from both parents.[3] Thus, healthy parents can have a child affected by a heritable disease in the same way that two brown-eyed parents can have a child with blue eyes.

3.4 Dominance, Selection and Evolution

We talked a lot about homo- and heterozygosity, diploidy and dominance. But what does this all have to do with evolution?

The dominance of an allele is not just an interesting quirk of nature that we can observe—it has significant implications for evolution. In particular, it greatly affects how selection acts on genotypes (alleles) over time.

Dominant alleles that influence fitness are always visible to selection. If they are beneficial, they will be selected for. Individuals with such alleles are more likely to have more viable offspring, and eventually, this beneficial allele spreads throughout the population. On the other hand, if the mutation is harmful, individuals who carry it will have, on average, fewer offspring, and the allele will eventually disappear from the population. If the harmful mutation is particularly severe, it is lost immediately, as the organism that carries it will die before leaving offspring

[3]This is the reason why children of related parents are at an especially high risk of inheriting a recessive disease—their parents have an increased chance of being carriers for the same disease.

and passing the mutation on to the following generation.

In contrast, recessive alleles are often shielded from selection by dominant alleles. A recessive allele with a lethal mutation can stay in the population for many generations. Only if an individual inherits two of such alleles does this show in its phenotype and is visible to selection. This happens fairly rarely if the recessive allele is rare in the population. Recessive harmful mutations can stay in the genome for ages, passed from generation to generation for thousands of years. In fact, most diploid organisms carry many lethal mutations that would kill them if they did not have the other working copy of that gene. This is one of the great advantages of being diploid.

However, while diploidy is a good thing for us, it is not such good news for breeders. For instance, imagine a breeder trying to eliminate a heritable disease from his livestock: he always breeds only healthy animals. If the disease is dominant, he will eliminate it after a single generation—all individuals that carry the dysfunctional allele will have developed the disease and will not be bred. However, the situation is more complicated if the disease is recessive. Even if the breeder only breeds healthy individuals, some of these animals are carriers, and thus, some of their offspring will inherit two dysfunctional copies and will be affected by the disease. This can go on for many, many generations.

So far, we have talked about traits influenced by single mutations and by one affected gene. However, there is another class of traits and diseases that are more complex. Take, for example, adult human height. People don't fall into tall and short categories depending on a single allele, and height can take any value on a continuous scale. Such traits are called, aptly, **quantitative** or **complex** traits. Do Mendel's inheritance patterns also apply to such traits?

Yes, they do. Though complex traits are affected by a large number of genes, each of these genes is inherited in a Mendelian manner. If a trait is only affected by one gene with two alleles, everyone in the population can have one of three possible allele combinations, AA, Aa or aa, giving only three different trait values (or two if there is dominance). But if the trait is affected by ten genes, there are $3^{10} = 59{,}049$ different combinations of alleles, and each combination can lead to a slightly different trait value. As a result, many traits vary almost continuously, even though individual genetic factors are distinct Mendelian alleles. But let's leave this important topic for Chap. 4!

DEFINITIONS

Carrier: an individual who possesses a recessive allele for a trait but does not express the trait phenotypically

Dominance: condition where an allele of a gene masks the expression of another allele at the same locus

Recessivity: condition where an allele's expression is masked by the presence of a dominant allele

Recombination: process by which genetic material is exchanged between homologous chromosomes, resulting in new allele combinations

Linkage: the tendency of genes located close together on a chromosome to be inherited together

SUGGESTED READING

- Mersha TB. From Mendel to multi-omics: shifting paradigms. Eur J Hum Genet. 2024 Feb;32(2):139–142
- Jain M et al. Nanopore sequencing and assembly of a human genome with ultra-long reads. Nat Biotechnol 36, 338–345 (2018).

3.5 Questions

1. **A man has sickle cell anaemia, while his wife is a carrier (but does not have the disease herself). Which of the following statements are true?**
   (A) All of their offspring will have the disease.
   (B) Any one of their offspring will either have the disease or will be a carrier.
   (C) The couple has two offspring. If one of these offspring has the disease, then the other offspring must be disease-free.
   (D) All their grandchildren must have the disease.
   (E) All their grandchildren will either have the disease or be carriers.
2. **Now consider a situation where the man has Huntington's disease, while the woman is disease-free. Which of the statements are correct now?**
   (A) All of their offspring will have the disease.
   (B) Any one of their offspring will either have the disease or will be a carrier.
   (C) All their offspring will have Huntington's disease, provided that the man is homozygous for the allele that causes the disease.
   (D) Any of their offspring will either have the disease or be carriers, in case the man is homozygous.
   (E) A grandchild of the couple has at least a 50% chance of having Huntington's disease, if the man is homozygous.
3. **Oculocutaneous albinism is a recessive disease, with the affected gene lying on chromosome 11. A healthy man, the son of two healthy parents, has a sister affected by this disease. He marries a woman with albinism. What can you say about this man's family?**
   (A) The man must be a carrier of albinism.
   (B) The parents of this man are carriers.
   (C) All of this man's children will have albinism.
   (D) All of his sister's children are carriers or have the disease.
   (E) All their children will be either carriers or have the disease.
4. **A breeder of rabbits decides that he wants to have all of his rabbits white (recessive trait) with floppy ears (dominant trait[4]). He selected all white rabbits with floppy ears and let them mate for two generations. Which of the following scenarios *are possible* in the following generations?**
   (A) In the first generation of offspring, all rabbits are white.
   (B) In the second generation of offspring, some rabbits are white, and some brown.
   (C) In the second generation of offspring, some rabbits have floppy ears, and some have normal ears.
   (D) In the second generation, the breeder could observe 9:3:3:1 ratio of different trait combinations.
   (E) It was impossible to observe any changes in population composition after such a short time.
5. **Look at the pedigree in Fig. 3.5. Female A had four daughters (C, E, F, H) and one son (J) with male B. What can you say by looking at this pedigree? Assume that no new mutations appeared in this family and the traits are either completely recessive or completely dominant.**
   (A) The trait is dominant.
   (B) Individual C is a carrier.
   (C) Individual G is a carrier.
   (D) We don't know if K and L are carriers.
   (E) Male B in the first generation is a carrier of the trait.

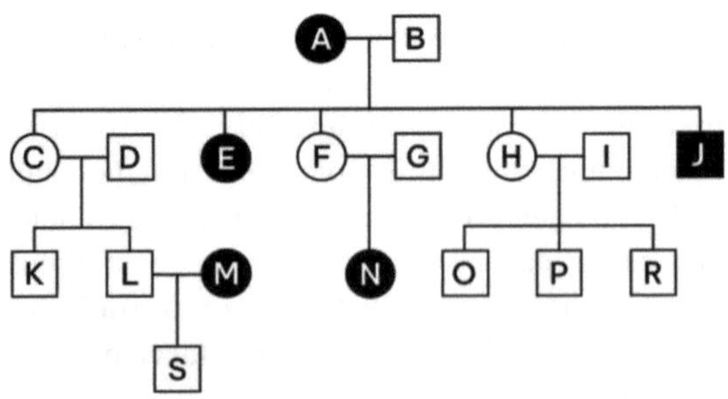

Fig. 3.5 Pedigree. Squares represent males, circles females. Black symbols represent individuals with a particular trait

[4] The true genetics behind bunny ears is actually more complex and not yet fully understood.

6. **Consider four genes, W, X, Y and Z, located on the same chromosome. Based on an experiment, scientists have estimated the probability of a crossover occurring between the pairs of genes as follows:**
   W - X ... 1.97%
   W - Y ... 6.02%
   W - Z ... 4.89%
   X - Y ... 4.15%
   X - Z ... 2.96%
   Y - Z ... 1.15%
   How are these genes located on the chromosome?
   (A) W—Y—X—Z.
   (B) Y—Z—X—W
   (C) Z—Y—X—W
   (D) Y—X—Z—W
   (E) W—X—Z—Y
7. **Fruit flies living in nature typically have long wings. Occasionally, mutants occur, with either short wings or curly wings. Suppose you cross a fly with short wings with a wild type with typical long wings. The F1 generation—the offspring of these two flies—all have typical long wings. Then, you let the F1 individuals interbreed and get 114 individuals of the F2 generation. If we know that the wing length is a trait following Mendelian inheritance, which of the following is correct?**
   (A) The mutation for short wings is dominant, and all F2 individuals will have short wings.
   (B) The mutation for short wings is recessive, and all F2 individuals will have short wings.
   (C) The long wings are coded for by a dominant allele.
   (D) The short wings are coded for by a recessive allele, and half of the F2 individuals will have short wings.
   (E) The allele coding for the wild-type (long) wings is dominant, and three-quarters of the F2 individuals will have long wings.
8. **Codominance means that in heterozygotes, both alleles are expressed, leading to a heterozygous phenotype, which is a combination of phenotypes of both homozygotes. An example would be blood groups. There are three alleles, A, B and 0, that code for four blood groups: A, B, AB and 0. Keep in mind that any human (being diploid) can carry only two alleles. Two types of interactions exist between these alleles: A is dominant to 0, and B is dominant to 0, but A and B are codominant. Peter, with blood group 0, is married to Ana, who has the AB blood group. Which blood groups can their children have?**
   (A) 0
   (B) A
   (C) B
   (D) AB
9. **Imagine that their daughter Maria is married to Jacob with the same phenotype for blood group as hers. What could be the phenotypes of their children?**
   (A) 0
   (B) A
   (C) B
   (D) AB

3.6 Project: Some Basics of Dragon Breeding

On the remote islands of Uramami, different dragon families live on two different islands. Dragons are ancient creatures, unfriendly to strangers, and usually breed only within the same island. However, recently, two specimens—one from each family—were captured by scientists and are now living in the small Zoo in Bratislava (Fig. 3.6).

Magnecia, a young female, is a small dragon with a shy and friendly nature. She is bright red with large, membrane-like wings that easily carry her lithe body. Despite her small size, she can be dangerous—her tail is equipped with sharp spikes. The curled horns on her head are small, with a decorative, rather than defensive, function. Magnecia is unable to breathe fire but belches large amounts of sulphur vapour.

Drakonikus is a young male with a large, bright yellow body. Despite his size, he is very friendly, curious and charming. His head is also equipped with small, curled horns, but his tail is spikeless. Drakonikus does not have any wings and cannot fly, but he can face any danger with his spectacular ability to breathe fire.

Feeling lonely and not knowing their family traditions, the two young dragons fell in love and started a family. There is a lot of excitement at the Bratislava Zoo: Magnecia has already laid four eggs. No one knows what the baby dragonlets will look like!

We don't know much about dragon genetics, but we know that they are **diploid**, all observed traits are either dominant or recessive, and both individuals are **homozygous** for all the observed traits. Table 3.1 provides a summary of the different dragon traits and alleles that encode them.

Part 1: Excitement at the Bratislava Zoo

1. From the information above, determine the genotypes of both dragons from the Bratislava Zoo (Table 3.2). Don't forget that both dragons are diploid and homozygous!
2. Write down possible gametes that each dragon can make. If there are several possibilities, write all of them down.
   Magnecia's eggs:
   Drakonikus' sperm:
3. How many possible offspring genotypes (different combinations of eggs and sperm alleles) are there?
4. Write down the genotype(s) of the first generation of offspring.
5. Describe the phenotype(s) of the first generation of offspring.

Part 2: Introducing the Swiss Dragon

Other zoos in the world are extremely interested in the events in Bratislava. A zoo in Zurich wants to buy the first female that hatches. They plan to call her *Stellaire* and cross her with their own dragon, *Ignitius*. We have only partial information about him. Ignitius is a small dragon with bright red colouring. With strong wings and a slim body, he can cross long distances quickly

Fig. 3.6 Magnecia and Drakonikus

Table 3.1 Description of alleles available for different genes and their phenotypes

Trait	Genetics
Body size	Allele B is dominant and responsible for large body size. Allele b, for small body size, is recessive
Colour	Dominant allele R makes red pigment, and r is non-functional. The absence of the pigment leads to yellow skin colour, while the presence of red pigment results in red skin
Wings	Having two recessive w alleles causes individuals to be wingless
Spikes	Dominant allele S is responsible for spike development
Horns	Dominant allele H leads to the development of large, straight and dangerous horns. The recessive allele h encodes small, curly horns
Fire	F is a dominant allele, giving the individual the ability to breathe fire. If fire is not produced, only sulphur comes out from the dragon's nostrils

Table 3.2 Genotypes of Magnecia and Drakonikus

Dragon name	Body size	Body colour	Wings	Spikes	Horns	Fire
Magnecia						
Drakonikus						

Table 3.3 Ignitius' genotype

Dragon name	Body size	Body colour	Wings	Spikes	Horns	Fire
Ignitius	... ...	RR	w ...	ss	Hh	f ...

and easily. Ignitius breathes fire and consumes his prey, whether grilled or cooked, never raw.

1. From the description above, fill in the missing alleles in Table 3.3 to complete Ignitius's genotype.
2. What can you say about his horns and spikes?
3. Think about what the offspring of Ignitius and Stellaire could have. Assume that they have 16 offspring, and answer the questions below. Hint: You can make a separate small Punnet square for each question.
   (a) How many offspring can you expect to be large and how many small?
   (b) How many of them will be red, and how many will be yellow?
   (c) How many of them will have wings?

Part 3: All About Wings and Spikes

1. Let's peek 200 years into the future! Stellaire and Ignitius have had 40 offspring. Researchers know that the genes for wings and spikes are on different chromosomes. How many offspring with the various genotypes and phenotypes (considering wings and spikes simultaneously) would you expect to see among these 40 dragons?
2. Researchers also know that the genes for wings and body size are on the same chromosome and, moreover, hypothesise that these genes are very closely linked. How many offspring of each genotype and phenotype would you expect if there were no recombination between these loci at all?
3. Three of the 40 offspring are small dragons without wings. How would these have arisen?

Part 4: Selection

1. In nature, the abilities to breathe fire and fly are extremely beneficial for the dragons. The majority of naturally occurring dragons are thus able to breathe fire and fly. From time to time, however, a sulphur-belching dragon appears, even if the parents could produce fire, and something similar may happen with the ability to fly as well. Explain how this is possible.

2. Large body size is fairly disadvantageous for dragons at present—they are easy to spot and are often captured by people. What would you expect to happen in nature with respect to the body size of dragons? Write a short scenario based on what you know about genetics and evolution to support your hypothesis.
3. In the past, many breeders wanted to have so-called "royal" dragons: yellow dragons that can breathe fire. They kept and bred only royal dragons. Which of the following statements are true?
   (A) After a single generation, the breeders got a clan of yellow dragons.
   (B) It was impossible to guarantee that all the dragons were yellow, as this is a recessive trait.
   (C) After a single generation, the breeders got a clan of dragons that could all breathe fire.
   (D) It is likely that some of the dragons in the new generation belched sulphur.
   (E) The breeders discovered that it was impossible to breed individuals for two traits simultaneously.

This project was developed by Barbora Trubenová and Kristína Hudáková.

Complex Genetics of Complex Traits

4

Michal Hledík, Himani Sachdeva, and Barbora Trubenová

► Many traits, such as human height or an individual's risk of developing type 2 diabetes, are known to be strongly influenced by genes, but geneticists have struggled to identify where in the genome the responsible genetic variants are located. These traits, which typically vary on a *continuous* scale, appear (at first glance) to be quite different from what Mendel had described. In this chapter, we will look at the inheritance and evolution of such "complex" traits and at the large-scale initiatives that aim to discover the genes that influence them.

Many of the phenotypes that we have encountered in previous chapters have a *simple* genetic basis. Take the example of seed shape in Mendel's peas or blood type in humans or rare diseases such as sickle cell anaemia: these are all controlled by mutations in a single gene or a very small number of genes. There is the "wrinkled seed allele" of the $SBE1$ gene in peas, three common alleles (the A, B and O alleles) of the ABO gene that code for blood type, or the "sickle cell allele" of the HBB gene. This may give the impression that for each genetically determined trait, there is *one gene* that encodes the phenotype. Examples where this is true are popular for a good reason—they nicely illustrate the basic principles of Mendelian inheritance. However, the genetics of traits actually gets more complex when we go beyond these classic examples. Is there a gene for being tall? A gene for high IQ? Or a gene for autism? We know that all three of these are highly heritable, meaning that they are, to a large extent, determined by our genetic material. But what do we know about the genes that influence them?

Let's talk about adult human height. People don't seem to fall into tall and short categories depending on a single allele. Height can take any value on a continuous scale and is therefore called a **quantitative trait** (because it is described by a number, not a category). Do Mendel's laws of inheritance even apply to such a trait? The short answer is yes. Traits such as height are influenced by a large number of genes, each of which is inherited in a Mendelian manner. For this reason, quantitative traits are also sometimes referred to as **polygenic** or **complex** traits.

M. Hledík
Institute of Science and Technology Austria, Klosterneuburg, Austria

H. Sachdeva (✉)
Faculty of Life Sciences and Faculty of Mathematics, University of Vienna, Vienna, Austria
e-mail: himani.sachdeva@univie.ac.at

B. Trubenová
D-USYS, ETH Zürich, Zürich, Switzerland

Eawag, Dübendorf, Switzerland

eduB, Bratislava, Slovakia
e-mail: barbora.trubenova@env.ethz.ch

B. Trubenová et al. (eds.), *Everyday Evolution: Unveiling Life's Hidden Patterns*,
https://doi.org/10.1007/978-3-032-07050-0_4

In this chapter, we will explore the inheritance and evolution of this rich class of traits. While we will mostly be talking about height as a classic example of a quantitative trait, everything we say also applies to a large class of so-called **complex diseases**. Despite not varying on a continuous scale (like height), these common diseases, such as autism, type 2 diabetes or schizophrenia, are also influenced by a large number of genes. In their case, we can think of the risk of developing the disease as an underlying quantitative trait.

4.1 Combining Contributions of Many Genes

How exactly does a large number of underlying genes lead to continuous variation in height across people? If a trait is affected by only one gene with two alleles, everyone in the population will have one of three possible allele combinations, aa, Aa or AA, giving at most three different trait values (or two, if one of the alleles is fully dominant). But if the trait is affected by ten genes with two alleles each, there are $3^{10} = 59{,}049$ different combinations of those alleles, and each combination could lead to a slightly different trait value. On top of that, most such traits are also affected by various non-genetic factors (diet, climate, lifestyle, etc.), which we refer to as the *environment*. As a result, many traits vary continuously, even though their individual genetic factors are distinct Mendelian alleles. Let's take a closer look at how this could work.

Imagine a hypothetical species, *Homo hypotheticus*, whose adult height is determined by ten genes with two alleles each: a and A at gene 1, b and B at gene 2 and so on to j and J at gene 10. In this hypothetical species, an individual carrying only the lower-case alleles at all genes ($aa\,bb\,cc\ldots jj$) would be 175 cm tall. Let's say Joe is such an individual,[1] and we call these the *reference alleles*.

[1] Joe might not actually exist in our population, but it makes things easier if we can compare everyone else to such a reference genome.

But most individuals in our population carry the *alternative* (upper-case) allele at at least some of the genes (in one or two copies), and this will make them taller or shorter than Joe. For example, each copy of the A allele of gene 1 makes a person 2.9 cm shorter, while each copy of the F allele of gene 6 makes them 3.3 cm taller. This is shown in Fig. 4.1.

This genetic variation leads to phenotypic variation in height. Most members of our population are around average, carrying some height-increasing and some height-decreasing alleles. But if someone happens to carry the height-increasing allele at many genes, he or she will be exceptionally tall. However, as we see in the histogram in Fig. 4.2, there will only be a few very tall (or very short) individuals.

Homo Hypotheticus vs. *Homo Sapiens*

There are several ways in which *Homo hypotheticus* differs from *Homo sapiens* or any other real organism.

One difference is that in reality, many traits are influenced by many more than ten genes—sometimes thousands of them. But more importantly, we don't know which genes they are or what their allelic effects are. We know that humans have millions of polymorphisms—places in the genome where some people are different from others. How can we even find the ones that are important for height or a particular disease? We will discuss this towards the end of the article.

Second, the height of real humans is also influenced by **environmental factors**. Environmental factors are even harder to pin down than the genetic ones: whereas the genotypes of different individuals can simply be sequenced and compared, measuring all potentially relevant environmental influences (including those before birth and in childhood) is virtually impossible. This is why we usually treat environmental effects as random (unpredictable)—for example, in the updated table in Fig. 4.3, we simply treat the trait value (height) of any individual as the sum of the genetic effects and a random environmental

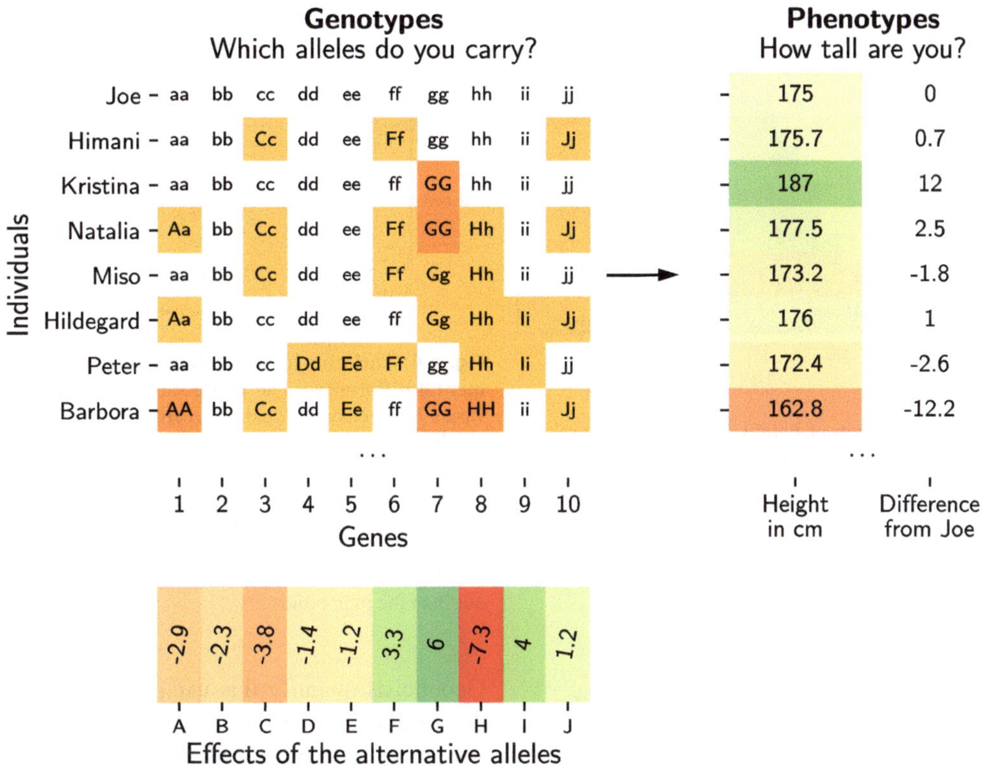

Fig. 4.1 **How genes determine height under the additive model.** The top left shows individuals with their genotypes: for example, Himani carries a single alternative (uppercase) allele at genes 3, 6 and 10, and Kristina carries two alternative alleles at gene 7. The top right shows how tall they should be if we consider only the effect of genes (first column) and how much taller or shorter they are than Joe (second column). We can predict anyone's height by taking Joe's height as the reference and adding to it the effects of all the alternative alleles that they carry. The allelic effects are at the bottom. For example, Himani's height is $175 - 3.8 + 3.3 + 1.2 = 175 + 0.7 = 175.7$. In this example, the reference alleles are more common in the population than the alternative alleles—but this does not have to be the case, as any allele can be chosen as the reference

effect (which may reflect the effects of diet or exercise). But despite the difficulties of measuring or even defining an individual's environment, we can at least quantify how important environmental effects are relative to the genetic effects in explaining differences between individuals—we will discuss this in a moment.

The third simplification we made is by taking the genetic component of height to be the *sum* of the effects of different alleles. This is called **the additive model**—because the phenotype is predicted by simply adding up the effects of all the variants that influence it. In reality, it can be more complicated. Take dominance: if an allele A is dominant, then carrying two such alleles (AA) has the same effect as carrying a single allele (Aa)—not twice the effect. If A is recessive, carrying a single allele would have no effect at all. Sometimes, the effect of an allele can depend on the rest of the genotype—this is called **epistasis**. For instance, it could be that B at gene 2 decreases height only if the individual is AA at gene 1. Luckily for geneticists, such complications are the exception rather than the rule and don't contribute much to variation in complex traits.[2] Thus, the additive model describes the relationship between genotype and phenotype quite well in most cases, providing useful predictions that we will discuss later.

[2]In contrast to Mendelian traits influenced by a single gene, where dominance can contribute significantly.

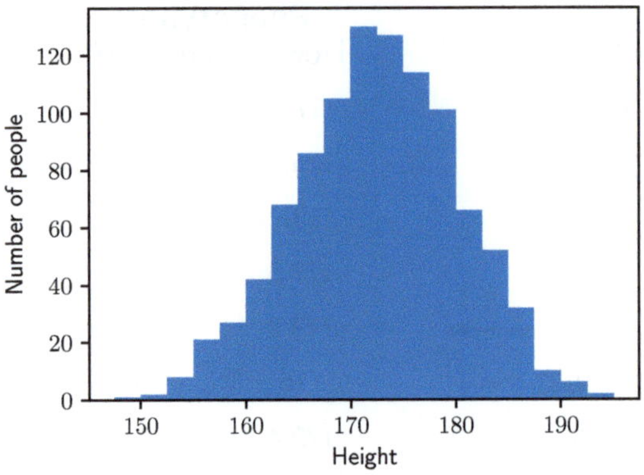

Fig. 4.2 Distribution of height across 1,000 members of the Homo hypotheticus population. Most individuals are around the average of about 172.5 cm, but the tallest and the shortest ones differ by more than 20 cm. There are, however, only a few of them

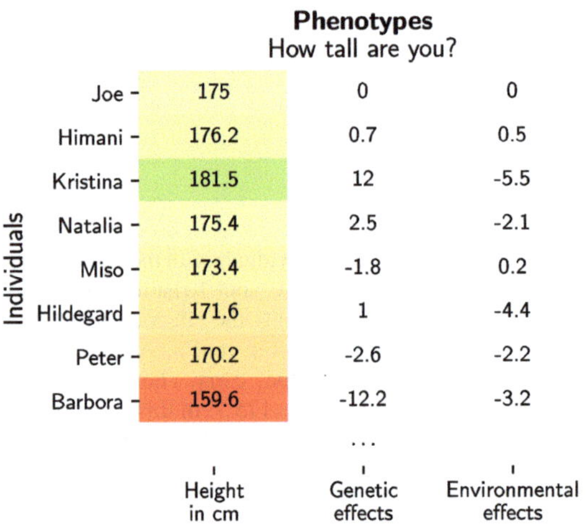

Fig. 4.3 Height of Homo hypotheticus including both genetic and environmental effects. The genotypes and allelic effects are the same as in Fig. 4.1, but now we add the environmental effects to the genetic effects. For example, Himani's height is that of Joe, plus the effects of all her alternative alleles, plus a random environmental effect: $175 + 0.7 + 0.5 = 176.2$

4.2 Nature vs. Nurture

To what extent are our traits determined by our genes, and to what extent by our environment?[3]

[3]Remember that environment has a broad meaning here—it includes factors such as lifestyle, diet, climate throughout life and anything else that is not genes.

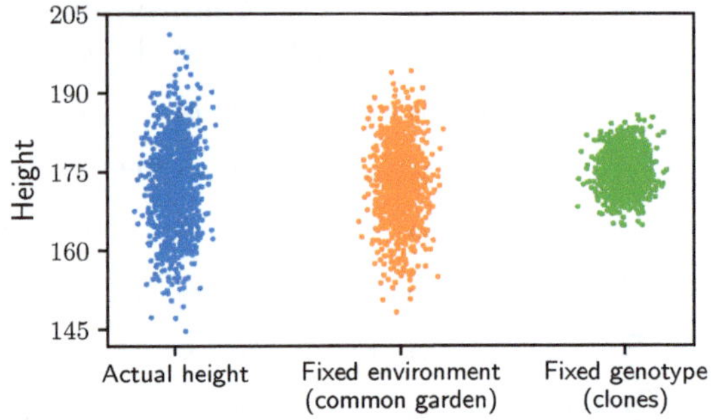

Fig. 4.4 Variation in height due to genetic vs. environmental differences. The figure shows how much height varies between individuals in reality (blue points), how much it would vary if the environment were fixed and all variation was due to genetic differences (orange points) and how much it would vary in a population of genetically identical clones that experience different environments (green points). In this case, heritability is high, $h^2 \approx 0.8$ or 80%, as the orange points are only slightly less spread out than the blue points

Geneticists quantify this using a quantity called the **heritability** of the trait. In the table above, height is determined as the sum of genetic and environmental effects, both of which vary across the population. Now imagine that all these individuals had grown up in *exactly* the same environment—eating the same food, playing the same sports, breathing the same air and so on. This would be hard to do with humans, but it is the basis of *common garden* experiments. In such experiments, scientists take seeds from different plants of the same species that normally live in different environments (e.g. some in the mountains and others near a river) and grow them in the same garden. If the offspring of the mountain and the river plants are different, even when grown together in the common garden, it indicates that these differences are at least partially genetic (Fig. 4.4).

In a perfect common garden, all environmental effects would be the same across individuals, and all differences between them would be genetic. These differences can be quantified using variance, which basically measures how much a bunch of numbers vary around their mean. More precisely, it is the average squared deviation from the mean. The heritability of a trait can then be calculated using the formula

$$h^2 = V_G / V_P$$

where h^2 is the heritability, V_P is the variance in the phenotype and V_G is the variance of the genetic effects (the variance that would remain in a perfect common garden).[4] If heritability is 0, then inherited genes play no role in phenotypic differences, while if it is 1, then the genome completely determines the trait, and the environment has no effect.

Estimating the heritability of a real trait in real humans is not easy since in practice, we do not know the genetic and environmental components of the trait value of each person (and hence, we cannot calculate the variances). Nor can we plant people in a common garden to see how much more similar they become. But there is a way around this problem, and it has to do with similarities between relatives, which we will discuss in the following section.

Scientists have measured the heritability for a number of traits—from height ($h^2 \approx 60\%$–80%) and body mass index (40%–60%) to diseases such as type 2 diabetes (between 20%–80%) to psychological traits such as IQ (around 50%). However, there are three things to keep in mind when talking about heritability.

First, a heritability of 80% means that 80% of the variation between individuals is explained by inherited genetic differences; it does *not* mean that an individual's height is influenced 80% by genes and 20% by the environment. Even if heritability is 1, environmental factors such as nutrition can still be involved, but they may simply affect everyone in the same way.[5]

Second, heritability is specific to the population in which it is measured. For example, estimates of the heritability of short-sightedness vary from 0.5 to 0.9 in different populations across the world because these have slightly different genetic variants and live in very different environmental conditions.

Third, even if someone carries many risk-increasing alleles for a highly heritable disease, a healthy lifestyle and medication can still help prevent or alleviate the disease.

Like Father, Like Son

Phenotypic similarity between relatives can be used to estimate heritability—but these similarities are interesting in themselves. Relatives share a part of their genome. Because of this shared portion, they will be similar in height. But not identical—there will be differences due to the genetic material that they don't share and due to different environmental influences. Since complex traits are influenced by a large number of genes spread across the genome, closer relatives will be more similar because they share a larger fraction of their genome. For example, individuals share their entire genome with an identical twin, 1/2 of it with parents or full siblings, 1/4 with grandparents or aunts or uncles and 1/8 with first cousins. This fraction is called relatedness (see Chap. 10).

In 1918, the statistician and geneticist Ronald Fisher calculated exactly how similar a pair of relatives should be, even without knowing exactly which genes are involved. For his calculation, he used the additive model and expressed similarity in terms of correlation coefficients. The **correlation coefficient** ρ measures the similarity between two variables (such as the height of a parent and the height of the offspring). It is a number between -1 and 1. If it is zero, the two variables are independent. If it is close to 1, one variable being high means that the other will tend to be high as well (taller parents tend to have taller children).

As a rule of thumb, the similarity between a pair of relatives is higher if the trait is more heritable and if they are more closely related. This can be expressed mathematically: the correlation coefficient between the trait values of two relatives is the heritability of the trait times their relatedness r, which is the fraction of the genome

[4]The fact that heritability is h squared and not just h is simply a convention.

[5]The opposite can also happen: some cancers such as leukaemia or lung cancer have a very low heritability ($h^2 \approx 0.08$ and 0.01 respectively), but genes are very much involved. These diseases are often triggered by novel mutations that are not inherited but occur during a person's lifetime.

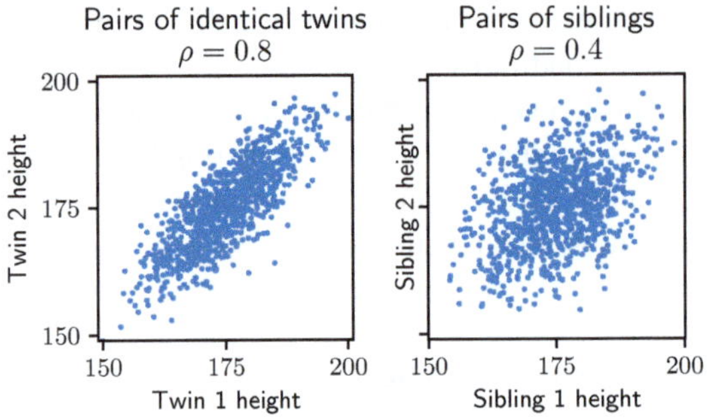

Fig. 4.5 Phenotypic similarity between relatives. Each blue point is a pair of identical twins (left) or siblings (right). The heights of the two twins (left) or siblings (right) are on the X and Y axes. Because of the relatively high heritability of height ($h^2 = 0.8$) and a completely shared genome ($R = 1$), identical twins are very similar in height (correlation $\rho = R\,h^2 = 0.8$), and the points lie close to the diagonal. This pattern is less pronounced but still present in siblings that only share half of the genome on average ($R = 0.5$), resulting in $\rho = 0.4$

they share, i.e. $\rho = h^2\,r$. Figure 4.5 illustrates this for identical twins and siblings.

Predicting phenotypic correlations gets a bit more complicated when relatives also have similar environmental influences, such as when siblings share not only genes but also a home and diet. Another possible complication is **assortative mating**—people tend to find mates who are similar to them. For example, tall people tend to marry other tall people. Thus, the offspring of a tall mother are likely to be tall—not only because of the alleles they inherit from her but also because of the height-increasing alleles they inherit from their father (who will be taller than the population average). As a result, children are more similar to each of their parents than the rule of thumb above would predict.

Fisher's predictions agree well with the observed correlations. By linking the heritability of a trait to the phenotypic similarity between relatives (which can be measured), Fisher demonstrated that Mendelian patterns of inheritance apply to complex traits and also established the additive model as a good approximation to the true relationship between genotype and complex traits. This also allows us to estimate heritability from the similarity between various pairs of relatives. This is most commonly done by contrasting the degree of similarity between identical twins ($r = 1$) and fraternal twins ($r = 0.5$). Both types of twins tend to be similar, both because of shared genes and because they grow up in similar environments. But if identical twins are even more similar than fraternal twins, it must be because they share more of their genome—and the difference in similarity can be used to calculate heritability.

4.3 Heritability and the Evolution of Complex Traits

Not only is heritability a useful measure of similarity between related individuals, but more fundamentally, it governs how quantitative traits evolve under natural or artificial selection. Consider a simple example: suppose a farmer has a crop of maize. She measures the oil content of the grains of each plant (e.g. by sampling a few seeds per plant). The average oil content is some value; let us call it X, but individual plants can vary significantly around this value: a histogram of the oil content of different plants might look something like the one in the top panel of Fig. 4.6. The farmer now selects the top 20% of plants with the highest oil content (those in the right tail of the histogram). The average oil content among these plants is some value Y, where $Y > X$. The farmer uses seeds from just these top 20% to produce next year's crop. Can she predict what the average oil content of next year's maize crop will be?

If oil content is determined entirely by environmental factors, e.g. the patch of soil a plant happens to find itself on, then we would expect the average oil content to revert back to X (since seeds will be planted at random across the same field). On the other hand, if oil content is determined entirely by genes, offspring would typically be very similar to their parents (from whom they inherit oil-increasing alleles), and the average oil content in the next generation would be close to Y, the average among the selected parents.

In general, how much any trait changes from one generation to the next depends on the

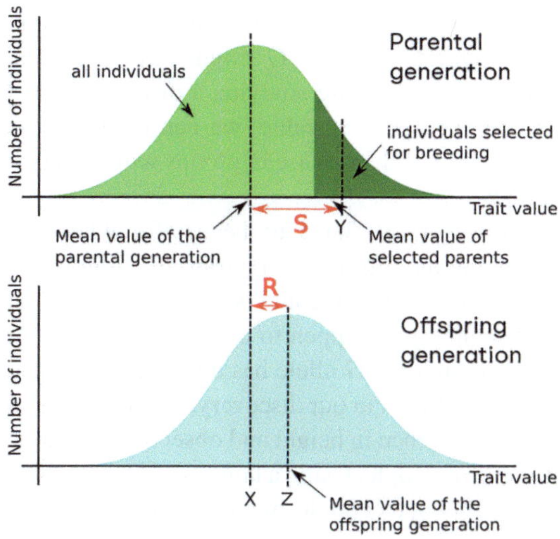

Fig. 4.6 Change in phenotype due to one generation of selection. The distribution of trait values changes from one generation (light green) to the next (blue) when individuals with high trait values (as shown in dark green) are chosen for breeding. The change in average trait value over one generation, R, is typically less than the parents' average (relative to the population average), S, as traits are only partially heritable

heritability of the trait. This is captured by the **breeder's equation**:

$$R = h^2 S$$

Here, R is the difference between the average oil content in this year's crop and the previous year's crop, S quantifies how different the selected parents are on average from the population average (in our example, $S = Y - X$) and h^2 is the heritability of the selected trait (for oil content in maize, h^2 is typically 60–70%). This is also illustrated in Fig. 4.6.

The breeder's equation is extremely powerful because it allows animal and plant breeders to estimate how any trait they are interested in (flower or seed size, milk or egg yield, even complex behavioural traits) would change under selection, even without knowing anything about the specific genes that influence the trait. All they need is an estimate of heritability, which they can obtain by measuring the correlation (similarity) between the trait values of relatives in a population. In fact, as we will see in the next section, finding the specific genes that are under selection is extremely difficult.

The breeder's equation predicts the effects of selection quite accurately over a few generations and is widely used in agriculture and animal breeding. It also describes how wild populations (trees or beetles in a forest, humans in a city) respond to natural selection, e.g. selection due to changing environments. However, it is a bit more complicated to use the breeder's equation in the context of wild populations, where it is not always obvious what particular traits are being selected for by the environmental shift and how these traits interact with each other. Nevertheless, the basic principle is the same: the extent to which traits change under selection, whether on a farm or in the wild, depends on how similar offspring are to parents, which in turn depends on how heritable selected traits are.

If we were to select for oil content by breeding maize over hundreds of (instead of a few) generations, could we still use the breeder's equation to predict the change under selection? The general answer is no—heritability, or more generally, the genetic variance of a trait itself changes over such long time scales. Why should this be? Some alleles are lost by chance, while some selected alleles that increase in frequency may eventually fix throughout the population. Further, new mutations that affect the trait also arise over long time scales. Thus, predicting the response to selection over hundreds of generations requires somewhat more sophisticated models than the breeder's equation.

However, there is an even more fundamental question here: can the oil content (or any other selected trait) even continue to increase under long-term selection over hundreds of generations? A partial answer is provided by the Illinois long-term selection experiment, where scientists have been selecting for high vs. low oil and protein content in different lines of corn since 1896. This experiment is still running, and the oil content has continued to increase (or decrease in the second line) at almost the same rate for more than a hundred years now! Scientists hypothesise that this kind of *long-term* response to selection in quantitative traits occurs because there is a *very large number of sites* on the genome which influence the trait. Thus, there is a lot of heritable genetic variation for such traits, which is also replenished by new mutations and is not exhausted even after a hundred years of selection. However, what factors ultimately limit the response to selection over very long time scales remains an important open question in evolutionary biology.

4.4 Can we Find the Genes That Influence Complex Traits?

We have discussed how complex traits are influenced by a large number of genes—which means that they have high levels of heritable variation where the extent of such variation (heritability) can be estimated from the degree of similarity between relatives. But today, we can go further by sequencing the genomes of large numbers of individuals and finding out who carries which alleles, thus identifying polymorphisms in the genome. Polymorphisms are sites in the genome where some people carry a different allele than others. Most of these differences are replacements of a single letter (nucleotide) in the DNA code: e.g. where some have a T, others have an A—these are called *single nucleotide polymorphisms* or SNPs. There are millions of such polymorphisms in the human genome. Can we find out which of these polymorphisms are actually responsible for the genetic variation in complex traits?

Geneticists are trying to do this in so-called **genome-wide association studies**—GWAS. The basic idea is simple: collect genotypes and trait values for many hundreds of thousands (or even millions!) of individuals. For each polymorphism, ask if people carrying different alleles have different average trait values. If yes, we say that the polymorphism is *associated* with the trait and could potentially have an effect on the trait.

For example, suppose we want to know which genetic variants make people taller. We focus on SNP with two alleles, where everyone has either the nucleotide T or G. If people who carry the G are, on average, taller than those who do not, this suggests that the G might have an effect on height.[6] If we are studying a disease, instead of the average trait value, we can compare the prevalence of the disease in groups with different alleles.

However, we have to be careful when announcing the discovery of a trait-affecting mutation. For example, it could be that taller people in our sample just happen to carry more G than T, even though the G allele has no effect on height. To be confident in our discovery, we need to see a large difference in height and observe it in a large sample of people. Using a larger sample allows us to reliably detect alleles with smaller effects.

Moreover, phenotypic differences between different genotype groups might have various causes. If we study the variation of height across Europe, we find that the Danish are, on average, taller than the Portuguese. The alleles that are common in Denmark but rare in Portugal will be associated with height—but this does not mean that these alleles actually make Danish people taller. Maybe they are taller simply because they have a different diet from the Portuguese.

For these reasons, association studies are only the first step in understanding the genetic basis of complex traits. Once polymorphisms associated with trait differences are discovered, biologists investigate whether and how these polymorphisms affect the properties of proteins and cellular processes such as metabolism. One can attempt to study these effects in specific tissues or in genetically modified bacteria or mice.

Since the early 2000s, scientists have been collecting genotypes and phenotypes from many thousands of people to search for associations between polymorphisms and traits or diseases. What have they found?

Take human height. In 2008, a GWAS on a cohort of over 60,000 people resulted in the discovery of 54 polymorphisms that, with high confidence, influence height. At that time, 54

[6] Since people are diploid, we actually have three genotype groups: TT, TG and GG. The polymorphism is associated with the trait if these three groups have different average trait values.

seemed to be a high number. But larger studies that followed discovered many more: in early 2018, a study with about 700,000 participants discovered 3,290 polymorphisms associated with height, while an even larger study with 5.4 million individuals in 2022 identified a whopping 12,111 polymorphisms!

How do all these different polymorphisms actually influence the phenotype? In Chap. 2, we saw examples of mutations in the protein-coding regions of the DNA that affect the *function* of the resulting protein molecules. But most polymorphisms identified by GWAS (over 90% of them) are outside of protein-coding regions of the genome.

While this remains an open question, scientists believe that these non-coding mutations alter how the DNA (near the mutation) coils and interacts with other molecules in the cell nucleus. As a result, nearby genes can become more or less active, changing the *amount* of proteins produced.

Autism

Autism is a well-known disorder with a wide range of symptoms such as learning deficit, impaired social interactions and communication, the prevalence of repetitive movements and behaviours and even sleep problems. These can be accompanied by a wide array of health issues, such as seizures or even gastrointestinal problems. The presence and severity of symptoms vary greatly, which makes it very hard to diagnose autism (or, as we now often refer to it, autism spectrum disorder). There is no simple clinical test, such as a saliva test, an MRI or a blood test, only behavioural observations, which can be extremely subjective and doctor dependent. This complicates matters for the scientists: even after years of research, we still don't know what causes this disorder.

What we do know is that autism has a strong genetic basis because monozygotic twins are much more likely to *both* be affected than dizygotic twins. This is quite unusual for such a symptomatically variable disorder. However, finding the genes responsible for autism has been difficult due to its high genotypic and phenotypic heterogeneity. Currently, more than 1000 genes are implicated in autism to varying degrees.

(continued)

In addition, a number of environmental factors have been closely linked to the occurrence of autism. These include the age of parents, viral infections during early development and complications during pregnancy. Although it is believed that environmental factors alone do not cause autism, they have often been found to increase the risk. However, one thing is clear: despite controversies over a possible link between childhood vaccines and autism spectrum disorders, many studies have shown that vaccines definitely *do not* cause or even contribute to autism.

GWAS have also discovered hundreds or thousands of polymorphisms associated with other traits such as body mass index (BMI) and diseases such as type 2 diabetes, schizophrenia and autism (see Box: Autism). Alleles in these polymorphisms slightly change the risk of developing the disease. Some alleles that increase the risk of type 2 diabetes decrease the activity of INS, a gene that produces insulin. Disease-associated alleles that are known to change the activity of a specific gene can help us understand the disease and potentially design treatments that reverse the effect of these mutations. For example, this has already happened with psoriasis—an autoimmune disease that affects the skin and joints. Following discoveries in GWAS, drugs have been developed to treat this disease.

The study of phenotypic variation has changed dramatically since the nineteenth century, from breeding pea plants to analysing DNA sequences of millions of people and studying how millions of polymorphisms shape who we are. Who knows where it will go next!

DEFINITIONS

Complex trait: a trait whose expression is influenced by multiple genes and the environment.
Additive Model: a simplified model of the relationship between genotype and traits, which assumes that an individual's trait value is the sum of the contributions of all trait-affecting genetic variants they carry and an environmental effect.
Heritability: the proportion of trait variance due to genetic differences between individuals.
Breeder's equation: a mathematical relationship that links the change in trait mean over one generation to trait heritability and the selection gradient (the deviation of selected parents from the population average).
Genome-wide association study (GWAS): a method that aims to identify the genetic variants underlying trait variation by measuring the difference between mean trait values of individuals carrying different genotypes at a given locus.

SUGGESTED READING

- Visscher PM, Hill WG, Wray NR. Heritability in the genomics era–concepts and misconceptions. Nat Rev Genet. 2008 Apr;9(4):255–66
- Grant PR, Grant BR. Predicting microevolutionary responses to directional selection on heritable variation. Evolution. 1995 Apr;49(2):241–251
- Mackay TF. Quantitative trait loci in Drosophila. Nat Rev Genet. 2001 Jan;2(1):11–20
- Abdellaoui A et al. 15 years of GWAS discovery: Realizing the promise. Am J Hum Genet. 2023 Feb 2;110(2):179–194
- Boyle EA, Li YI, Pritchard JK. An Expanded View of Complex Traits: From Polygenic to Omnigenic. Cell. 2017 Jun 15;169(7):1177–1186

4.5 Questions

1. **Barbora and Peter from the table in Figure 1 in the article have children. One of their children has ordered a genetic test from a company, but the results got mixed up with those of other people. Based on the table in the article, which of these genotypes could possibly belong to this child? Assume there have been no new mutations.**
   (A) Aa bb cc Dd ee ff Gg hh ii Jj
   (B) Aa bb cc Dd Ee ff Gg Hh ii Jj
   (C) Aa bb Cc Dd EE ff Gg HH Ii jj
   (D) Aa bb Cc DD EE FF Gg Hh Ii jj
   (E) AA bb Cc dd EE Ff gg Hh Ii jj
2. **For a very long time now, breeders of dragons have known that the heritability of the length of a dragon's tail is around 0.8. What are the implications of this?**
   (A) Parents with tails about 1 m long will have offspring that will, on average, have tails about 80 cm long.
   (B) About 80% of the variability in the tail length can be explained by genetics.
   (C) With the best possible nutrition and care, a breeder can extend a dragon's tail by a maximum of 20%.
   (D) About 20% of the variability in the tail length can be explained by environmental factors.
   (E) Identical dragon twins will always have identical tails.
3. **Scientists are interested in understanding the variation in the number of flowers produced by different individuals of a common flowering plant. They find that the variance in the number of flowers per individual is** 12 **in a natural population but only** 4 **for plants grown in a common garden (using seeds taken from the natural population), as shown in Fig. 4.7. Which of the following statements are true?**
   (A) Most of the differences in the number of flowers per plant can be explained by environmental factors.
   (B) Heritability of the number of flowers produced by plants is about 0.33.

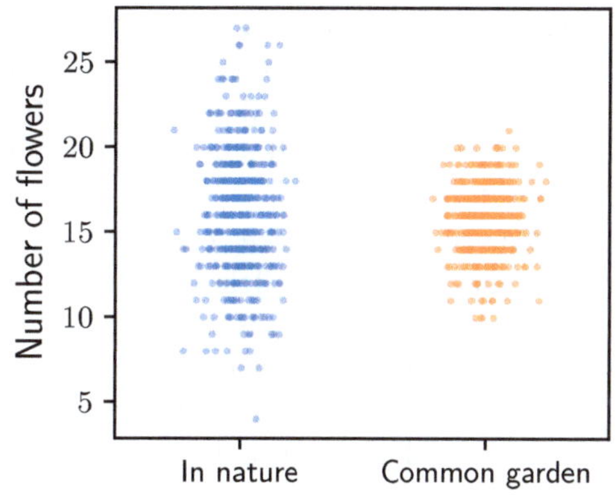

Fig. 4.7 **Variation in the number of flowers**

   (C) The correlation coefficient between the number of flowers in sibling plants will be around 0.33.
   (D) Plants in nature produce thrice as many flowers on average as plants in the common garden.
   (E) Plants in nature can respond faster to selection than plants in the common garden.
4. **People from the Netherlands are taller on average than those from Slovakia. What are the possible reasons for this difference?**
   (A) Different genetic composition of Dutch people and Slovaks.
   (B) The Dutch diet and lifestyle make Dutch people taller than Slovaks.
   (C) Dutch people have more genes for height than Slovak people do.
   (D) Slovak people have fewer height-increasing alleles on average.
   (E) Many Dutch people carry a mutation (or multiple mutations) that makes them taller than the average Slovak.
5. **After years of hard work, scientists finally managed to sequence all the dragons on the islands of Uramami. They noticed that dragons from Ur island carry LL, DD alleles and have, on average, 2 cm longer horns than dragons from Am island with $\ell\ell$, DD alleles. Further, the horns of the Am island dragons are, on average, 4 cm longer than the horns of dragons from Ami island, which carry $\ell\ell$, dd alleles. (Remember, all dragons on the islands are homozy-**

gotes.) What can we conclude from scientists' observations? Assume that the length of dragon horns is described by the additive model and is NOT influenced by the environment.

(A) Contribution of each L allele is 1 cm, compared to the ℓ allele.
(B) Contribution of each L allele is 2 cm, compared to the ℓ allele.
(C) Contribution of each d allele is -2 cm, compared to the D allele.
(D) A dragon heterozygous on both loci would have horns of the average size of its homozygous parents.
(E) Crossing an $LLdd$ dragon with $LLDD$ dragon would result in a dragon with horns 1 cm longer than a cross of $\ell\ell DD$ with $LLdd$.

6. **Scientists performed breeding experiments on a species of fly and measured a variety of traits—shape, size and traits related to life history—for all individuals in the experiment. They found that full sisters (who had the same mother and father) had more similar leg lengths than half-sisters (who had the same mother but different fathers) on average. However, scientists observed a different trend for other traits—e.g. the age of sexual maturity (at which female flies first produced eggs) was similar between full sisters but only to the same extent as between half-sisters. Based on this information, which of the following conclusions can you draw?**
(A) Leg length is more heritable than the age of sexual maturity.
(B) An experiment that selects for flies with a younger age of sexual maturity would be more successful (over multiple generations of breeding) than an experiment which selects for flies with shorter leg lengths.
(C) There is higher overall variance for leg length than for age of sexual maturity in the population.
(D) We expect to see a positive correlation between the leg length of mothers and daughters in the population.
(E) Leg length is not influenced by the environment of the fly.

4.6 Project: On Breeding of Creatures

Part 1: Puzzling Lengths or Lengthy Puzzles

All unicorns have one horn each. They are very proud of them and compare them constantly. In Table 4.1, you find the genotypes of five unicorns and the lengths of their horns. The length of a unicorn's horn is determined entirely by genetics, i.e. it is unaffected by the environment.

1. What are the contributions of the individual alleles A, B, C and D?
2. What is the genotype of the smallest possible horn?
3. What about the longest?
4. How many different values can the length of a unicorn's horn have?
5. What length does a completely heterozygous unicorn have? (Write down their genotype as well.)
6. What length do completely homozygous unicorns have? (Write down their genotypes as well.)

Part 2: Dragon Tails Again

Disclaimer: In this project, we will assume that the tail length in a dragon is entirely determined by their genetics, even though we know this is not true!

The shortest dragon tail ever observed was only 2.4 m. The length of the dragon's tail is determined by five genes, each of which can have two possible alleles. Each allele labelled with a capital letter changes the tail length by a specific amount, shown in Table 4.2 (ignore the coin part for now).

1. How long is the longest dragon tail? Again, assume that the environment does not affect the length.

Do you remember Stellaire and Ignitius? Scientists already have the genotypes of both dragons, shown in Table 4.3.

2. Look at the genetic sequences of Stellaire and Ignitius, and determine the lengths of their tails.

When they were introduced to each other, both dragons fell in love immediately. They have happily settled in the Zurich Zoo and are ready to start their own family. Just yesterday, Stellaire laid her first six eggs!

Let's think about their future offspring. Take coins of the denominations described in Table 4.2—they represent the different genes. Throw them to determine which alleles (first or second) make it to the gamete that Stellaire produces. Do this six times to produce six eggs.

Then, do it again to create six sperm that Ignitius produces and then pair the eggs and the sperm to create six offspring.

3. For each offspring, write the genotype and the expected length of their tails.
4. What is the longest tail, and what is the shortest tail *you observed*? Write the genotypes and the length.
5. What are the longest and the shortest tails *you could possibly get*? Write the genotypes and the length.

A long time ago, scientists believed that the length of dragon tails was determined by a single gene. Imagine that there was just a single

Table 4.1 Unicorns' genotypes and horn lengths

Name	Genotype	Horn length
Annion	aabbCcdD	11 cm
Beatrones	aabbCCdD	13.5 cm
Claudion	aAbbCCdD	14.5 cm
Darius	AABBCCdD	11.5 cm
Eristoteles	aaBBCcDD	5.5 cm

Table 4.2 Length contributions of different alleles

Allele	K	L	M	N	O
Contribution [cm]	17	31	41	51.3	52.3
Coin (Euro)	10c	20c	50c	1	2

Table 4.3 Genotypes of Ignitius and Stellaire

Ignitius	Kk	Lℓ	mm	Nn	Oo
Stellaire	KK	LL	mM	nn	Oo

gene with two possible alleles, with each T allele adding 192.6 cm to tail length. Two parents, both with Tt genotypes and 432.6 cm long tails, and have six offspring.

Flip a coin six times for each parent to decide which allele is used to make a gamete. Make six gametes from each parent, and then pair them to produce six offspring.

6. What are the lengths of their tails?
7. What is the longest tail, and what is the shortest tail *you observed*? Write the genotypes and the length.
8. What are the longest and the shortest tails *you could possibly get*? Write the genotypes and the length.

Think about the differences between both scenarios (with five or with one gene determining tail length) and about your answers. Then go to our online simulator https://en.edub.sk/everyday-evolution and look for *random-dragon-generator* to create 100 of offspring at once.

9. Make histograms of the tail lengths for both scenarios.
10. What do you observe? How many longest or shortest possible tails did you get in each case?
11. When was the variance in tail length larger?
12. When was the length range wider?

This project was developed by Barbora Trubenová and Kristína Hudáková.

Part II

Evolution and Humans

Out of Africa

5

Barbora Trubenová, Himani Sachdeva, and Kristína Hudáková

▶ Have you ever tried to construct your family tree to see if you have any historical celebrities in your family? Or have you ever attempted to unravel the complex pedigrees of dynasties like the Habsburgs or the Manchus in an effort to better understand history? If yes, you are probably familiar with family pedigrees. However, today we will go further back than this and try to reconstruct (a part of) the evolutionary history of *all* humans living on Earth today.

The question of human origin has fascinated us for ages. Where did we come from? Who were our ancestors? How did they spread across the Earth, and when? While some people are content to go back only a few generations, others are interested in more ancient history, tracing their roots back hundreds of thousands of years. In this chapter, we will learn how modern science enables deeper insights into human evolution than ever before, provides powerful tools to search for our ancestors and answer fundamental questions about human origin.

Constructing your own pedigree can be fun. You have two biological parents, who also have two parents each—your grandparents, who, again, have two parents each. With every generation further back in your history, the number of your great-great-great-grandparents doubles, whether you know them or not. However, if you continue this line of thought, at some point, you will have more ancestors in a single generation than people who have ever lived on this planet! What happened? How did this contradiction arise?

While the doubling rule usually works well in the beginning (3–4 generations), your ancestral lines will soon start to merge. Some of your great-great-great-grandparents were likely first or second cousins who shared a grandparent or were related to each other in some other way. Not too long ago, the pool of available partners was very limited; people did not travel very far and tended to marry someone within their village or living nearby. So, to find our ancestors, we often employ a completely different approach. You can pick any person on this planet, and be sure that both of you share a common ancestor! How deep into the past you have to go to find this ancestor often depends

B. Trubenová (✉)
D-USYS, ETH Zürich, Zürich, Switzerland

Eawag, Dübendorf, Switzerland

eduB, Bratislava, Slovakia
e-mail: barbora.trubenova@env.ethz.ch

H. Sachdeva
Faculty of Life Sciences and Faculty of Mathematics, University of Vienna, Vienna, Austria
e-mail: himani.sachdeva@univie.ac.at

K. Hudáková
eduB, Bratislava, Slovakia
e-mail: kristina.hudakova@edub.sk

B. Trubenová et al. (eds.), *Everyday Evolution: Unveiling Life's Hidden Patterns*,
https://doi.org/10.1007/978-3-032-07050-0_5

on how far apart you live. If you pick someone from your home town or that of your parents, you probably don't have to go back too many generations to find a common ancestor. However, if you consider people from two different long-separated populations, such as Australian Aborigines and an Amazonian tribe, then you may need to go back a few thousand years to find their common ancestor and use a different approach: looking at their genetics.

5.1 Our Common Ancestors

How do we find the common (shared) ancestors of a group of individuals? If we had perfect family records for everyone in the world, then we could find these common ancestors by tracing our ancestry backwards in time and simply stopping at the first individual who is found in the genealogical tree of every person. Unfortunately, as you might have guessed, there are no family records that go back thousands of years. Moreover, as we will see below, defining common ancestors is not straightforward!

Another way to learn about ancestry and the relationships between different people is by examining their DNA. As explained earlier, you inherit half of your genetic material from each of your parents. The DNA sequences you inherit are almost identical to theirs, since very few mutations occur in one generation. The same is true for your siblings. Statistically speaking, you share about half of your genetic material with your full brother or sister and a quarter with your first cousins, with very few differences caused by mutations. Thus, the more similar the DNA of any two people, the more recently they shared common ancestors.

However, tracing our genetic history using DNA is a bit more complicated than this. As mentioned earlier, genes and chromosomes inherited from the mother and the father are reshuffled during sperm (or ovum) production. This means that your father may pass on to you the gene for eye colour from your grandmother and the gene for hair colour from your grandfather. Thus, if you trace the ancestry of different genes or stretches of DNA backwards in time, you will find that different genes originate from different ancestors in your family tree. Moreover, if you compare your genome with those of your siblings or cousins, you will find that some segments trace back to a very recent common ancestor (e.g. a parent or grandparent), while others are very distantly related. This would be the case for those segments that you have inherited from your grandmother, while your sibling has inherited from your grandfather. Thus, to find the common ancestor of these, you may need to go back thousands of years to an individual who is a genetic ancestor of both your grandmother and grandfather.

Many of these complications can be avoided if we look at segments of DNA that do not undergo recombination or shuffling. This is particularly useful when trying to find the common ancestors of all humans. Luckily, there are two special DNA molecules that avoid getting reshuffled, and consequently, there are two ancestors that we can find relatively easily: Mitochondrial Eve (mt-Eve) and Y-chromosomal Adam (Y-Adam).[1]

5.2 mt-Eve and Y-Adam

Both parents contribute equally to the nuclear genome through their ovum (egg) and sperm. However, mothers also provide the zygote with its cytoplasm and with mitochondria that contain *their own DNA*. Thus, while our nuclear DNA is a mixture of the DNA of our four grandparents, our mitochondrial DNA (mt-DNA) is inherited only from the mother and is *an almost exact copy* of our maternal grandmother's mitochondrial DNA, with just a few copying errors—mutations.

Consider all humans alive on Earth today. All these humans have mothers, and all these mothers have their own mothers. If we try to find the mothers of all living humans and their mothers, then their mothers and so on, the number of women we are looking at keeps decreasing as we go back in time, as some women (sisters)

[1]Many scientists now prefer to use "mt-MRCA" and "Y-MRCA" to avoid confusion.

Genealogical and Genetic Ancestors

Your family tree or pedigree—also called **a genealogical tree**—is based on tracing your family ancestors and includes all individuals that were involved in your origin. However, it does not mean that you share genes with all of them! You inherited your genes from your biological parents. These genes came packaged in 46 chromosomes, 23 from each parent, who also have their own 46 chromosomes. This means that each parent passes on only half of their genetic material to you; the other half is lost (or your siblings might inherit some of it). Your parents also inherited half of their genetic material from each of their parents, so you share approximately one-quarter of your chromosomal DNA with each of your grandparents. However, due to the chromosomal shuffling and recombination, it is quite likely that you inherited slightly more from one of your grandparents than from the other.

If you go back several generations, you encounter many ancestors who were simply not lucky enough to pass any of their genetic material on to *you*. It doesn't mean they weren't important for you to be born. It only means that *you* do not share any of their DNA. By looking only at those of your ancestors who contributed to your DNA, you can construct another type of tree—**a genetic tree**—which is a subset of your genealogical tree.

will have the same mother. This is because some women have no daughters, while others have many. Thus, eventually, we will arrive at a single female who is matrilineally connected to all of us. This lady is Mitochondrial Eve (referred to as mt-Eve in short), our matrilineal Most Recent Common Ancestor (MRCA). Her mitochondrial DNA is present in *all* humans currently living on Earth.

The concept of mt-Eve is illustrated in Fig. 5.1. Note how mt-Eve (in the circle) is connected to each woman in the bottom row via an *unbroken line of daughters*. Such a line consisting only of your female ancestors is called a **matrilineal line**, or **lineage**. All the other matrilineal lines either died out or had at least one generation in which no daughters, only sons, were born. Of course, each of us has many ancestors who do not belong to the matrilineal line—your father's mother is an example of an ancestor who is not matrilineal to you. While you most likely inherited some of her chromosomal genes, you don't have any of her mitochondrial DNA.

Interesting Fact

Mitochondria have a fascinating history themselves: their ancestors were unicellular organisms—free-living bacteria. Approximately 1.5 billion years ago, long before multicellular organisms evolved, some of them began living symbiotically within other cells, gradually losing their ability to exist as independent organisms. Today, even though they still contain parts of their own DNA, they have lost many important genes that are required to survive and reproduce independently (see Chap. 15). However, they are vital for cellular respiration in all multicellular organisms!

Mt-Eve was named after Eve from the Genesis story in the Bible, but she has no actual connection with the biblical Eve. Mt-Eve was certainly not the first woman on Earth, nor was she the only female living in her time (see Fig. 5.1). She had ancestors who are also our common ancestors. The name Mitochondrial Eve just refers to the *most recent* of these ancestors—the woman who is the common great-great-grandmother to all of us via matrilineal inheritance. She also had many female contemporaries—researchers believe that as many as 20,000 individuals of mt-Eve's (our) species may have lived at the same time as her, but mt-Eve was the only "lucky mother" who produced an *unbroken* line of daughters that persists to this day. This means that only mt-Eve's mi-

Fig. 5.1 Finding mt-Eve. Some women leave more daughters, and some leave none. In this example, all women in the last generation belong to a single matrilineal lineage (in red) and descend from one woman—mt-Eve (in the circle). All other matrilineal lineages (blue, green, etc.) have died out

tochondrial DNA is carried by all humans living today, while that of her contemporaries has been lost.

Is there a male counterpart of Mitochondrial Eve? Yes, he is called Y-chromosomal Adam, and he is the patrilineal human MRCA. The "Y-chromosomal" in his name comes from the male Y-chromosome, which is passed exclusively from father to son.

Reproductive cells—gametes—are produced via meiosis, which involves chromosomal shuffling and recombination (see Chap. 12). During meiosis, the two homologous chromosomes exchange segments of DNA to create two recombined chromosomes. Each gamete then receives—at random—one of these recombined chromosomes from each pair.

However, Y chromosome, only found in males, is more or less exempt from the recombination process. It looks quite different from its partner, the X chromosome, so the two recombine very rarely, only at their tips. This allows us to trace its origin by following the line of fathers—the patrilineal line—back in time all the way to one single man. Again, his existence does not imply that he was the first man or the only man on Earth. None of Y-Adam's contemporaries, however, have a direct male line that continues to the present day. Either the line died out or the line had at least one generation in which no sons, only daughters, were born.

5.3 Tracing Our Ancestors

Scientists have even been able to estimate when these ancestors lived using the **molecular clock technique**. Based on the molecular clock, they estimate that Y-Adam lived more than 300,000 years ago (though a few studies suggest that it may have been as recent as 120,000–156,000 years ago). Mt-Eve seems to be much more recent, with studies estimating that she lived about 150,000 years ago. This means that Y-Adam and mt-Eve *themselves never met*. We also know that both of them lived in Africa and that their descendants started to wander out of Africa around 60,000 years ago, replacing other human populations such as *Homo erectus*.

Interesting Fact

We can find the genealogical MRCA of all humans alive today (or the genetic MRCAs of specific segments of the genome). However, this is not the same as the MRCA of humans alive at some point in the remote past or future. As people die, some lineages may become extinct, and a more recent individual may become the most recent common ancestor of all humans living at that time. The extinction or rediscovery of a small isolated group living deep in a jungle can drastically change the estimates of when our MRCA lived! Therefore, you may find many different estimates of when Y-Adam or mt-Eve lived, depending on when the article you are reading was written.

But how do we know all this? Our DNA again provides the clues. From the genetic analysis of mitochondrial DNA and Y-chromosomes in many individuals from different ethnic groups and tribes, as well as archaeological finds, we

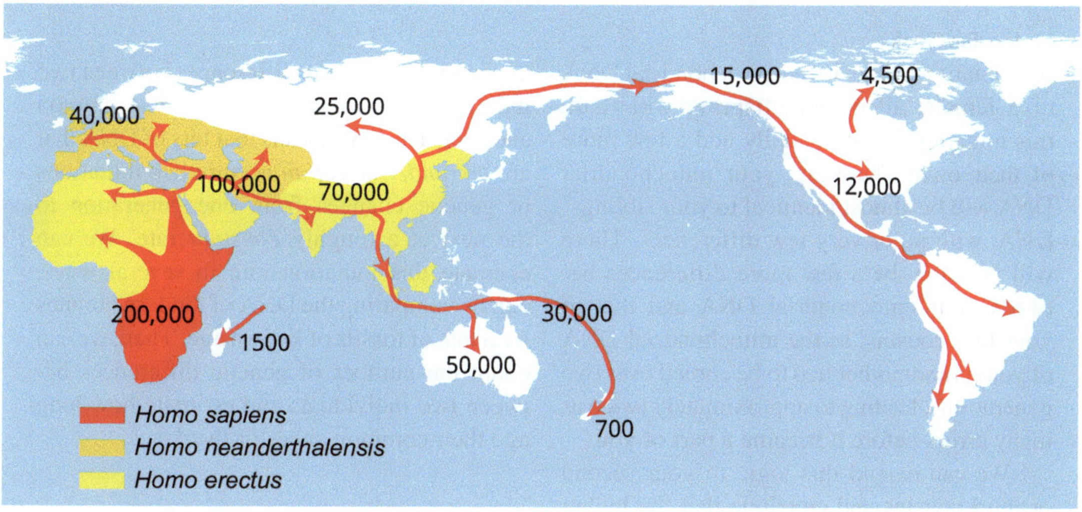

Fig. 5.2 Estimated migration routes of humans from Africa. The number on each arrow shows how many years ago each migration most probably occurred. (NordNordWest, Public domain, https://creativecommons.org/public-domain/ viaWikimediaCommons)

can reconstruct how our ancestors left Africa and populated various continents (Fig. 5.2). We can even investigate how humans adapted to different natural environments!

We can divide the human population into several groups based on the similarity of their mitochondrial DNA sequences. These groups are called **haplogroups**; each haplogroup has certain typical mutations that distinguish it from other haplogroups. Each of these mutations can be traced to a single female in whom it first occurred.

By identifying which mitochondrial DNA sequences are common in a region and assessing how similar or different they are to the DNA sequences in other regions of the world, we can trace the migration routes of our ancestors as they spread across the Earth (Fig. 5.2). As an example, mutations found in DNA sequences of modern-day Native Americans are most similar to the ones found in Asia—this allowed scientists to infer that ancient humans most likely migrated to the Americas via Siberia.

You may wonder what happens when we go even further back in time. At this timescale, different subspecies, then species, start to merge. You can find the most recent common ancestor between *Homo sapiens* and Neanderthals, between hominins and chimpanzees and between all apes and any other species. The principle is the same: some species are more closely related, sharing more features and more similar DNA, while others are more distantly related, showing less similarity. Going further and further back in time, we can construct the biggest, the most complete family tree of all—the Tree of Life.

Molecular Clock

If a mutation occurs in the mitochondrial DNA of a female, all of her offspring will share this mutation and potentially add a few more of their own. Therefore, your mitochondrial DNA will be almost identical to your siblings' DNA, with no or very few differences. There will probably be a few more differences between your mitochondrial DNA and that of your first cousins, as the mitochondrial DNA of your grandmother had to be copied over two generations, leading to approximately twice as many errors before it became a part of you.

We can extend this logic to your second or third cousins and conclude that the higher the number of genetic differences between two individuals, the longer ago they had a common ancestor. The key point is that bits of DNA that are *not under selection* accumulate mutations or genetic changes from one generation to the next at a roughly *constant* rate. We can estimate this mutation rate in several ways, e.g. by comparing the DNA of modern humans with that of fossils of known age. Then, we can count the number of genetic differences between two individuals and estimate how long ago their common ancestor lived.

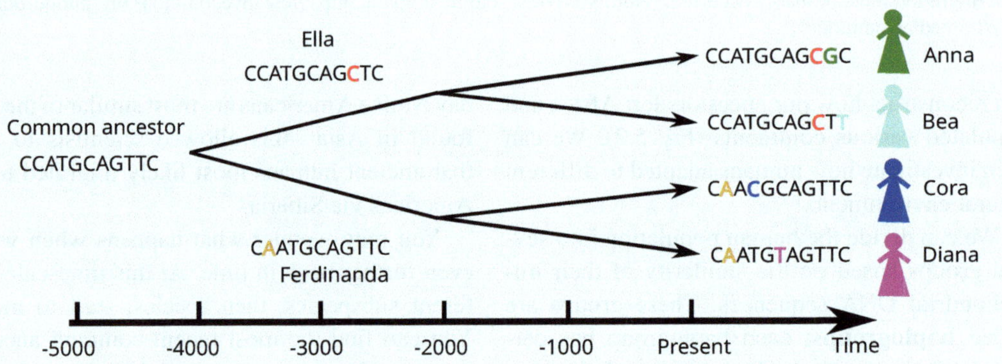

Calibrating molecular clocks. Different mutations occurred in different females (Ella and Ferdinanda) who shared a common ancestor. Each of these females passed her respective mutation to her daughters and granddaughters, some of whom also eventually acquired their own mutations and passed them on, creating distinct lineages (Anna, Bea, Cora and Diana).

Genetic Diversity in Different Human Populations

The sequencing of millions of human genomes has revealed some interesting facts about human genetic variation. More genetic diversity is found in Africa than in the rest of the world combined. In fact, the farther away a human population is from Africa, the lower its genetic diversity. This pattern is a very important piece of evidence in support of the "Out of Africa" hypothesis.

Generally, larger populations tend to be more diverse. Any small sub-group of a large population carries only a small fraction or subset of the mutations present in the whole population. Therefore, each small sub-group that migrated out is less diverse than the source population, giving rise to the patterns in human genetic diversity that we see today. Of course, in the last hundred years or so, more and more humans have been migrating longer distances. It will be interesting to see how this alters geographic patterns of diversity in the future!

DEFINITIONS

Molecular clock: a method that estimates the time to the most recent common ancestors of individuals or populations based on the number of genetic differences between them and using information about mutation rates.

MRCA: The genealogical Most Recent Common Ancestor is the most recent individual from whom all currently living humans are directly descended. The genetic MRCA of a specific genomic region is the most recent individual ancestral to all living humans for that region; different genomic regions have different MRCAs. However, MRCA can also refer to the most recent common ancestral population of different species, such as the Last Universal Common Ancestor (LUCA) for all life or the Last Eukaryotic Common Ancestor (LECA) for all eukaryotes (see Chaps. 15 and 9).

Y-Adam: a single male who is patrilineally connected to all living human beings, the patrilineal most recent common ancestor.

mt-Eve: a single female who is matrilineally connected to all living human beings, the matrilineal most recent common ancestor.

SUGGESTED READING

- Mendez FL et al. An African American paternal lineage adds an extremely ancient root to the human Y chromosome phylogenetic tree. Am J Hum Genet. 2013 Mar 7;92(3):454–9.
- López S, van Dorp L, Hellenthal G. Human Dispersal Out of Africa: A Lasting Debate. Evol Bioinform Online. 2016 Apr 21;11(Suppl 2):57–68
- Cann RL. Y Weigh In Again on Modern Humans. Science. 2013; 341, 465–467
- Hunter P. The genetics of human migrations: Our ancestors migration out of Africa has left traces in our genomes that explain how they adapted to new environments. EMBO Rep. 2014 Oct;15(10):1019–22

5.4 Questions

1. **Which of the following statements about mitochondrial DNA are true?**
   (A) Mitochondrial DNA codes for some proteins which are necessary for survival.
   (B) Siblings of the same sex always have identical mitochondrial DNA, while siblings of different sexes always differ in their mitochondrial DNA.
   (C) Mitochondrial DNA is found in all organisms living on Earth today.
   (D) When a human zygote is formed, the sperm contributes only nuclear DNA while the egg contributes only mitochondrial DNA.
   (E) We could easily survive even if we had our mitochondrial DNA completely removed.
2. **Identify all correct statements about Y-Adam.**
   (A) A man living in Africa 200,000 years ago had three sons. This individual could have been the Y-Adam only if at least one of his sons also had male descendants.
   (B) A new discovery of an isolated tribe may shift our estimates of the Y-Adam closer to the present.
   (C) All humans living today can trace at least one of their sex chromosomes back to the Y-Adam.
   (D) All primates living on Earth today share a common ancestor. If this ancestor was a male, then it can be called the Y-Adam of all primates.
   (E) Y-Adam is the most ancient common ancestor of all people living today.
3. **Which of the following statements are true?**
   (A) Epicanthic fold is a skin fold that covers the inner corner of the eye, most commonly found among East Asians and Southeast Asians. As this feature is not usually found among Africans, it disproves the "Out of Africa" hypothesis.
   (B) Any two sisters share exactly half of their genome.
   (C) A woman with a specific beneficial mutation in her nuclear DNA had four sons, but didn't leave any female offspring. The absence of daughters is the reason why this mutation will eventually be lost.
   (D) If two distinct living species have a common ancestor, this means that they can interbreed and produce fertile offspring.
   (E) Male mitochondria cannot be used in ancestral DNA analysis, because mitochondrial DNA is inherited only matrilineally.
4. **Jerry, whose father suffered from a rare disease, characterised by muscle weakness and gradual loss of muscle coordination, marries his cousin Amy. The disease is caused by dysfunctional mitochondria. Which of the following can be true about the children of this couple?**
   (A) If Amy's mother is a sister of Jerry's mother, the children will be healthy.
   (B) If Amy's mother is a sister of Jerry's father, all the girls will have the disease, and the boys will be healthy.
   (C) Regardless of whose sister Amy's mother is, all the boys will have the disease, and the girls will be healthy.
   (D) All the children will have the disease if Amy's mother is a sister of Jerry's father.
   (E) If Amy's mother and Jerry's father are half-siblings, then the children must be healthy.
5. **To estimate the time to the most recent common ancestor of two species, we use the following information:**
   (A) The age of the Earth
   (B) How similar they are in appearance
   (C) The number of genomic sites at which two species differ
   (D) The rate of mutation
   (E) All of the above
6. **Figure 5.3 shows the number of changes that have accumulated in a genetic sequence versus time in three different graphs. Which of these graphs would make a good candidate for a molecular clock? Choose only one answer.**
   (A) Graph A, because the curve becomes flat after some time.
   (B) Graph B, because it is a straight line.
   (C) Graph C, because the change in sequence is the most rapid.

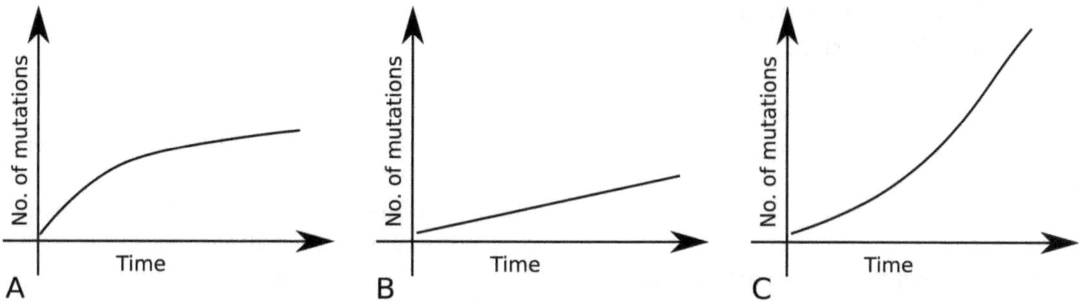

Fig. 5.3 **The number of changes that have accumulated in a genetic sequence over time**

(D) Any of the three genetic sequences will make a good molecular clock because genetic changes occur in all of them.
(E) None of the three, as there should be no genetic changes for a molecular clock to work.

7. **The "family tree" of Mitochondrial Eve's descendants was created by:**
(A) Calculating the time at which humans and apes shared a common ancestor
(B) Comparing the DNA of fossilised humans with that of living humans
(C) Analysing the genealogical trees of humans alive today and then extrapolating back to an initial ancestor
(D) Analysing the similarities of mitochondrial DNA of a large and diverse sample of living humans to create related groups
(E) Reconstructing Eve's complete genome

8. **Which of the following statements are true?**
(A) Any two humans living on the Earth share more DNA with each other than with any other species.
(B) Asians and Europeans have several genetic mutations not found commonly in Africans. This disproves the "Out of Africa" hypothesis.
(C) There were multiple human-like (hominin) species, of which only one (*Homo sapiens*) survives today.
(D) The maximum amount of genetic diversity in human mitochondrial DNA is found in European females.
(E) Modern-day Tibetans have several genetic mutations that help them survive at high altitudes. This is an example of a recent human adaptation to an environmental challenge.

5.5 Project: How Can We Find Our Ancestors?

In this activity, you will try to trace back your virtual ancestors using the methods explained in the this chapter.

Part 1: Finding Y-Adam and mt-Eve

Look at the scheme on the right (Fig. 5.4). Each horizontal row of couples represents one generation. We assume that each couple has children only with each other; lines connect the pair to their offspring. In this example picture, the first pair in the first generation had one son and one daughter, who got married to the daughter and the son of the other couple.

Your Task

Using the knowledge you gained from reading the chapter, look at Fig. 5.5 depicting one small population, and answer the questions below.
In your answers, you can either mark the individuals in the picture or identify them using their sex (male/female), pair (the column label) and generation number (the row number), e.g. "female from pair A in the 1st generation".

Questions

1. Who is the matrilineal MRCA of the blue and the purple woman?
2. Who is the patrilinear MRCA of the orange and the green men?
3. Is there also a purely matrilineal or purely patrilinear MRCA of all four coloured (purple, orange, green and blue) individuals? If yes, who is it? If not, why not?
4. Is there also the matrilineal MRCA of all **women** living in the 9th generation? If yes, who is it? If not, why not?
5. Is there also a mt-Eve—matrilineal MRCA—of **all humans** in the 9th generation? If yes, who is it? If not, why not?
6. If you imagine the following generation (10th generation, not shown), is there mt-Eve of all the people of the 10th generation? If yes, who is it? If not, why not?
7. Would there be Y-Adam of people living in the 10th generation? If yes, who is it? If not, why not?
8. Which couple represents the **first meeting** between the Y-Adam and mt-Eve lineages?
9. Can you find the matrilineal or patrilineal MRCA of all four coloured individuals? If you can't find the MRCA of all four, can you find the MRCA of any three of them?

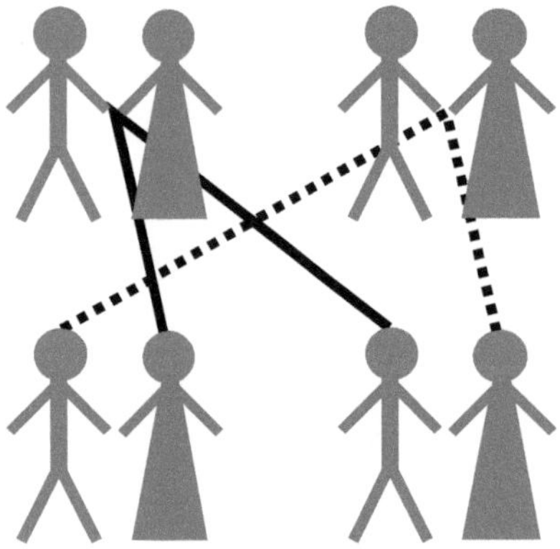

Fig. 5.4 Parents and offspring

Part 2: Tracing Back Migration Patterns

Imagine that you know that your great-great-great-grandmother came from a group of small, remote populations in the mountains (Fig. 5.6). You just don't know which one. These populations used to be quite isolated from each other, and it took your ancestors several thousand years to travel through the region and establish small villages. Every new village was colonised by a small number of closely related individuals who had the same mitochondrial DNA sequence.

Luckily, you know your own mitochondrial DNA sequence, as well as the mitochondrial DNA sequences from the teeth of females who

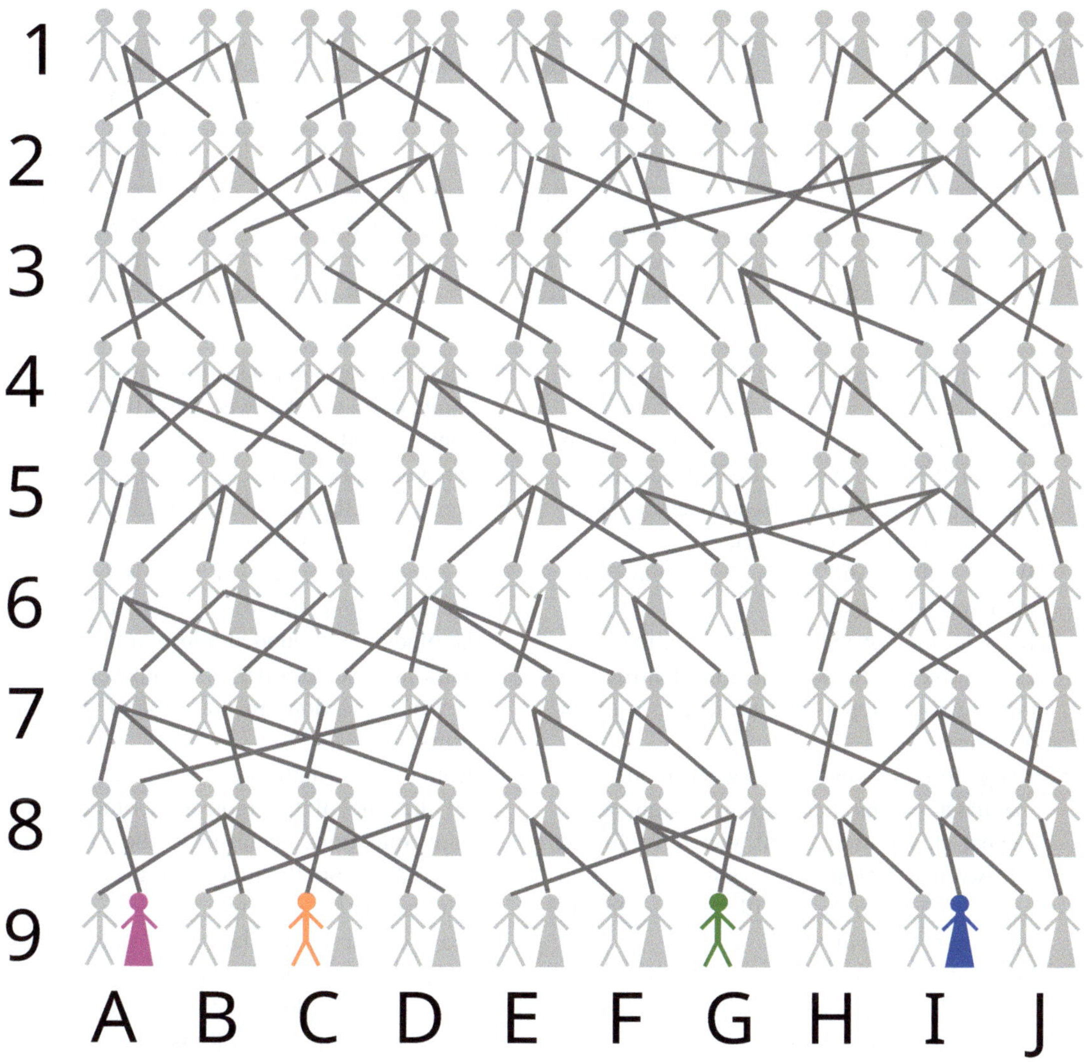

Fig. 5.5 Ten generations of a small population

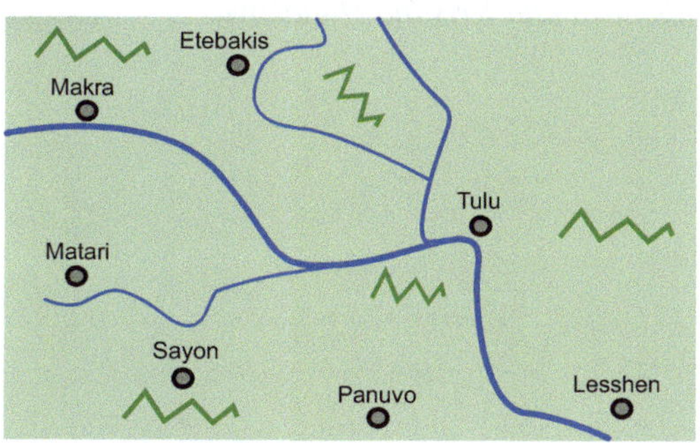

Fig. 5.6 Map of the region

lived in these villages thousands of years ago (representing different haplogroups).

Your Task

You want to draw a map of your ancestors' migration across the region by drawing arrows between the villages. The instructions and questions below will help you do this.

Instructions

1. Study the sequences in Table 5.1 carefully, and highlight locations (columns) where different mutations occurred over thousands of

Table 5.1 Your mitochondrial DNA and that of the individual haplogroups

Individual	Place	Genome
You	Your hometown	A T G C A A T C G C C T A T T C T
Athena	Matari	A T G G A T T C G C C T A C T C T
Mokosh	Sayon	A T G G A T T C G C C T A T T C T
Juno	Makra	A T G G A T T C G A C T A C T C T
Gaia	Tulu	A A G G A T T G G A C T A C T C T
Freyja	Etebakis	A A G G A T T C G A C T A C T C T
Kali	Panuvo	A T G G A A T C G C C T A T T C T
Atabey	Lesshen	A A G G A T T G G A C T A C T A T

years in the mitochondrial DNA of your ancestors.

2. Can you find the sequence that is most similar to yours? Which woman does it belong to? This woman was your closest ancestor—your great-great-great-grandmother.
3. Which village did your great-great-great-grandmother come from?
4. Keep looking at the table. Who is the closest relative of the woman you identified in the previous step? Link these two women together.
5. Keep looking and linking women with the most similar haplogroups (sequences). Repeat this process until you have linked all women. Note that sometimes two haplogroups can be equally similar to the one you are considering!
6. Without knowing anything about the age of the fossils, can you decide which haplogroup is the oldest one?
7. If two villages have very similar haplogroups (which differ by just one letter), then it is very likely that people from one of the villages colonised the other. Use this information to connect villages that were directly colonised from each other.
8. Can you draw a direction of migration between the villages at this stage? If yes, do so; if not, explain why not.
9. The recent discovery of fossil records and artefacts shows that the village of Makra is the oldest one, founded approximately 3000 years ago. Reconsider your answers to the previous questions in the light of this new information!
10. From the information you have, can you find out the rate of the molecular clock? How many mutations occurred in the mitochondrial DNA from the time the region was first colonised to the present day?
11. Draw a scheme similar to the one in Box: Molecular Clock in the chapter.
12. Can you now draw the complete map of how the region was colonised and villages founded? If yes, do it!

This project was developed by Barbora Trubenová and Kristína Hudáková.

6 Evolution, Antibiotics, and Us

Hildegard Uecker

▶ Antibiotics are an integral part of modern medicine, so widely used that healthcare without them has become inconceivable. But what do antibiotics look like from the perspective of bacteria? How does antibiotic treatment change their environment? How do bacteria adapt to this dramatic change? And what happens to the "good" bacteria in our body during treatment? In this chapter, we will make an excursion into the fascinating world of bacteria and have a closer look at the evolution of antibiotic resistance.

Before the era of antibiotics in which we live today, many diseases were nearly untreatable. Even minor cuts, if they became infected, could lead to amputated limbs or even death. Bacteria—the cause of these infections and of many diseases, such as bacterial pneumonia, or tuberculosis—are single-celled organisms (see Fig. 6.1 for pictures of some species). They reproduce asexually: a bacterium divides into two daughter cells that are almost exact copies of the original, up to a few mutations. How fast bacteria divide depends on both the species and the environment, i.e. the site of infection. The division time of bacteria in the human body is on the order of hours. Thus, a few days (or even just one) are often sufficient for bacteria to reach a population size large enough to make us ill.

In the second half of the nineteenth century, several researchers made an intriguing discovery. They observed that laboratory animals injected with both **pathogenic** (i.e. disease-causing) and certain other types of bacteria would not develop the dangerous infectious diseases, such as anthrax or cholera, normally caused by the pathogenic bacteria. One of these protective bacteria was *Pseudomonas aeruginosa* (Fig. 6.1), at the time called *Bacillus pyocyaneus*, and it turned out that it secreted a substance that prevented the growth of several pathogens and could be used to treat the associated diseases. This secretion, known as pyocyanase, became the first antibiotic ever used in a hospital! Unfortunately, its curative effects were inconsistent, and it was quite toxic to the patient.

The next antibiotic was discovered in 1909 by a systematic search for a "magic bullet" that would specifically target the pathogen and no other cells, and thus be harmless to the patient. Paul Ehrlich had observed that dyes stained some cells, but not others, which made him believe that they could constitute such a magic bullet. The 606th substance that he and his collabora-

H. Uecker (✉)
Institute of Integrative Biology, ETH Zürich, Zürich, Switzerland

Department of Theoretical Biology, Max Planck Institute for Evolutionary Biology, Plön, Germany
e-mail: uecker@evolbio.mpg.de

B. Trubenová et al. (eds.), *Everyday Evolution: Unveiling Life's Hidden Patterns*, https://doi.org/10.1007/978-3-032-07050-0_6

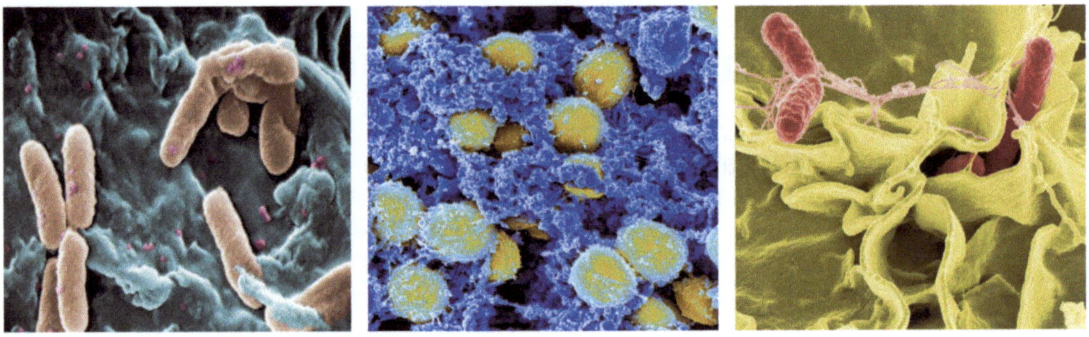

Fig. 6.1 **Various bacterial species.** The pictures show from left to right *Pseudomonas aeruginosa*, *Staphylococcus aureus*, and *Salmonella* species. (Images: (1) Centers for Disease Control and Prevention/Janice Haney Carr, (2)+(3) National Institute of Allergy and Infectious Diseases)

tors tested was a success, resulting in the antibiotic Salvarsan to treat syphilis. The screening of synthetic dyes for antimicrobial effects was continued by other researchers and led in 1935 to the discovery of Prontosil, which was used to treat a number of bacterial infections. Unlike pyocyanase, Salvarsan and Prontosil were not natural compounds but were produced synthetically. The real breakthrough in antibiotic discovery was achieved with the development of penicillin (see Box: The Discovery of Penicillin), which was very effective and had no or only weak toxic effects.

The risk of bacteria becoming resistant to antibiotics was realised early on. Resistant bacteria—unlike their sensitive relatives—can multiply and grow in spite of the presence of the antibiotic. The specific drug is thus ineffective and cannot be used to treat the infection. In his Nobel prize lecture, Alexander Fleming (one of the "fathers" of penicillin), who had observed resistance to penicillin in the lab and in patients, warned:

It is not difficult to make microbes resistant to penicillin in the laboratory by exposing them to concentrations not sufficient to kill them, and the same thing has occasionally happened in the body. The time may come when penicillin can be bought by anyone in the shops. Then there is the danger that the ignorant man may easily underdose himself and by exposing his microbes to non-lethal quantities of the drug make them resistant.

Today, antibiotic resistance has become one of the greatest dangers to healthcare worldwide. The European Centre for Disease Prevention and Control estimates that resistant bacteria are responsible for more than 35,000 deaths every year just in the European Union, Iceland, Liechtenstein, and Norway.[1]

Where do these resistant bacteria come from? What makes them different? And what does evolution have to do with it? These are some of the questions we will explore in the following sections.

6.1 Antibiotic Resistance

Let us consider one example of how an antibiotic works and how a cell can become resistant to it (for other examples of antibiotic action, see Box: How Do Antibiotics Affect Bacteria?). Before a cell divides, its DNA needs to be replicated, and the cell possesses a whole toolkit of molecules for this purpose. When the antibiotic agent in our example enters the cell, it attaches itself to one of these tools, preventing the tool from working properly. Instead of performing its task in DNA replication, the tool destroys the DNA! We have achieved our goal: the bacterial cell dies.

[1]Based on infection data from 2016 to 2020, European Centre for Disease Prevention and Control. Assessing the health burden of infections with antibiotic-resistant bacteria in the EU/EEA, 2016–2020. Stockholm: ECDC; 2022.

The Discovery of Penicillin

Before Alexander Fleming left St. Mary's Hospital in London for his summer vacation in 1928, he left some Petri dishes with *Staphylococcus* bacteria on his lab bench. When he came back in September, either he or his colleague Merlin Pryce—history is vague on this point—spotted something interesting: One of the plates had been contaminated by a mould, and around this mould, there was a bacteria-free zone! Fleming said "That's funny". And with these words, one of the most influential discoveries in the history of medicine began. Further investigation revealed that the mould belonged to the bread mould family *Penicillium*, and it secreted a substance that could kill bacteria of various species.

Alexander Fleming preserved some of this specific mould and wrote an article about his findings. This article was found 10 years later by Ernst Chain, who was a researcher in Howard Florey's team at Oxford University. This team began testing the secretion for use in medical treatment. One of the major obstacles turned out to be how to purify and produce enough of the antibiotic substance. This problem was solved primarily by Norman Heatley, another member of Florey's lab. He not only succeeded in purifying the penicillin out of the "mould juice", but he also observed that the mould was growing best in the bedpans borrowed from an infirmary and, based on this knowledge, designed the setup to grow large amounts of *Penicillium* mould. It still took enormous efforts before penicillin finally arrived on the market. Its use became widespread in the early 1940s and saved millions of lives during World War II.

Only a few years later, in 1945, Alexander Fleming, Howard Florey, and Ernst Chain were awarded the Nobel Prize in Physiology or Medicine. Norman Heatley was, sadly, forgotten.

However, among the many bacteria, there may be some with a mutation in their DNA that causes the tool to have a slightly different shape. In this case, the antibiotic cannot attach to it, like a key that does not fit a lock. Cells with this altered tool are not affected by the antibiotic and can replicate normally. They pass the life-saving mutation on to their daughter cells, which again replicate and leave descendants. After a while, all the normal cells are dead, and the population consists entirely of mutated, resistant cells. This is Darwin's natural selection among bacteria: the fittest ones thrive and take over. For us, this means that the infection is back.

There are other mechanisms through which a cell can be resistant to an antibiotic. For instance, a cell might be able to pump the antibiotic out faster than normal cells or be able to destroy antibiotic molecules, which can occur inside or outside the cell. If sufficient amounts of the antibiotics are destroyed by resistant cells, the neighbouring sensitive cells are protected and can survive as well.

Let us take a closer look at the fitness of bacteria with and without the resistance mutation. As we already know, the fitness of any organism depends on its environment. An antibiotic can be a very significant component of a bacterium's environment. Fig. 6.2 shows how the growth rate (i.e. the difference between the replication and death rates) of sensitive and resistant bacteria depends on the antibiotic concentration. The growth rate is one measure of bacterial fitness. We can see that when the antibiotic concentration is close to zero, resistant cells actually grow more slowly than sensitive cells. This is because resistance mutations are often (but not always) *costly*, i.e. cells with this mutation divide less frequently or die earlier than mutation-free cells. Thus, in the absence of antibiotics, natural selection tends to eliminate the resistant cells, and only a few of them are present in the bacterial population. (However, antibiotic resistance can also persist for a very long time in the absence of antibiotics if costs are small or absent or if the genes are maintained through horizontal spread, as explained in the next section.)

How Do Antibiotics Affect Bacteria?
To understand how antibiotics attack bacteria, we need to briefly consider what a bacterium looks like and how it lives. The outermost layer of a bacterial cell is the cell wall. Under the cell wall, there is the cell membrane that regulates which molecules go in and out. Inside the cell, there is a complex machinery made up of molecules such as proteins that are responsible for the "work" inside the cell. A cell's DNA stores the information for the manufacturing of this machinery. When a cell divides, its DNA is replicated and distributed to the two daughter cells.

Antibiotics can attack bacteria in various ways. Some antibiotics destroy the cell wall of the bacteria or the cell membrane. Other antibiotics inhibit DNA replication or the production of proteins. Depending on their mechanism of action and their concentration, antibiotics can decrease the replication rate or increase the death rate of bacteria.

A given antibiotic is not effective against all bacterial species; some bacterial species will have **intrinsic resistance**, and will not be affected by certain antibiotics. An antibiotic that is effective against many different bacterial species is called a **broad-spectrum antibiotic**.

When we are sick and take medication, antibiotics only do part of the job of clearing the infection. The rest is done by our immune system.

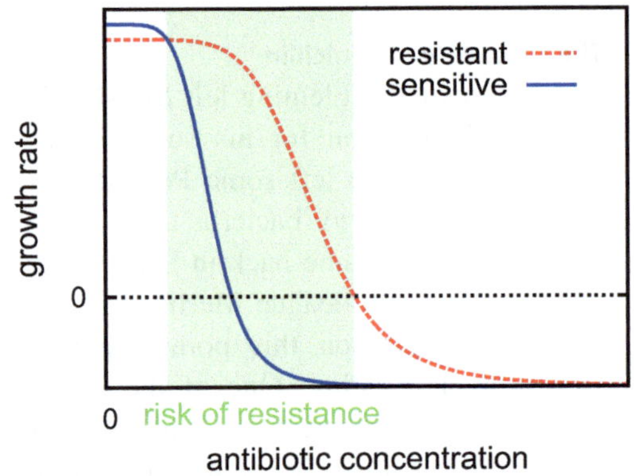

Fig. 6.2 Bacterial growth in the presence of an antibiotic. Growth rate of sensitive and resistant bacteria as a function of the antibiotic concentration. Within the green region, a risk of resistance evolution exists: resistant bacteria have a positive growth rate and grow better than sensitive ones

In contrast, in the presence of a sufficiently high concentration of antibiotics (starting at the left margin of the green region in Fig. 6.2), resistant bacteria grow faster than sensitive bacteria. As the concentration of antibiotics increases, the sensitive bacteria eventually may not be able to grow at all (the solid blue curve goes below the dotted line, corresponding to no net growth). However, even resistant bacteria can only tolerate a certain amount of the antibiotic: if in our example from above many antibiotic molecules attempt to attach to a mutated tool, chances are that one of them will succeed and break it. Thus, at very high antibiotic concentrations (to the right of the green region), the resistant bacteria also show negative growth, i.e. they die out.

When bacteria are exposed to the concentrations of the green region in Fig. 6.2, we face the risk of resistance evolution: resistant bacteria are fitter than sensitive bacteria (the dashed red curve is higher than the solid blue curve), and they are able to grow (the red curve is above the dotted zero line).

The level of antibiotic concentration a resistant cell can tolerate depends on the specific mutation. There are many mutations that confer resistance to low levels of antibiotics, whereas only a few make the cell resistant to high concentrations. Cells with several resistance mutations can usually withstand higher drug pressure than cells with just one mutation. For example, if the DNA that specifies the construction of the replication tool has two mutations, then the resulting tool would be altered to a greater extent, making it even harder for the antibiotic to attach to it.

Resistant bacteria may exist in the population prior to treatment, or they can appear by chance through mutations during treatment. Resistance is particularly likely if the drug dose is too low. (Remember Alexander Fleming's warning in his Nobel prize lecture!) In order to prevent the evolution of resistance, one ideally gives doses that are high enough that a single mutation in the DNA is not sufficient to achieve resistance. Acquiring two mutations that together provide resistance to these high doses is much less likely than gaining one. If we think of rolling two dice, getting two sixes is much less likely than getting just one six. There are, of course, more considerations entering the decision about antibiotic dosing; especially the dose should not be toxic to the patient.

Another option is to administer two antibacterial drugs simultaneously. If a different mutation is required for each, the bacteria would need two mutations to survive. This strategy has been used to treat tuberculosis from very early on. In many other cases of bacterial infection, combination therapy is not applied. The reasons are diverse and include an increased risk of side effects and higher monetary costs—or maybe also habit. Moreover, although the idea of combination therapy is very compelling, we do not know whether it is really always better than monotherapy. Exploring the benefits and risks of combination therapy is an active field of research.

6.2 Plasmids: Vehicles of Antibiotic Resistance Genes

Bacteria have a single chromosome that contains their genetic information. The essential genes that a bacterium needs, i.e. to grow, metabolise "food", and replicate, are all located on this chromosome. But in addition to the chromosome, bacterial cells can contain DNA in a second format called **plasmids**. Plasmids are (most often circular) DNA molecules that are separate from the chromosome and carry additional genes. During cell replication, they are also copied and transmitted to the daughter cells (just like chromosomal DNA).

There are many different types of plasmids. No one fully understands how or why all these plasmids persist as they are not essential for the cell. They are sometimes depicted as parasites (but read on). And now, here is something really outlandish: some plasmids are able to transfer and "infect" a second cell, similar to a parasite transferring from one host to another (see Fig. 6.3).

But plasmids have genes that influence traits. In other words, traits can be passed on from one cell to another. To understand the impact of this, let us imagine we shake hands with our neighbour, and suddenly we have their eye colour! Moreover, this transfer is possible between different bacterial species: imagine our cat rubs against our leg, and we grow a tail![2]

Back to antibiotic resistance: antibiotic resistance genes are often located on plasmids. For example, the plasmid can carry genes for "pumps" that transport the antibiotics out of the cell. Thus, a bacterial cell can become resistant not only by acquiring a mutation in the chromosome (as discussed above) but also by receiving a resistance plasmid. Resistance genes on plasmids can spread easily across bacterial communities, making their way from one species to another. This alone makes resistance on plasmids particularly dangerous, and what is more, plasmids often accumulate resistance genes to several antibiotics. If a cell receives such a plasmid, treatment with all of these antibiotics becomes ineffective. Thus, in the presence of antibiotics, such plasmids change from a burden into a great benefit for the bacterial cell.

6.3 Antibiotic Resistance in Commensal Bacteria

Each of us carries approximately as many bacterial cells as our own human cells (of the order of 30 trillion, i.e. $3 \cdot 10^{13}$). These bacteria reside in our gut, on our skin, and in other body parts. We call them *commensal bacteria*. The word *commensal* derives from medieval Latin, where "commensalis" means "sharing a table". These bacteria are normally harmless or even beneficial inhabitants of our body and interact with us in many complex ways that are far from being fully understood. For example, some of them help us digest food that we would otherwise be unable to process. They also protect us from

[2]Of course, this is a simplification. The process of plasmid transfer is more complicated than shaking hands or a cat rubbing against a leg.

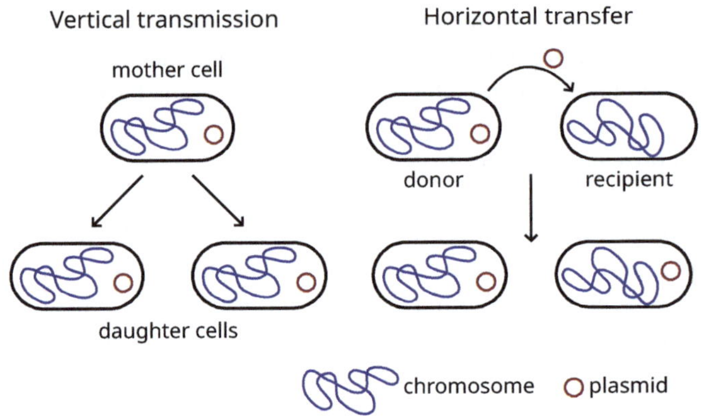

Fig. 6.3 The propagation of plasmids. On the left, we can see how a plasmid is passed from the mother cell to both daughter cells during cell division. The right side depicts the horizontal transfer of the plasmid: a copy of the plasmid is transferred from a plasmid-carrying to a plasmid-free cell. As a result, both cells carry the plasmid

pathogens: incoming pathogens need to compete for resources with our harmless commensal bacteria, which makes it harder for them to become abundant and cause an infection.

When we are sick and take antibiotics, we expose not only the pathogen (that we want to attack) but also part of our commensal flora—the community of our commensal bacteria—to the drug. This has two effects. First, by killing "good" bacteria, the flora is disturbed. One consequence of this disturbance is that the "bad", toxin-producing bacterium *Clostridium difficile* can become abundant and cause diarrhoea, a common side effect of antibiotic treatment. Alterations of the commensal flora due to antibiotic therapy might persist for years and may have unknown long-term consequences. Some researchers have suggested a link to an increased risk of certain diseases, such as Crohn's disease. Second, the antibiotics select for resistance in the commensals: drug-sensitive commensals are suppressed, and drug-resistant commensals spread. In fact, even without recent antibiotic treatment, we are very likely to harbour antibiotic-resistant bacteria in our commensal flora, which we may have acquired from contaminated food or from other people or which are just there (see Box: Antibiotic Resistance without Antibiotics).

Normally, we do not need to worry about resistant commensal bacteria in our bodies—they are just as harmless as their drug-sensitive relatives. However, commensal bacteria can cause nasty and life-threatening diseases if they end up in places where they do not belong, e.g. in the bloodstream or in the lung. Many infections acquired during hospital stays are caused this way, e.g. *Escherichia coli* that are usually found in the gut or other commensal bacteria can get into the bloodstream through catheters, causing sepsis. Treatment of these diseases can be extremely complicated when the disease-causing bacteria are antibiotic resistant. Methicillin-resistant *Staphylococcus aureus*, or MRSA, has caused problems for just this reason: it is a commensal bacterium that can turn pathogenic and cause bloodstream infections, pneumonia, or infections at surgical sites. MRSA are often resistant to multiple antibiotics of various classes, making infections hard to treat.

Resistance in the commensals can possibly turn problematic for a second reason. If the resistance genes are located on plasmids, they might be transferred from commensals to pathogens. In fact, researchers have indeed detected cases where *E. coli* and *Klebsiella* passed on their resistance genes to *Salmonella* within the human gut during treatment. In that case, the commensals are not themselves culprits but do serve as reservoirs of resistance genes.

Bacteria are complex organisms with great adaptive potential, and they are many. Whenever we use antibiotics, we exert selection pressure on bacteria that could lead to the evolution and

Where Does Selection for Resistance Occur?

There are three key areas where the selective pressure for antibiotic resistance is particularly strong:

Medicine: Whenever we take antibiotics (or give them to our pets), we promote the evolution of resistance. Antibiotics are, of course, often a life-saving medicine. However, antibiotics are used very widely, including for harmless infections that our immune system can usually handle on its own. Moreover, antibiotics are sometimes even taken if the patient has viral infection, which is not only useless (antibiotics do not act against viruses) but also imposes unnecessary selection pressure on our commensal bacteria (see main text) and hence increases the probability that they will become resistant. In many countries, antibiotics can be bought without a doctor's prescription.

Agriculture: Huge amounts of antibiotics are used in agriculture. Intensive livestock breeding requires high levels of treatment, and depending on the country and its regulations, antibiotics are also used preventatively and as growth promoters. Agriculture is a serious driver of antibiotic resistance, and resistant bacteria from livestock can be transferred to humans through contaminated food, manure on fields, etc. Therefore, many countries are increasingly tightening their regulations for antibiotic use in agriculture. For example, in both the EU and the USA, it is not permitted to use antibiotics as growth promoters.

Environment: Antibiotics (and antibiotic-resistant bacteria) are introduced to the environment through wastewater from households, agriculture, and pharmaceutical companies. In one study, researchers followed the course of a river in Canada and found that the amount of antibiotics and resistant bacteria detected along the river matched the type of land use at these sites. Remember, even low levels of antibiotics that do not kill the sensitive bacteria may select for resistance.

spread of resistance. Even if we use high doses or combinations of drugs, a risk of resistance remains. This is evolution happening in our bodies, in the bodies of our farm animals, and in our wastewater. And it is rapid. When a new antibiotic is released onto the market, it usually only takes a few years before the first resistant cases are detected. The development of new antibiotics cannot keep up with the speed of evolution.

So what can we do? The best way to slow down the evolution of resistance is to reduce the selection pressure, i.e. to use fewer antibiotics, and, when we do, to use them intelligently. But what does "intelligent use" entail? To really know that, we still need to learn more about bacterial evolution.

DEFINITIONS

Pathogenic bacteria: bacteria that cause disease

Commensal bacteria: bacteria that we carry in and on our bodies. Harmless under normal circumstances, they can also turn into pathogens

Broad-spectrum antibiotic: antibiotic that is effective against many different bacterial species

Plasmid: DNA element that is separate from the chromosome; antibiotic resistance genes are often located on plasmids. Some plasmids can "infect" other cells on cell-to-cell contact

SUGGESTED READING

- Hutchings MI, Truman AW, Wilkinson B. Antibiotics: past, present and future. Curr Opin Microbiol. 2019 Oct;51:72–80
- Andersson DI, Hughes D. Selection and Transmission of Antibiotic-Resistant Bacteria. Microbiol Spectr. 2017;5(4):10.1128/microbiolspec.mtbp-0013-2016

Antibiotic Resistance Without Antibiotics: A Soldier in World War I and an Amazonian Tribe

In 1915, a young British soldier, Ernest Cable, was hospitalised in a French hospital for dysentery. Dysentery is a diarrhoeal disease caused by bacteria of the *Shigella* species. It cost many the lives during World War I and continues to do so in developing countries. Unfortunately, Ernest Cable did not survive. However, the bacteriologist of the hospital isolated the bacteria, and they were deposited in the UK National Collection of Type Cultures. About 40 years later (in the 1950s), the bacteria were tested for antibiotic resistance. It turned out that they were resistant to penicillin and erythromycin. Almost 100 years after Ernest Cable's death, researchers sequenced the genome of the bacteria that had killed the soldier, and they found the genes conferring resistance to these two antibiotics. Now look at the dates: penicillin was first discovered in 1928 and only widely used much later, and yet, some bacteria were already resistant to it in 1915!

In 2015, researchers investigated the commensal bacteria of a remote Amazonian tribe, the Yanomami, for antibiotic resistance. As far as it is known, the Yanomami have never had contact with Western people before and have never taken antibiotics. And yet, their bacteria carry antibiotic-resistance genes!

Let us consider the origin of antibiotics. Most antibiotics are natural compounds (remember pyocyanase and penicillin from the introduction) or derived from natural compounds, i.e. natural compounds with some modifications (semi-synthetic). Only very few are fully synthetic, meaning that they are entirely designed and built synthetically without using any natural compound in the process.[a] That means bacteria have been confronted with many of these compounds long before we started using them as medicines, leading to the evolution of bacterial defence mechanisms. It is thus perhaps not surprising that we find resistance genes to those antibiotics even without any drug exposure. However, the Yanomami also harbour resistance genes to fully synthetic antibiotics. One possible explanation could be that these genes provide cross resistance to some natural and to these synthetic drugs.

[a]Examples include sulfonamides and fluoroquinolones.

- Clemente JC et al. The microbiome of uncontacted Amerindians. Sci Adv. 2015 Apr 3;1(3):e1500183
- San Millan A. Evolution of Plasmid-Mediated Antibiotic Resistance in the Clinical Context. Trends Microbiol. 2018;26(12):978–985

6.4 Questions

1. **When we talk about antibiotic resistance, we mean:**
   (A) The human body not being able to respond to the antibiotic treatment
   (B) The viruses surviving in the human body during the antibiotic treatment
   (C) The bacteria causing the disease not surviving in the human body during the antibiotic treatment
   (D) The disease-causing bacteria surviving in the human body despite antibiotic treatment
   (E) The human body resisting a bacterial disease in the presence of antibiotics
2. **You have the flu. You should take:**
   (A) A broad-spectrum antibiotic
   (B) An appropriate antibiotic at a high dose
   (C) No antibiotics at all
   (D) Antibiotics in combination
   (E) A low dose of an antibiotic
3. **Which of the following statements are true?**
   (A) Bacteria and other microbes (such as fungi) naturally produce antibiotics.
   (B) Resistant bacteria are always resistant to all available antibiotics.
   (C) Low antibiotic concentrations can give a selective advantage to resistant bacteria.
   (D) Antibiotic-resistant bacteria can spread to humans through contact with a live animal, food, or water carrying antibiotic-resistant bacteria.
   (E) Within the human body, antibiotic resistance can sometimes spread from one bacterial species to another.
4. **Commensal bacteria:**
   (A) Should urgently be cleared from our body
   (B) Can cause infections
   (C) Are harmless under normal circumstances even if resistant
   (D) Cannot evolve resistance to antibiotics
   (E) May act as a reservoir of resistance genes
5. **Plasmids:**
   (A) Can carry antibiotic resistance genes
   (B) Are separate from the bacterial chromosome
   (C) Cannot be passed on to the daughter cells
   (D) Might be transferred to other cells on cell-to-cell contact, but only when both cells belong to the same species
   (E) Carry the genes necessary for cell division

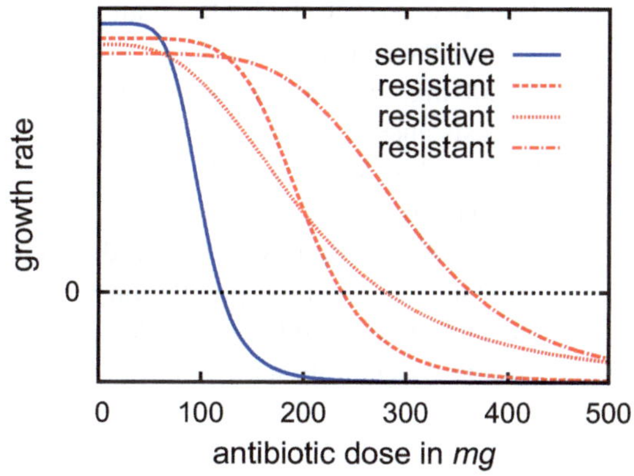

Fig. 6.4 **Growth rate of sensitive and resistant bacteria as a function of the amount of antibiotics administered**

6. **Figure 6.4 shows the growth rate of sensitive and resistant bacteria as a function of the amount of antibiotics administered. The three red curves correspond to three different resistant mutants. All of these mutants differ from the sensitive strain by a single mutation. No other mutations are known that—on their own—can confer resistance to the drug. The optimal drug dose should (1) suppress the growth of the sensitive strain, (2) make the evolution of resistance by a single mutation impossible, and (3) be as low as possible to minimise toxic side effects. Which of the following doses best fulfils all conditions?**
   (A) 50 mg
   (B) 100 mg
   (C) 250 mg
   (D) 400 mg
   (E) 500 mg
7. **Looking at the figure from the previous question, which of the following statements are true?**
   (A) Above doses of 300 mg, there is no risk of resistance.

(B) The dose needs to be higher than 200 mg to clear the sensitive strain.
(C) At 200 mg, three different mutations confer resistance.
(D) Doses below 50 mg do not select for resistant bacteria.
(E) The relative proportions of the different resistant strains in the population will be the same at both 100 mg and 200 mg antibiotic concentrations.

8. **Which of the following measures can reduce the spread of resistant bacteria?**
(A) Regular handwashing and other hygiene measures when sick
(B) Vaccination against bacterial infections
(C) Antibiotic treatment of all family members if one person is sick
(D) Use of leftover antibiotics when you feel ill the next time
(E) Staying at home when ill

6.5 Project: Modelling the Evolution of Antibiotic Resistance

In this activity, you will simulate a bacterial infection and observe how the dynamics of different bacterial types change under various antibiotic treatment strategies, including the impact of patient decisions and the spread of resistance through horizontal gene transfer. This project will help you understand the complexities of antibiotic resistance, a major global health threat.

Material

- Many dice (~100). You will need dice of at least four distinct colours (e.g. green, yellow, orange, red).
- Alternatively, Internet access to a virtual dice roller: www.random.org/dice
- Paper and pencil for recording data
- A flat surface with margins for rolling dice

Basic Procedure: Understanding Your Bacterial Population

1. The dice represent the bacteria in a bacterial infection. Each dice represents a population of one million bacteria of a specific type. Some of the bacteria are resistant to some antibiotics from the beginning of the simulation, and some of them may even share their resistance with others.
    - **Sensitive bacteria (S):** Sensitive to both antibiotic A and antibiotic B. Represented by **green** dice
    - **Resistant to A bacteria (RA):** Resistant to antibiotic A but sensitive to antibiotic B. Represented by **yellow** dice
    - **Resistant to B bacteria (RB):** Resistant to antibiotic B but sensitive to antibiotic A. Represented by **orange** dice
    - **Multi-resistant bacteria (MR):** Resistant to both antibiotic A and antibiotic B. Represented by **red** dice
2. Each day in the simulation represents a 24-hour period during which either an antibiotic dose is applied or, in the absence of antibiotics, bacteria have a chance to reproduce. One day is simulated by rolling the dice once.
3. When dice are rolled, the number shown on each die indicates the fate of that particular bacterial population (reproduce, survive, or die). This depends on the bacterial type and whether an antibiotic is present.
    - In the *absence* of antibiotics, three bacterial types (Sensitive, Resistant to A, and Resistant to B) reproduce on a roll of 2, 4, or 6 and survive on a roll of 1, 3, or 5. Multi-resistant bacteria reproduce only on a roll of 6 and survive on a roll of 1 to 5.
    - Table 6.1 describes what happens to the different bacteria in *presence* of two types of antibiotics.
4. Reproduction is represented by adding one more dice of the same colour to the population, death by removing the dice from the population, and survival by keeping the dice in the population.
5. What is more, resistance can spread among bacteria even without reproduction, through horizontal gene transfer (HGT)! This process occurs when bacteria are in close contact and transfer genetic material. After the dice have been rolled, observe the remaining dice (surviving bacteria): If two dice representing different bacterial types end up physically touching each other after the roll, an HGT event occurs between them.
    - When a Sensitive (S) bacterium is in close contact (touching) with a Resistant to A (RA) bacterium, the Sensitive bacterium **changes to Resistant to A (RA)**. Replace the green dice with a yellow dice.
    - When a Sensitive (S) bacterium is in close contact with a Resistant to B (RB) bacterium, the Sensitive bacterium **changes to Resistant to B (RB)**. Replace the green dice with an orange dice.
    - When a Sensitive (S) bacterium is in close contact with a Multi-resistant (MR), the Sensitive bacterium **changes to Resistant**

Table 6.1 Changes in populations of different types of bacteria in the presence of two types of antibiotics

	Antibiotic A			Antibiotic B		
Bacterial type	Reproduction	Survival	Death	Reproduction	Survival	Death
Sensitive (S)	–	6	1–5	–	6	1–5
Resistant to antibiotic A (RA)	6	2–5	1	–	6	1–5
Resistant to antibiotic B (RB)	–	6	1–5	6	2–5	1
Multi-resistant (MR)	–	1–6	–	–	1–6	–

Table 6.2 Number of bacteria alive after each day in the simulation

	Number of bacteria alive after each day									
Bacterial type	day 1	day 2	day 3	day 4	day 5	day 6	day 7	day 8	day 9	day 10
S										
RA										
RB										
MR										

to B (RB) or A (RB)—your choice. Replace the green dice with a dice of a suitable colour.

- When a Resistant to A (RA) bacterium is in close contact with a Resistant to B (RB) bacterium, **both of them change to Multi-resistant (MR)**. Replace the yellow and orange dice with two red dice.
- No HGT occurs between identical types (e.g. S and S) or if one of the partners is already MR.

Simulation

A person is sick, and the bacterial population causing their infection is represented by 88 green dice (S), 3 yellow dice (RA), and 9 orange dice (RB). Initially, there are 100 dice (representing 100 million bacteria) in total.

Part 1: Initial Treatment (Days 1–5)

1. Doctor's Initial Choice: The doctor does not know the exact composition of the bacterial population's resistance. Therefore, he or she randomly chooses Antibiotic A or Antibiotic B to treat the infection. In your simulation, perform a coin toss (e.g. Heads for Antibiotic A, Tails for Antibiotic B) to determine which antibiotic is initially chosen.
2. Apply Treatment: For the next 5 days (Days 1–5), apply the chosen antibiotic daily.
3. Record Data: At the end of each day, note the number of dice of each bacterial type (S, RA, RB, MR) into Table 6.2.

Part 2: Patient's Decision and Doctor's Re-evaluation (Days 6–10)

At the end of Day 5, the patient may already feel much better—actually, they may feel so well that they may decide to stop the treatment immediately. However, even if they feel well, they still may go to keep their appointment with the doctor to evaluate the treatment.

4. **Patient's choice:** Observe the total bacterial count at the end of Day 5, and decide for the patient: Will they stop the treatment on their own, or will they go to the doctor? If the patient decides to stop the treatment: Continue the simulation for five more days (Days 6–10) with *no antibiotics applied.*
5. **Doctor's action:** The doctor re-evaluates and switches the antibiotic. If Antibiotic A was used for Days 1–5, use Antibiotic B for Days 6–10. If Antibiotic B was used for Days 1–5, use Antibiotic A for Days 6–10. If the patient decides to see the doctor: Continue the simulation for five more days (Days 6–10) with the *newly chosen antibiotic applied daily.*
6. Create a line graph showing the population dynamics of each bacterial type (S, RA, RB, MR) and the total bacterial population over the

10 days of the simulation. You may use the provided graph template in Fig. 6.5 or create your own.

Analysis

After completing the 10-day simulation and recording all your data, address the following questions:

1. Describe the state of the bacterial population at the end of the entire 10-day treatment period.
2. Based on your simulation, was the overall treatment successful in eliminating the infection? Can we consider this person healed? Explain.
3. If any bacteria remain, identify their type(s), and discuss why they persisted.
4. How did the decision on Day 5 impact the bacterial population dynamics during Days 6–10?
5. If you had chosen the alternative path at Day 5, what do you hypothesise the outcome might have been? Briefly explain your prediction.
6. In which aspects does this dice-rolling simulation effectively represent the reality of antibiotic treatment and bacterial evolution?
7. In which aspects is this simulation an oversimplification of biological reality?
8. Would using *both* Antibiotic A and Antibiotic B at the same time (combination therapy) from Day 1 likely result in a faster and more complete clearing of the infection in this specific scenario? Explain your reasoning.

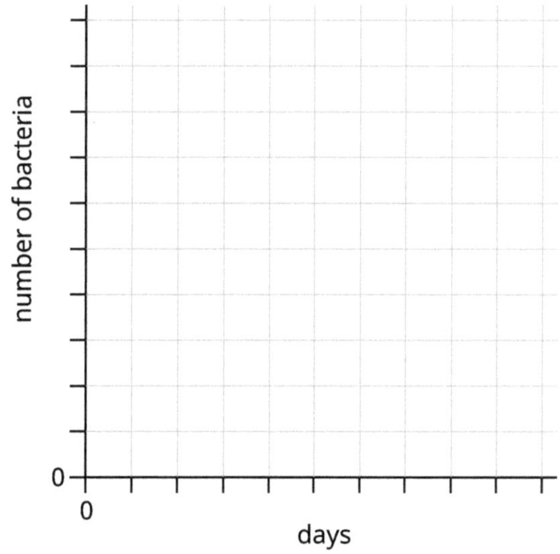

Fig. 6.5 **Population dynamics of each bacterial type**

9. What would be the primary disadvantage(s) of routinely using combination therapy for every bacterial infection in a real-world clinical setting?
10. What did you observe regarding the Multi-Resistant (MR) bacteria? How did their presence and behaviour affect the overall outcome of the infection, particularly after the antibiotic switch (if applicable)?
11. Considering the persistent challenge posed by multi-drug resistant bacteria (often called "superbugs") as illustrated by your simulation, research and describe at least two major strategies that are currently used to combat their spread.

This project was developed by Kristína Hudáková. The use of dice to model stochasticity in biological processes is an adaptation of established educational simulations.

A Glimpse into the World of Human Viruses

7

Hildegard Uecker and Barbora Trubenová

> ▶ Viruses are tiny biological agents that possess a genome but cannot replicate on their own. In order to replicate, they need to infect host cells and hijack the host's machinery. Therefore, one can argue that they are not really living organisms. Alive or not, viruses are very successful at tricking and hijacking cells. There are an estimated 10^{31} virus particles on this planet, and viruses infect all forms of life. They play an important role in many ecosystems and shape life on Earth. In this chapter, we will have a look at viruses that infect humans. But don't forget—no form of life is safe from a viral attack!

Viruses are classified as **DNA viruses** and **RNA viruses** depending on whether their genetic material is stored as DNA or as RNA. About 70% of all viruses are RNA viruses. As we already know (see Chap. 2), RNA is less stable than DNA, and RNA replication is more prone to errors. Thus, RNA viruses have a higher mutation rate than DNA viruses.

Outside of host cells, the genetic material of a virus is packed within the **capsid**, a shell made up of proteins. The entire unit (genetic material and capsid) is called a **virion**. When a virus infects a host cell, the viral DNA or RNA is released from the capsid into the cell, where it is replicated by the molecular tools of the host cell. Finally, viruses need to spread from cell to cell within an organism and from one organism to another (see Box: Viral transmission: How do viruses spread).

We are all infected by viruses. Luckily, many of them do not cause symptoms and remain unnoticed. They may have negative long-term effects, be entirely harmless, or even provide benefits—it is often hard to know. We only notice viruses when they cause diseases.

7.1 Viral Diseases

Viruses are responsible for many human diseases, including the common cold, influenza, measles, chickenpox, rubella, hepatitis, AIDS, Ebola, Zika, and others. Some viruses can even cause cancer (Peyton Rous was awarded the 1966 Nobel Prize in Medicine for this discovery).

H. Uecker (✉)
Institute of Integrative Biology, ETH Zürich, Zürich, Switzerland

Department of Theoretical Biology, Max Planck Institute for Evolutionary Biology, Plön, Germany
e-mail: uecker@evolbio.mpg.de

B. Trubenová
D-USYS, ETH Zürich, Zürich, Switzerland

Eawag, Dübendorf, Switzerland

eduB, Bratislava, Slovakia
e-mail: barbora.trubenova@env.ethz.ch

B. Trubenová et al. (eds.), *Everyday Evolution: Unveiling Life's Hidden Patterns*,
https://doi.org/10.1007/978-3-032-07050-0_7

Most (if not all) human viruses derive from animal viruses. For example, the measles virus most likely evolved from an ancestor of rinderpest, a virus that infects cattle. Normally, viruses that are well adapted to an animal host are not well adapted to humans. This is because viruses need to interact with their host's proteins to complete their life cycle. These proteins are not always identical in animals and humans, which means the viral tools cannot be used on the human version of a protein. However, if the virus is at least not entirely unfit in the new environment of the human body and also has some chance to spread between humans, it might accumulate adaptive mutations and evolve into a well-adapted human pathogen. Some animal viruses already have the potential to spread in a human population and just need an opportunity to make the jump. Infectious diseases caused by pathogens that infect both humans and animals are called zoonoses. Some viruses become purely human viruses over time, such as the measles virus.

The emergence of new human viruses from animal viruses is facilitated through close contact between animals and humans (providing the virus with many chances to infect humans) and large, dense communities (allowing the virus to spread between humans more easily). Changes in human lifestyle can therefore contribute to the emergence of new diseases. For example, researchers hypothesise that the evolution of the measles virus has occurred around 2500 years ago, when humans started to live in communities that were large enough to maintain the virus. (However, there is much uncertainty in the estimates both for the emergence of measles and that of large settlements.) Just a few years ago, we witnessed how the jump of a virus from bats to humans sparked a global pandemic. We will talk more about the COVID-19 pandemic later.

Even after the successful transition from an animal to a human pathogen, viruses continue to be challenged by the human immune system. On the one hand, the genes coding for the components of our immune system can evolve through mutation and recombination and selection for variants that are better at fighting common (or less common) pathogens. This is a process happening across human generations. On the other hand, the immune system of a single person itself is adaptive and can—once an infection occurs—respond and produce new tools tailored to fighting the specific virus attacking the person (however, some viruses manage to hide from our immune system).

One of these tools is **antibodies**. The production of antibodies is triggered by surface proteins of the virion (which belong to the capsid or to an envelope covering the capsid). Substances that trigger an immune response are called **antigens** (antigen = *anti*body *gen*erator). The antibodies produced by the immune cells inactivate the virus by binding to these antigens. Antibodies are antigen specific, i.e. they can only bind to the antigens they are made for. Another immune system entity that fights viruses are killer T-cells that circulate through the body to find and kill infected cells. Killer T-cells recognise infected cells from the viral antigens present on their surfaces. However, just like antibodies, killer T-cells are antigen specific: infection by a particular virus triggers the production of specialised killer T-cells that can act against it. Following an infection, we keep producing the specific antibodies and killer T-cells for a long time, sometimes for decades. With many diseases, we can nevertheless become infected recurrently throughout our lives. Whether this is because immunity wanes over time or because the virus changes (as discussed further below), or a combination of both, is a subject of research in each specific case.

Our immune system has many more components. In particular, it also has non-adaptive parts that act rapidly and are not specific to a given pathogen. For example, we also have natural killer cells that are not antigen specific and constantly "patrol" our body to kill any infected cell as a quick first response to invading pathogens. Another important defence mechanism of our immune system is that cells, when infected, may secrete molecules that serve as warning signals to other cells. After receiving these signals, cells become more selective about which substances they let in—making infection less likely—and more ready to degrade nucleic acids—destroying the virus if it manages to enter.

Viral Transmission: How do Viruses Spread?

Viruses can pass to new cells in two ways—either *horizontally* or *vertically*.

Horizontal transmission refers to the spread of the virus from an infected to an uninfected cell of the same generation. In order for horizontal transmission to take place, new virions must be produced. To do this, the virus makes use of the molecular machinery of the host cell to build a capsid and assemble the virion. Then, to reach uninfected cells, the new virions are usually released into the liquid surrounding the cell. This release can happen all at once, killing the host cell, or, more gradually, leaving the host cell intact. Free virions cannot move by themselves. Rather, they rely on luck—on random encounters with new host cells. Virions can also reach uninfected cells through *cell-to-cell spread*, i.e. by moving directly from an infected to an adjacent uninfected cell without passing through any liquid. This direct transmission is conducive to viral spread within an organism because it is fast and helps the virus escape the individual's immune system.

With **vertical transmission**, the virus is transmitted from the mother cell to the daughter cells during cell division. Some viruses incorporate their genetic material into the host genome so that it gets replicated together with the host DNA. RNA viruses that use this strategy are called retroviruses. Upon infecting a cell, they reverse-transcribe their RNA into DNA. The viral DNA is then integrated somewhere in the genome of the host cell. We will learn more about an important retrovirus—the HIV virus—below.

Transmission between individuals can occur either horizontally or vertically. Horizontal transmission between different individuals can, for example, happen through free virions in saliva droplets. Other viruses use *vectors*—some outside agent—to spread from one individual to another. For instance, most plant viruses use insects to get from one plant to another. Vertical transmission can occur from the mother to the foetus. This happens with the rubella virus, which causes German measles. Normally, German measles is not a serious disease. However, if a pregnant woman passes the virus to her foetus, this can cause serious harm to the child's health, even severe birth defects. Furthermore, viruses can be passed on to the next generation by infecting cells of the germ line (egg/sperm). Viruses that are transmitted via the germline are usually not very harmful to their host—otherwise, they would compromise their own survival. We will talk more about such transmission and its consequences later.

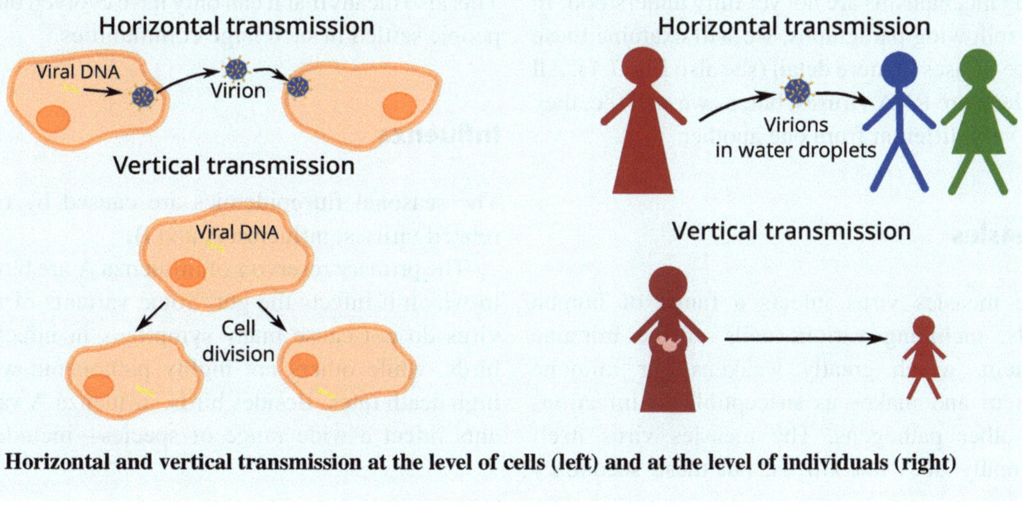

Horizontal and vertical transmission at the level of cells (left) and at the level of individuals (right)

If a virus gains mutations that alter its antigens, the old antibodies and killer T-cells will not work anymore, and our immune system must produce new, suitable ones to fight the evolved virus. There is great variation in the patterns of antigenic evolution across viruses. Some viruses evolve antigenic changes very slowly. Once we have manufactured the antibodies (and other tools) to fight them, we are immune for the rest of our lives (provided immunity does not wane) since the virus does not change a lot over decades. An example is the measles virus. Other viruses change more quickly. While they might not escape the immune system during the course of infection, a slightly changed variant might come back one or 2 years later and be able to infect us anew. For instance, influenza viruses do not evolve much during the time they are in a person, but they evolve a great deal from one season to the next. Yet other viruses perpetually escape our immune systems, leading to chronic infections. The reason can be that their antigens evolve very rapidly or that we cannot eliminate the virus quickly, giving it sufficient time to change. One example is the human immunodeficiency virus (HIV), which evolves a lot within every single patient and repeatedly escapes our immune system.

Why viruses are so variable in their antigenic evolution is an important topic of ongoing research. Researchers have gained some insight into why the measles, influenza, and human immunodeficiency viruses adapt to the pressure of our immune systems to varying extents, but the underlying mechanisms are not yet fully understood. In the following paragraphs, we will examine these three viruses in more detail (see also Fig. 7.1). All of them are RNA viruses, but, as we will see, they are very different from one another.

Measles

The measles virus infects a range of human cells, including various cells of the immune system, which greatly weakens our immune system and makes us susceptible to infections by other pathogens. The measles virus itself normally does not kill us, but these secondary infections can be life-threatening. In countries with good healthcare, about 0.1% of infected people die. However, in low-income countries several percent of patients die following a measles infection, and even more in particularly vulnerable groups such as among displaced children with some estimates being as high as 30%. Further, while measles is usually cleared fully by our immune system, in rare cases, it can infect our central nervous system, persist there, and ultimately lead to a fatal neurological disease.

Fortunately, once we recover from measles, we have acquired lifelong immunity. The protein that serves as an antigen in measles is needed for entry into the host cell. Research suggests that the specific structure of this protein cannot easily change without compromising the virus' ability to attach and infect host cells. The vast majority of mutations that change the antigen hence make the virus unable to complete its life cycle. Moreover, the measles virus has multiple antigenic sites on one of its proteins, and those are targeted by different antibodies. It would therefore need to accumulate several mutations to escape our immune response, which is rather unlikely to happen.

The lifelong immunity acquired through infection implies that the measles virus can only persist if the community size is large enough that sufficiently many new susceptible individuals are born every year. Otherwise, the virus fails to find new victims fast enough and dies out at some point. The measles virus requires a community size of around 250,000–500,000 inhabitants to persist. This also means that it can only have evolved once people settled in such large communities.

Influenza

The seasonal flu epidemics are caused by two related viruses, influenza A and B.

The primary reservoir of influenza A are birds, in which it infects the gut. Some variants of the virus do not cause many symptoms in infected birds, while others are highly pathogenic with high death rates. Besides birds, influenza A variants infect a wide range of species—including

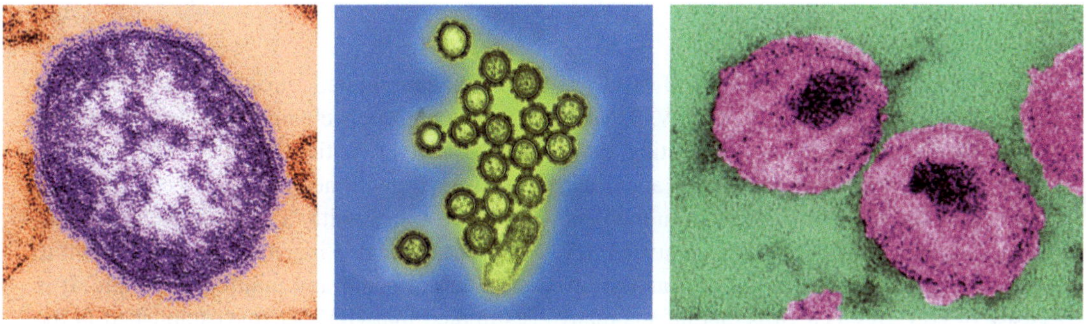

Fig. 7.1 **Various viruses.** Colorised transmission electron microscopic images of the virions of measles, influenza, and HIV (from left to right). (Images: (1) CDC/Cynthia S. Goldsmith; William Bellini, Ph.D. (2) National Institute of Allergy and Infectious Diseases (3) CDC/A. Harrison; Dr. P. Feorino)

humans. In humans, it infects the epithelial cells[1] of the respiratory system. Influenza B is an almost exclusively human pathogen.

Influenza viruses possess two proteins on their surface: hemagglutinin (H) and neuraminidase (N). These two proteins are crucial for the entry of the virus into the cell and for its release from the cell. They are also the main antigens of influenza. For influenza A, there are 18 subtypes of hemagglutinin and 11 subtypes of neuraminidase, which exist in many combinations. Influenza A viruses are classified on the basis of their H and N subtypes, e.g. H1N1 or H3N1. The subtypes differ considerably from each other, and not all of them work on human cells.

Unlike measles, the influenza antigens are flexible, allowing for mutations to occur without detrimental effects to the virus. The process of antigenic changes due to mutations is called *antigenic drift*. Influenza A has yet another way to evolve. The RNA of the influenza virus is not a continuous strand but is divided into segments (eight for influenza A). When two different strains (or *variants*) infect the same cell, these segments can be shuffled. This is called *antigenic shift*. Due to its broad host range, co-infection of cells by two different strains is relatively common for influenza A (unlike influenza B). This co-infection could, for example, happen in pigs that have contact with both birds and humans: the pig might be simultaneously infected by an avian and a human influenza A strain. Sometimes, these newly combined viruses are able to infect humans, and since they are different from previously encountered strains, we may not have pre-existing immunity. Newly emerged strains, whether they are a mix of other strains or directly jumped from birds to humans, can cause pandemics with many infected patients. One such pandemic was the Spanish flu, which started in 1918 and lasted 3 years, causing 50–100 million deaths.

HIV

The human immunodeficiency virus (HIV) is a retrovirus, i.e. it integrates its genetic code into the host genome. It targets various cells of the immune system that carry the protein CD4 on their surface, for example helper T-cells. HIV, therefore, harms the immune system. Without treatment, it causes so many helper T-cells to die over time that the immune system breaks down, resulting in AIDS (acquired immunodeficiency syndrome). People then become easy targets for other pathogens, leading to very poor health and, ultimately, death. HIV thus kills indirectly. Unlike influenza and measles, HIV does not get transmitted through everyday contact. Infection occurs through unprotected sex, contaminated blood transfusions, needles shared between drug addicts, or from mother to infant at birth.

It is estimated that one HIV-infected helper T-cell produces on the order of 10,000 new viri-

[1] Epithelial cells are the cells at the surfaces of organs or cavities like the lung.

The Evolution of HIV

It took scientists a great deal of effort to uncover the evolutionary history of HIV. By sampling blood from monkeys, collecting thousands of samples of faeces from apes, and comparing the DNA sequences of human and simian viruses, they eventually identified SIV (Simian immunodeficiency virus) as the ancestor of HIV. SIV is a virus that infects many non-human primates in Africa. While we do not know with certainty how it jumped over to humans, the accepted view is that hunters caught the virus from the blood of killed monkeys and apes. The infection of hunters with SIV has probably happened many times, but either the foreign virus could not proliferate well in the new host or could not be transmitted to other people. In a few cases, however, the transition was successful, and the virus accumulated mutations that made it better adapted to existence within and transmission between humans. Transmission was, however, not only facilitated through viral adaptation but also through changes in human lifestyle in the early twentieth century. Many people in Africa transitioned from small communities to large cities with a high population density, dramatically increasing the contact rate between infected and uninfected people. It is estimated that HIV became established in the human population in the 1930s. Since people also started to travel more and farther, the virus could spread widely. One transition of SIV to humans was particularly successful. It occurred in chimpanzees and led to the HIV strain that is called HIV-1 Group M, where "M" stands for "major". This group of HIV viruses reached the USA by the 1970s and is responsible for the worldwide HIV epidemic. Other groups of HIV are less prevalent: group O, for instance, is found almost exclusively in West Central Africa, where it has infected about 100,000 people.

ons. It is possible that the cell bursts to release the virions, but usually it stays intact. How else do helper T-cells die in the process? At various points of the viral life cycle, viral proteins may trigger molecular pathways that ultimately lead to the death of the infected cell. Or the cell may be identified as infected and killed by killer T-cells. However, the vast majority of helper T-cells that die in an HIV infection are not infected. They either destroy themselves following failed incomplete infection or upon receiving signals from infected cells.

Not all infected cells start producing new virions right away. Some of them just continue to divide, passing the virus in their genome on to the daughter cells. These cells are called latently infected cells. They are the reason it is impossible (at present) to completely remove HIV from the human body. The virus hides within these cells from all currently available treatments and from our immune system—no antigens are presented at the surface of these cells—until, at some point, the production of new viruses starts again.

Our immune system produces antibodies and killer T-cells against HIV, but it cannot keep up with the virus for several reasons. First, HIV destroys precisely those cells that are required for the immune response. Our body produces new helper T-cells, but at some point, it gets tired, and production slows down. Second, the virus evolves within our body during the infection, escaping our immune response—mutation and recombination within the viral population result in new variants that the already-present antibodies and killer T-cells cannot bind to. HIV has an extremely high mutation rate, which, combined with a very high replication rate, leads to the rapid emergence of mutants. These mutants can proliferate undisturbed until our body recognises them and can produce suitable new antibodies and build new killer T-cells. It is an arms race between the virus and our immune system that our immune system finally loses. Luckily, nowadays, we have very good drugs available that very efficiently suppress the virus and allow patients a nearly normal life.

Interesting Fact

A very small fraction of people—probably less than 1%—are resistant to HIV. They carry a mutation that alters the shape of one of the surface proteins on their helper T-cells, preventing the virus from entering the cell.

7.2 Vaccines

Vaccines work by introducing viral antigens into the body, thereby triggering an immune response similar to the one triggered by the pathogen, leading to the production of antibodies. Vaccines may, for example, contain dead viruses or modified live viruses that, for instance, are unable to replicate at body temperature ("live vaccines"). Depending on the virus, vaccines either prevent disease entirely or reduce symptoms, providing protection from severe disease. Furthermore, similar to natural infections, vaccines for some diseases provide us with long-lasting immunity, while for others, immunity wanes over time. The potential for antigenic evolution of a virus strongly influences how easily we can develop vaccines against a specific disease and how effective they are. As we have seen in the previous section, the speed of antigenic evolution differs greatly between the measles virus, the influenza virus, and HIV.

The measles virus is so invariable that vaccines derived from a strain isolated in the 1950s still provide protection today. Since the measles virus changes so little and the vaccine (similar to infection) elicits a lasting immune response, we only need to get vaccinated once to possess lifelong immunity. This has a very important implication. Above, we have learned that the measles virus can only persist if the community size is large enough. Through vaccination, we can effectively reduce the community size. This, in turn, means that we could drive the measles virus to extinction if sufficiently many people are vaccinated (measles is a purely human pathogen without an animal reservoir). This strategy has been successful with its close relative, the rinderpest virus. In 2011, it was officially declared to be eradicated. This has been the second time in human history that a virus has been driven to extinction through vaccination. The first virus to be eliminated by a global effort was the highly dangerous smallpox virus, which was declared extinct in 1980. (In both cases, vaccination programs were complemented by other measures such as close surveillance and quarantining.)

The First Vaccine

The cowpox virus mainly infects rodents but also other animals, including cows (the reason for its name) and humans. In humans, it leads to a rather harmless disease that resembles a mild form of smallpox, which in turn is a severe human disease with high mortality. By the second half of the eighteenth century, it had been realised that people who had overcome cowpox (e.g. milkmaids who had contracted the disease from cows) would not fall ill of smallpox. The physician and scientist Edward Jenner (1749–1823) pursued this further, about a 100 years before viruses were even discovered. In 1796, he inoculated an 8-year-old boy with pus from the cowpox blisters of a milkmaid. Two months later, he inoculated him with the pus of smallpox blisters—the boy did not turn ill. Edward Jenner's development of the very first vaccine was, at the same time, the first step on the way to the eradication of smallpox. The word "vaccination" is derived from the Latin word 'vaccinus', which means "coming from a cow".

Influenza, unlike measles, requires a shot with an updated vaccine every year. Normally, the vaccine contains antigens of three different influenza variants. In order to choose the variants to include, researchers predict which strains are most likely to spread in the next season. No vaccine has yet been found for HIV, but scientists are working on it. One difficulty is the huge genetic diversity of the HIV virus. Another is that we do not know what a successful immune response to

Herpes Viruses: Hidden But Ready to Come Back

Herpes simplex 1 (HSV 1) is the culprit behind the blisters on our lips, while its cousin—HSV 2—causes the more serious sexually transmitted diseases. Chickenpox, a very common childhood disease, is also caused by a herpes virus (the varicella-zoster virus).

While different herpes viruses cause very different diseases, most of them have something in common: once we have them, they are almost impossible to get rid of. This is because, during infection, small numbers of viruses hide within cells and escape the immune system. In this way, the virus can persist in the host for the host's entire life. Most of the time, it does not cause any symptoms. Only from time to time do these hidden viruses reactivate and cause another round of infection—another blister on the lip. Herpes viruses are extremely widespread among humans, and a latent form of some herpes virus can be found in almost everyone.

Interestingly, the varicella-zoster virus that causes chickenpox the first time it infects a person causes a different disease with different symptoms when it is reactivated. Once a person has overcome chickenpox, the virus persists in the nerve roots near the spinal cord. In times of weak immunity—for instance, as the result of other diseases, stress, or simply old age—the virus can reactivate and travel from the nerve root to the endings in the skin, producing painful blisters—a disease called shingles. Unlike the rash in chickenpox, the blisters usually only occur in a small area of the skin. One cannot contract shingles from another person. However, a person with shingles spreads the varicella-zoster virus, which can cause chickenpox in someone who has not had it yet.

HIV even looks like—nobody has it. Luckily, we have developed some very potent HIV drugs that control the virus extremely well. Moreover, some drugs are available as pre-exposure prophylaxis (PrEP). A recent breakthrough is a new PrEP medication that only needs to be injected once every 6 months.

Interestingly, vaccines can actually trigger viral evolution. First, viral strains with mutated antigens are favoured by natural selection. Unlike the ancestral strains, mutants may replicate not only in unvaccinated but also in vaccinated hosts. Thus, they can spread more easily through the host population. However, the evolution of vaccine resistance is luckily rare, and most vaccines to human diseases, such as the measles or the polio vaccine, have remained effective through time. Viruses that continuously escape from natural immunity simultaneously also escape from vaccine-induced immunity (such as the influenza virus or SARS-CoV-2). Knowing how much of this escape is driven by each type of immunity requires careful research in each case. Second, viruses might evolve to become more virulent in response to vaccine-acquired immunity. For instance, the virus causing Marek's disease in poultry has evolved higher virulence in response to a series of vaccines. Third, attenuated viruses used in live vaccines can, in rare cases, acquire mutations that turn them pathogenic. This is, for example, a risk in the oral poliovirus vaccine. However, overall, while vaccine-induced viral evolution is possible, the risk is very low, and the huge benefits of vaccines largely outweigh them. One cannot stress enough how much vaccines benefit us.

Importantly, through vaccination, we protect not only ourselves but also others in our community, provided the vaccine prevents infections (or at least reduces onward transmission). If we do not develop an infection, we cannot transmit the virus to people around us, including those who cannot be vaccinated because their immune system is compromised or because they are too young to be vaccinated. If everyone around them is vaccinated, however, these people are protected as well. This is called **herd immunity**.

7.3 The COVID-19 Pandemic

On December 31, 2019, Chinese authorities informed the WHO of cases of pneumonia for which the causal pathogen was unknown. The pathogen was identified as a coronavirus on January 7, and on January 10, its genome sequence was published. Coronaviruses are a group of related RNA viruses that infect a large number of mammals and birds. The newly emerged one is the seventh (known) coronavirus that infects humans. The first four have been circulating in the human population for a long time. They usually cause mild symptoms and are responsible for 15%–30% of cases of the common cold each year. The other two—SARS-CoV-1 (last cases detected in 2004) and MERS-CoV (Middle East respiratory syndrome-coronavirus)—emerged in the twenty-first century, and both have very high mortality rates. Analysis of the RNA sequence of the newly emerged virus revealed that it is closely related to SARS-CoV-1, giving it its name SARS-CoV-2 (see Fig. 7.2 for an image). The disease caused by the new coronavirus was named COVID-19, which stands for "**co**rona**vi**rus **d**isease 20**19**". In many people, the virus causes symptoms similar to those of a bad cold or the flu. Some people develop no symptoms at all. For others—especially for the elderly and for those whose health is already compromised for other reasons—it can be life-threatening.

Starting with the local outbreak in Wuhan, the virus quickly spread around the world. On January 30, the World Health Organization (WHO) declared the outbreak a public health emergency of international concern; on March 11, it was declared a pandemic. By that time, 114 countries were affected, almost 120,000 cases had been identified, and more than 4,000 people had died. Six weeks later, the number of detected cases had reached more than 2.5 million, and more than 190,000 people had died of COVID-19. In most countries around the globe, governments demanded that people keep physical distance from each other, entire countries were in lockdown, borders were closed, and flight connections were halted.

Crossing Species Boundaries

Where did this new virus that shook the world so badly come from? In Sect. 7.1, we have learned that human viruses often originate from animal viruses, and this is also true for SARS-CoV-2, which made its way from bats to humans, although we still do not know how exactly that happened. As far as we know, the other two coronaviruses that caused epidemics in the twenty-first century—SARS-CoV-1 and MERS-CoV—also have their origin in bats, which are a common asymptomatic host of coronaviruses. For SARS-CoV-1, palm civets likely acted as intermediate hosts, while MERS-CoV (Middle East respiratory syndrome-coronavirus) jumped from bats to dromedary camels, from which it is transmitted to humans.

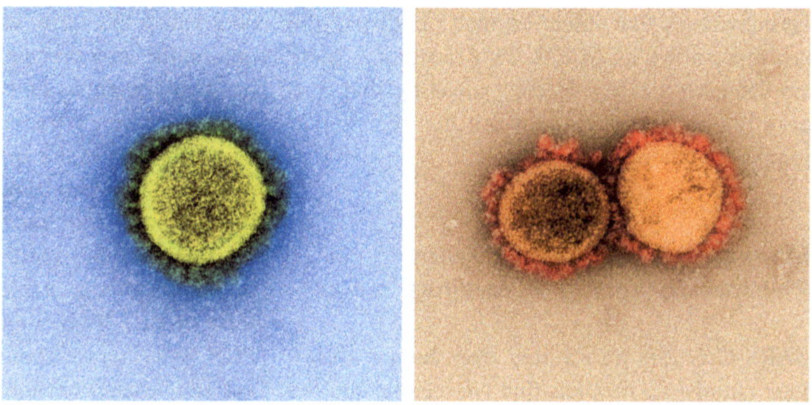

Fig. 7.2 **SARS-CoV-2.** Colorised electron microscope images of SARS-CoV-2. The "crown" gives coronaviruses their name. (Image: NIAID, License: CC-BY-2.0 (https://creativecommons.org/licenses/by/2.0))

Coronaviruses are known to cross species boundaries quite easily. What makes them so evolvable? First, their genomes are large. There is thus a lot of opportunity for mutations to occur. Yet, their per-nucleotide mutation probability is actually lower than that of other RNA viruses since they possess repair mechanisms, which most RNA viruses do not. Beyond mutations, there is another way in which the genome of coronaviruses can change. When two different coronaviruses infect the same cell, there is a high probability that they recombine with each other—i.e. swap part of their DNA. This creates a new virus that may have quite different features. Thus, due to their overall abundance, plus ample opportunities for mutating and recombining, coronaviruses have high genetic diversity. This makes it likely that at least some lineages will evolve to match a new host well enough to somehow proliferate.

A second factor contributing to the high evolvability of coronaviruses is the viral protein which they use to attach to the host cells and enter them. Host cell attachment and entry are mediated by the so-called spike (S) glycoprotein that is part of the viral envelope. The coronavirus spike (S) glycoprotein is very robust to changes, which means that the corresponding gene can acquire many mutations without compromising the viability of the virus. The existing variation in this glycoprotein means that there likely exists a variant that can attach to a new host. Further adaptation to the new host is then possible.

Finally, even though distinct strains of coronaviruses may use different receptors for the same host, the two SARS coronaviruses (and related coronaviruses that infect animals) both use the receptor known as ACE-2. This receptor is quite similar across many species, meaning that it is easy for coronaviruses that use this receptor to jump to another host.

Managing the Pandemic: From Physical Distancing to Vaccines

SARS-CoV-2 is transmitted from human to human mainly via droplets that we release into the air while breathing and speaking and especially when we cough or sneeze. The size of these droplets varies. Large ones fall to the ground quickly. The water in small droplets evaporates before it reaches the ground, leaving behind minerals, proteins, and viral particles. These tiny water droplets and their remains are called aerosols and can remain in the air for a long time. People infected with COVID are infectious even if they do not experience symptoms yet or stay entirely asymptomatic. This makes the spread of the virus so difficult to contain: Recognising and isolating people with symptoms is relatively easy. Recognising and isolating people who are infected but perfectly healthy, in contrast, is hard. What tools were available to manage the pandemic?

Perhaps the most straightforward way to avoid infections is for everyone to keep a safe physical distance from others and not gather indoors. Physical distancing is, of course, no long-term solution. However, it was pivotal at various points in the pandemic to avoid overburdening the healthcare system. In the absence of any interventions, a virus that enters a susceptible population grows exponentially—not even the best healthcare system can keep up with such an explosion of patients. Spread through the air can also be prevented through masks. While masks were in many places in short supply at the beginning of the pandemic, production ramped up over time, offering an alternative to physical distancing.

Another important tool are tests that allow us to tell whether a person is infected with SARS-CoV-2. The person can then quarantine until they are no longer infectious. Through contact tracing, people who met this person and may have contracted the virus can be identified and can quarantine as a precaution as well. (Later in the pandemic, many countries developed smartphone apps for contact tracing: if a person registered their infection in the app, all recent contacts were informed automatically.) Tests furthermore make it possible to monitor the developments of case numbers: are they increasing, warranting caution, or are they at a stable low level, and we can relax a bit? Two important types of tests are PCR (polymerase chain reaction) tests and rapid antigen tests. For both, swabs are taken from the nose or the throat. PCR tests detect the viral RNA and are very sensitive, i.e. they detect even very small amounts of virus. The disadvantage is that the sample has to be sent to a laboratory, and the tested person therefore does not obtain the result immediately. Antigen tests detect viral antigens. They are less sensitive and may miss cases, but they are much cheaper and very fast: the result is available within 15 minutes. Antigen self-tests can easily be done by everyone at home. PCR tests were performed right from the start of the pandemic on symptomatic patients, but to use them as a management tool for prophylactic testing (e.g. before boarding planes), the capacities first needed to be built. Antigen tests were developed later and were mostly important from late 2020 on. PRC and antigen tests were not the only tests in use. So-called LAMP (loop-mediated isothermal amplification) tests detect the viral RNA itself and are very accurate, similar to PCR tests. However, they do not require analysis by a laboratory and give rapid results, and self-test kits for use at home became available at similar times as antigen self-tests. They are more expensive than antigen tests though and not quite as simple to use.

Over time, antiviral drugs to treat COVID-19 infections became available. The real game changer, however, were vaccines. Vaccines against COVID-19 were developed at record speed, and the first vaccines were approved as early as December 2020. Take a moment to appreciate what an amazing achievement that is! For COVID-19, along with other vaccine types, a new type of vaccine was used for the very first time, so-called mRNA vaccines. Unlike classical vaccines, these vaccines do not contain any viral antigens. Instead, they contain the template—the mRNA—to build them. As the mRNA enters our cells, we produce the viral proteins ourselves for our immune system to react. Irrespective of the type of COVID vaccine, immunity to infection unfortunately wanes quickly. However, protection against severe disease persists for longer. Vaccines, therefore, eventually allowed us to transition back into normal life. The public health emergency was declared over on 5 May 2023. The virus is here to stay, though.

The control measures implemented by governments differed widely between countries and the different phases of the pandemic. In spring 2020, many governments implemented rules for physical distancing, for example, limiting the number of people who were allowed to meet. Events such as concerts were, of course, cancelled, restaurants could not open, people had to work from home when possible, and in many countries, schools were temporarily closed. Later, alternative measures complemented or replaced physical distancing. For example, wearing a mask became mandatory in public transport, or having an official negative test result was needed to board planes, attend concerts, or go to a restaurant. Various restrictions continued in many countries for about 2 years.

Managing a pandemic is not only about curbing viral spread—the social and economic needs of people need to be taken into account as well. For example, school closures are a measure which can have severe consequences, especially for children from poor socio-economic backgrounds. The relative importance of the pros and cons of certain measures shifts as case numbers go up or down, as new knowledge about the virus accumulates, and as immunity in the population builds up, requiring constant adjustment of interventions. Unsurprisingly, people can have very diverging opinions about what constitutes an appropriate measure, leading to non-adherence and protests against the government rules and social frictions between different groups in a country's population. Pandemics thus have severe effects on societies in numerous ways.

The Evolution of SARS-CoV-2 During the Pandemic

With so many infections all around the globe, it is no surprise that SARS-CoV-2 has evolved a great deal from early 2020 to now. A virus may evolve in various traits; the most important ones are perhaps its transmissibility, its escape from pre-existing immunity, and its virulence. The WHO has labelled five variants as so-called

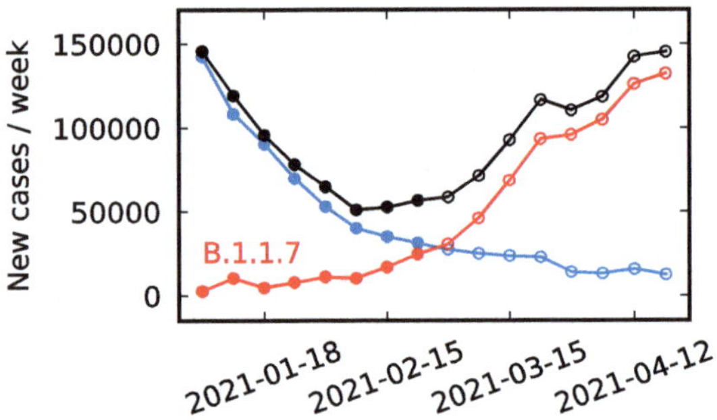

Fig. 7.3 Spread of COVID-19 in Germany in early 2021. The black curve shows the weekly number of cases. The red curve shows those caused by the Alpha-variant (B.1.1.7), and the blue curve those caused by all others. (Figure reproduced with modifications from Traulsen et. al (2022) The Covid-19 pandemic: Basic insights from basic mathematical models. NAL-live 2022.3, v1.0. Licence: CC BY 4.0, https://creativecommons.org/licenses/by/4.0/)

variants of concern as they differed in important ways from the previous ones—the Alpha, Beta, Gamma, Delta, and Omicron variants.

The first new COVID variant—the Alpha variant—was detected in September 2020. It turned out to be much more transmissible than the original virus. As Fig. 7.3 illustrates, the great lockdown that was in place in Germany since mid-December was sufficient to suppress the spread of the original strain, but the Alpha variant could spread despite control measures, leading to a new wave of infections. With more variants arising, the virus became even more transmissible.

The Omicron variant, which was first detected at the end of 2021 and is still the dominant variant today, was able to infect people with pre-existing immunity. By the time it appeared, many people had been infected or vaccinated already, which gave it a great selective advantage over the other circulating variants. New subvariants continue to show antigenic drift. Luckily, the virus overall evolves slowly enough that our vaccine updates can keep up with it.

A lot of people believe that viruses always evolve to become less virulent—i.e. less harmful to their hosts—over time since not killing their hosts helps them persist. This is, unfortunately, a too simplistic view of evolution. However, we were lucky: Omicron is indeed rather mild. In combination with the immunity that we have now almost all gained through vaccination and/or infection, the virus has become less dreadful than it was. Nevertheless, it can still cause severe disease and even death. Furthermore, even after mild infections, some people have serious, long-lasting symptoms, suffering from so-called long COVID.

7.4 Our Genome Marked by Viruses

Interactions with viruses have shaped our genome in two fundamental ways. The first is based on the fact that viruses are pathogens from which we must defend ourselves. Most obviously, viruses and other pathogens impose selection on the genes associated with our immune system. A particular drastic example of a disease that imposed selection on immune-related genes is the Black Death, a pandemic of plague in the fourteenth century. Researchers estimate that individuals who were homozygous for a specific allele had a 40% survival benefit over those homozygous for the disadvantageous allele. Less obviously, viruses also put selection pressure on many other genes. This is because viruses interact with many different kinds of proteins—e.g. those on the surfaces of cells—not just with the ones related to immune responses.

A second way in which viruses have shaped our genome is more direct. As explained above, some viruses can integrate into the genome of their host. Normally, they are in somatic cells, and in order to proliferate, they must infect other cells of the same type and eventually of another host. Sometimes, however, they end up in cells of the germline (eggs, sperm). Then, they are passed on to the children of that individual, who then carry the virus in every one of their cells. A virus that made it into the germline and is passed on vertically to the following generations is called an endogenous virus. Initially, endogenous viruses can still escape from host cells to infect other hosts. Over time, however, mutations accumulate in the viral genes until the virus can no longer produce viable virions and leave the host cell. They remain in our DNA as viral fossils. About 8% of our DNA consists of these viral relics. And now, it gets exciting: some of these viral genes give rise to new human genes that perform fundamentally important tasks in our body.

Interesting Fact

One example of a viral fossil is the gene *Syncytin-1*, which is expressed exclusively in the placenta, where it is crucial for fusing the placenta to the uterus, thus securing the transport of nutrients from the mother to the foetus. The same gene found in humans is also present in other primates. Similar genes are found in many other mammal species, such as mice, cats, and dogs. However, these genes do not derive from the same virus as our *Syncytin-1* gene. What does this mean? It means that there have been several infection events in different mammal lineages with different viruses whose genes have been co-opted by different hosts to fulfil the same function! Not all mammals have *Syncytin*-like genes. Actually, there is great diversity in how exactly the placenta functions across species, and one of the reasons is thought to be infection (or non-infection) by a range of retroviruses across species since the placenta (and hence live birth) evolved ca. 130 million years ago.

7.5 Making Use of Viruses

We have talked a lot about how viruses make us ill. In this section, we will see how we can purposefully use them to *treat and cure* diseases. With scientific knowledge, we are turning our enemy into our friend.

Some mutations in our genes can cause diseases. For instance, sickle cell anaemia, a very severe and painful disease, is caused by a single-nucleotide mutation in the haemoglobin gene. This mutation distorts the shape of the red blood cells, impairing oxygen transport in the blood. Patients often need blood transfusions.

What if we could replace or complement this gene with a functional, unmutated copy? We do not need to change it in all cells in the body, only in the source of red blood cells—the bone marrow, where the hematopoietic stem cells are found. These cells constantly divide and produce all kinds of blood cells. If we could replace at least part of the stem cells of a patient who has sickle cell anaemia with stem cells that have a healthy gene, that would permanently cure the person. One way to achieve this is to transplant bone marrow from healthy individuals but this requires a suitable donor and is risky. It would be much better if we could take the patient's own stem cells and modify them. And this is where viruses enter the scene.

As we mentioned above, retroviruses integrate into the genome of the host cell. This property can be exploited for **gene therapy**. In order to develop a therapy for sickle cell anaemia, researchers engineered an artificial virus that carries a functional β-globin gene in its genome. This artificial virus is based on the human immunodeficiency virus, but the researchers made so many changes to it that it cannot harm us (it cannot even replicate within cells). To treat patients with sickle cell anaemia, doctors extract blood stem cells from the patient and let the artificial virus infect the extracted stem cells. This does not harm the cells, but they now carry a healthy copy of the gene. The doctors then implant the infected cells back. This therapy was used for the first time in 2014 to treat a 13-year-old boy. It was a success—even after several years, the boy still had enough healthy red blood cells that he no longer needed blood transfusions! In 2023, such gene therapy was approved as a regular treatment for sickle cell anaemia in the USA (see Chapt. 17 for another approved gene therapy).

Viruses have been omnipresent throughout human history and evolution, and they remain omnipresent in our lives. They have adapted to proliferate within our bodies and spread from human to human, and they keep evolving. In this chapter, we have only glimpsed a small portion of their world. There is so much more yet to discover about this ubiquitous form of life—or not life.

How Our Immune System Fights and Remembers Pathogens

The human immune system is very complex, involving many cell types, molecules, steps, and pathways. Here, we explain the basic principles of a very important component of our adaptive immune system, the production and maintenance of antibodies.

Antibodies inactivate the virus by binding to its antigens. They are antigen specific, i.e. a given antibody can only bind to specific antigens. When we are infected by a new virus that we have not previously encountered, we do not have the right antibodies readily available. However, our immune system is able to manufacture them, and this is why we call this part of it "adaptive". How does this work?

In the production of antibodies, two important cell types of our immune system are involved—*B-cells* and *helper T-cells*. The "factory" for the production of B-cells is in the bone marrow, whereas helper T-cells are produced in the thymus. We constantly produce new B- and helper T-cells for several decades (later in life, the production wanes). High rates of somatic recombination during their production lead to an enormous diversity in their cell pools. It will become clear in a moment why this diversity is crucial.

B-cells circulate in our body to "search" for pathogens. They possess receptors at their surface that can bind to viral antigens. A given B-cell can only bind specific antigens. Yet, thanks to the immense diversity in the pool of B-cells, it is nevertheless very likely that we possess B-cells that can bind the present antigens, at least weakly. Once the antigen has been bound to the B-cell, the cell engulfs it, chops it into little pieces, and presents some of the pieces on its cell surface.

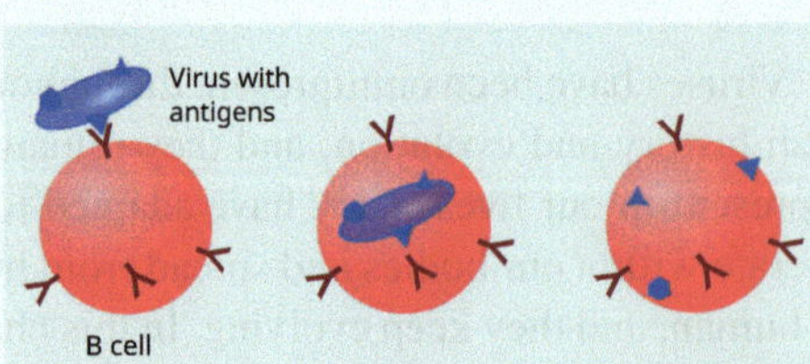

Acquisition and presentation of viral antigens by a B cell. If the receptors on the surface of a B-cell match the antigens of a virus, they bind the antigen (left), engulf it (middle), and present pieces of it on their surface (right)

Then, a helper T-cell (circulating in the body as well) attaches to the antigens presented by the B-cell. This "activates" the B-cell. Again, a given helper T-cell can only bind to specific antigens, and the diversity in the helper T-cell pool is important.

Once activated, the B-cell starts producing antibodies. However, these first antibodies often only bind weakly to the viral antigens (they very much resemble the B-cell receptors). The B-cell, therefore, undergoes several rounds of replication with very high mutation rates and selection for variants that bind the antigens more strongly. It then creates a lineage of two different kinds of B-cells, plasma B-cells and memory B-cells. The plasma B-cells are the cells that produce the perfect antibodies.

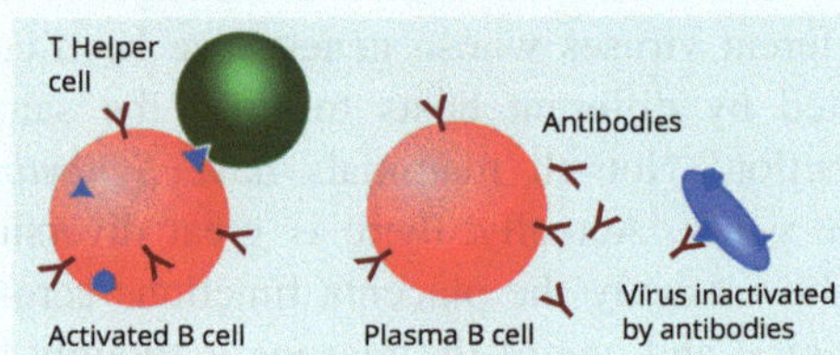

Inactivation of viruses by antibodies. If a helper T-cell attaches to the antigens presented on the surface of the plasma B-cell, the plasma B cell is activated and produces antibodies, which bind to antigens on the surface of the viruses and inactivate them.

Plasma B-cells and memory B-cells can live for long periods of time. The plasma B-cells continue to produce antibodies, and when the pathogen attacks us again, we are prepared and can fight it off quickly—before it can cause an infection. If the existing antibody concentration is insufficient to clear the pathogen, the memory B-cells can be reactivated to produce new plasma B-cells. For some viruses, we keep immunity for a long time in this way. For others, immunity wanes over time (even if the virus does not change). It is not well understood which factors influence the duration of immunity. One of them is how strong the immune response was in the first place.

DEFINITIONS

Capsid: the protein shell that encloses the genetic material of a virus
Virion: a complete, infectious virus particle outside of a host cell
Antigen: a molecule that can trigger an immune response in the body
Antibody: a protein that our immune system tailors to bind to specific antigens and inactivate the pathogen
Herd Immunity: when a sufficient percentage of a population is immune to a disease, making its spread unlikely and offering protection to those who are not immune
Gene therapy: a medical technique that involves introducing, removing, or changing genetic material in a patient's cells to treat or prevent disease

SUGGESTED READING

- Mushegian AR. Are There 10^{31} Virus Particles on Earth, or More, or Fewer? J Bacteriol. 2020 Apr 9;202(9):e00052-20
- Longdon B, Brockhurst MA, Russell CA, Welch JJ, Jiggins FM. The evolution and genetics of virus host shifts. PLoS Pathog. 2014;10(11):e1004395.
- Woolhouse ME, Haydon DT, Antia R. Emerging pathogens: the epidemiology and evolution of species jumps. Trends Ecol Evol. 2005;20(5):238–244
- Düx A et al. Measles virus and rinderpest virus divergence dated to the sixth century BCE. Science. 2020 Jun 19;368(6497):1367–1370
- Kennedy DA, Read AF. Why does drug resistance readily evolve but vaccine resistance does not? Proc Biol Sci. 2017;284(1851): 20162562
- Weiss RA, Stoye JP. Virology. Our viral inheritance. Science. 2013 May 17;340(6134): 820–1

7.6 Questions

1. **Which of the following statements is (are) true?**
   (A) Viruses can replicate outside of host cells.
   (B) The common cold is a bacterial disease.
   (C) Influenza A only infects humans.
   (D) HIV evolved from SIV.
   (E) It is possible to eradicate certain viruses through vaccination.
2. **Picture the following scenario: A country introduces a rubella vaccination. Before this, no one in the population had been vaccinated. From then on, 40% of all children receive a rubella vaccine at 15 months of age. Few people get vaccinated later in life. What are the likely consequences?** *(Information: Infection with rubella confers lifelong immunity.)*
   (A) Over the years, the average age at which unvaccinated people are infected increases.
   (B) The virus spreads much more rapidly than before.
   (C) Fewer children get the disease.
   (D) If a rubella epidemic happens 25 years after the introduction of the vaccine, a large number of children with birth defects due to the viral infection of the foetus will be born ("large" compared to a population without vaccination).
   (E) The virus infects more infants (< 15 months) than before.
3. **Cowpox is an infectious viral disease. The virus mainly infects rodents but also other animals, including cows (the reason for its name) and humans. In humans, it leads to a rather harmless disease that resembles a mild form of smallpox, which in turn is a severe human disease with high mortality. By the second half of the eighteenth century, it had been realised that people who had overcome cowpox (e.g. milkmaids who had contracted the disease from cows) would not fall ill with smallpox. How can this be explained?**
   (A) The cowpox virus occupies the same cells that the smallpox virus needs to infect to cause disease. Since they are already occupied by the cowpox virus, the smallpox virus cannot infect them anymore.
   (B) Cowpox and smallpox have similar antigens, and antibodies developed against cowpox are also effective against smallpox. This is unusual.
   (C) When the cowpox and the smallpox virus infect the same cell, their genomes recombine. The recombined virus is harmless. Since so many cells are coinfected by both viruses, no pure, harmful smallpox virions will be assembled.
   (D) The cowpox virus stimulates certain bacteria on our skin to change their surface proteins. As a consequence, the smallpox virus is able to infect the bacterial cells and prefers those over the human cells.
   (E) The cowpox virus kills the cells that the smallpox virus needs to infect to cause disease. These cells are only renewed very slowly, meaning that the smallpox lacks cells to infect for up to a year following a cowpox infection.
4. **Which of the following statements are true?**
   (A) People are resistant towards RNA viruses, because cells cannot make copies of RNA.
   (B) If we cannot observe any HIV antigens on a human cell's surface, that means this cell is not infected by this virus.
   (C) If you get vaccinated against influenza, you will be protected against flu for at least the next 3 years.
   (D) Viral antigens can evolve by different mechanisms, and so the virus may become resistant to antibodies developed by the host organism.
   (E) Antibiotics development is important to fight viral infections.
5. **Which of the following factors would make it more likely that viruses cross species boundaries, i.e. jump from one host species to another?**
   (A) High levels of genetic diversity within the virus population
   (B) Recombination within the virus population

(C) Sudden close contact between two species that have been evolving separately for some time
(D) High levels of antibiotics in the environment
(E) High levels of habitat fragmentation within the host species

6. **8% of the human genome appears to be of viral origin. Which of the following statements best describe this part of the genome?**
   (A) The main function of this part of the genome is to provide defence against viral pathogens.
   (B) This part of the genome cannot be copied by DNA polymerase.
   (C) This part of the genome is unique to humans and distinguishes them from their closest living relatives, the chimpanzees.
   (D) Genomic regions of viral origin are a result of ancient infections in which a sperm or egg cell (in our ancestors) got infected.
   (E) Mutations in these regions are more likely to be deleterious.

7.7 Project: The Importance of Vaccination

Some people cannot be vaccinated. This concerns, for example, young babies, pregnant women, or people whose immune systems are severely compromised. A weakened immune system can be due to diseases such as cancer but also to immunosuppressive drug treatment (after organ transplantation, for instance, people need to take medication that suppresses their immune responses such that their body does not reject the foreign organ). The immune systems of immunocompromised individuals may not respond to the vaccine, resulting in a lack of immunity. Second, while the risk of complications is very low in healthy people, life vaccines (e.g. the measles vaccine) can be very dangerous to people with a weakened (or not yet fully developed) immune system. At the same time, the people who cannot be vaccinated are often the ones for whom the disease is most dangerous.

While the people themselves cannot be vaccinated, those in their surroundings who are healthy can be. By doing so, they protect the unvaccinated from infectious diseases as well. This is called *herd immunity*. Let us look at this phenomenon more closely.

Material and Preparation

In this project, you will simulate the spread of a measles virus in a population of individuals, some of whom are vaccinated, while others are not. Initially, two people are infected with measles. They have contracted the disease outside of the community we are looking at. You will examine three different populations of 100 individuals, each with a varying number of vaccinated individuals: 0, 50, and 95. You will need:

- Red, blue, and white paper[2]
- A piece of green paper
- Scissors
- Small box or other container

Cut about 100 paper squares of each colour. These squares represent healthy but unvaccinated (white), vaccinated (blue), and infected (red) individuals. Using these squares, you will create different populations for the different experiments. Cut two more squares from the green paper—these represent individuals who cannot be vaccinated due to one of the reasons listed above and for whom the disease is particularly risky ("high-risk individuals").

Simulation Steps

Read the steps of the simulation below carefully. You will follow them to simulate three different populations (A, B, and C).

1. **Let's mix:** four individuals go to a party, another four to a shopping mall, another four to a restaurant and to many other places. Close your eyes and pick 40 squares from the box, and place them in **10 different piles of 4 squares each**. These ten piles represent different groups of people coming into close contact, which could spread the disease if any of them are infectious.
2. **Infection:** Now look at the piles. If any of the piles contain an infected individual, all the other unvaccinated individuals in the group are infected with measles: replace all the white squares in this pile with red ones. Vaccinated individuals (blue squares) are not infected (if there are any—in Simulation A, there are no vaccinated individuals).
3. Are there any susceptible individuals (green squares) in the infected pile? If yes, mark this event in the table, and replace them with red ones.
4. **Counting:** Count the number of infected and healthy individuals, and write these numbers into the corresponding tables below.
5. Place all 100 squares back into the container (make sure they are of the correct colour—some of them are new, some old). Shake thoroughly.

[2]Of course, you can use different colours.

6. Repeat steps 1–5 six times to simulate the spread of the disease over time, and use your data to fill in corresponding tables.

Before you start the simulations, write a clear and testable hypothesis for how the number of vaccinated individuals affects the spread of the disease.

Simulated Scenarios

A. No One Is Vaccinated

In the first simulation, no one is vaccinated. Take 96 white squares (unvaccinated individuals), 2 red squares (individuals with measles), and 2 green squares (high-risk individuals), and shake them thoroughly in a small container. Follow the steps described in "Simulation Steps" above, and fill in Table 7.1.

B. 50 % of the Population Is Vaccinated

In the second simulation, only 50 % of individuals are vaccinated. Take 46 white squares (unvaccinated individuals), 50 blue squares (vaccinated individuals), 2 red squares (individuals with measles), and 2 green squares (high-risk individuals), and shake them thoroughly in a small container. Follow the steps described in "Simulation Steps" above and fill in Table 7.2.

Table 7.1 Spread of measles in an unvaccinated population

Time point	0	1	2	3	4	5	6
Infected	2						
Healthy	98						
High-risk individuals affected [y/n]							

Table 7.2 Spread of measles in a partially (50 %) vaccinated population

Time point	0	1	2	3	4	5	6
Infected	2						
Healthy	98						
High-risk individuals affected [y/n]							

Table 7.3 Spread of measles in a partially (95 %) vaccinated population

Time point	0	1	2	3	4	5	6
Infected	2						
Healthy	98						
High-risk individuals affected [y/n]							

C. 95 % of the Population Is Vaccinated

In the last simulation, almost everyone, except for very few individuals, is vaccinated. Take 1 white square (unvaccinated individual), 95 blue squares (vaccinated individuals), 2 red squares (individuals with the measles), and 2 green squares (high-risk individuals), and shake them thoroughly in a small container. Follow the steps described in "Simulation Steps" above and fill in Table 7.3.

Analysis

1. Make a graph (or three different graphs) showing the changes in the numbers of healthy and sick individuals over time in all simulated populations.
2. What do you observe? In a few short sentences, summarise the most important results of the simulations.
3. What happened with the high-risk individuals in your simulations? Were they affected in all four scenarios?
4. If not, why not? If yes, when?
5. Compare your hypothesis and your results. Do your results confirm or contradict your hypothesis? Explain.
6. What do you think would happen if more people went out every time?
7. In order to achieve herd immunity, doctors recommend vaccinating 90 to 95 % of the population. Why is the percentage so high?
8. You only ran the simulation once for each scenario. Is this enough to make scientific conclusions?
9. Go to https://en.edub.sk/everyday-evolution and look for Vaccination simulator there. We have prepared a computer simulation for you that runs the same simulations as you just

did by hand. It runs 1000 times and gives you the average numbers. Play with the number of vaccinated individuals. What do you observe? Are the results consistent with the results of your simulation? If not, can you explain?

This project was developed by Kristína Hudáková, Hildegard Uecker and Barbora Trubenová

Genetics and Nature Conservation

8

F. Gözde Çilingir

- Did you know the Earth's evolutionary history has witnessed at least five mass extinction events? In these incidents, species went extinct due to geological, climatic, or external factors such as an asteroid impact. Now, fast forward to the present era. Due to the increasing human population, overconsumption, and environmentally unfriendly policies, the current species extinction rate is 100 to 1000 times higher than the background extinction rates revealed in the fossil records from the times between mass extinction events! Scientists warn that we are in the middle of a sixth mass extinction, the only event of this kind caused solely by a single species: humans.

Biological diversity, or biodiversity, refers to the variety of life on Earth. Scientists have shown that we are losing biodiversity at an unprecedented rate as a result of several factors, most significantly habitat destruction, overexploitation of organisms, global climate change, pollution, and invasive species. Biodiversity can be observed at the level of **ecosystems**, **species**, and **individuals** (Fig. 8.1). If we want to preserve the diminishing biodiversity, we must protect it at all levels and consider all the factors responsible for its destruction. In other words, we cannot save ecosystems without saving the species that inhabit them, and we cannot save species without preserving the diversity between individuals, most importantly their **genetic diversity**. Therefore, I will mainly focus on genetic diversity in this chapter.

Genetic diversity is the variation in genetic composition among individuals in a population, a species, an assemblage, or a community. It is shaped by millennia of evolutionary history, a product of many years of trial and error in evolutionary strategies. Genetic diversity affects the fitness of an organism, a population, or a species.

As an example, let us think about pandas, a mammalian species that is vulnerable to extinction. Imagine that scientists cloned these animals, and eventually, hundreds of cloned pandas were reintroduced into their natural habitats. Now, suppose the pandas' environment changes: for instance, a disease hits the population. Because these clones are genetically identical, their immune systems are also mostly the same. If these clones lack the specific gene variants that can help them fight the disease, they will all die. In contrast, if these panda populations were not composed of identical clones, there would be greater genetic diversity. Thus, there is a higher chance that some individuals carry gene variants

F. G. Çilingir (✉)
Eidgenössische Forschungsanstalt für Wald, Schnee und Landschaft (WSL), Birmensdorf, Switzerland

Department of Evolutionary Biology and Environmental Studies, University of Zurich, Zürich, Switzerland
e-mail: goezde.cilingir@wsl.ch

B. Trubenová et al. (eds.), *Everyday Evolution: Unveiling Life's Hidden Patterns*,
https://doi.org/10.1007/978-3-032-07050-0_8

Fig. 8.1 **Various levels of biodiversity.** Diversity can be observed at different levels

that help them survive. The pandas that manage to survive will have the chance to mate, reproduce, and become the ancestors of future generations.

In other words, genetic diversity provides the evolutionary raw material that increases a population's chance of adapting to its changing environment. Therefore, maintaining genetic diversity is crucial for populations to endure. Yet, genetic diversity depends on several internal factors, such as the size of a population, and external conditions, such as environmental effects.

8.1 Measuring Genetic Diversity

To better understand the processes that maintain or reduce diversity, we first need to have good measures of genetic diversity. One such measure is the level of **heterozygosity** within a population.

As we already know, if a diploid individual carries exactly the same alleles (gene variants) at the very same position (locus) on a particular chromosome, then that individual is **homozygous** for that allele; if it carries two different alleles at the locus, then it is **heterozygous** at that locus. The "observed" heterozygosity at a locus, denoted by H_O, is then the proportion of individuals that carry two different alleles at that locus.

In order to quantify inbreeding, we typically compare the observed heterozygosity with the level of heterozygosity we would expect in a hypothetical randomly mating population with no inbreeding. This is also known as the "expected" heterozygosity, denoted by H_E, and can be calculated from the allele frequencies at any locus (see Box: Measuring Genetic Diversity). Sometimes, we are interested in estimating the (observed or expected) heterozygosity across the entire genome, rather than at specific loci. This can be calculated by averaging over H_O (or H_E) calculated for different loci across the genome.

In populations with high levels of inbreeding, the observed heterozygosity is usually much lower than the expected heterozygosity. This means that there are fewer heterozygotes than one would expect based on allele frequencies at various loci. Since deleterious (harmful) alleles are often recessive, i.e. they decrease fitness (e.g. by causing diseases) only in their homozygous form, lower levels of heterozygosity are generally associated with lower fitness.

Interesting Facts

For example, let's look at Scotland's red-billed choughs (a bird species in the crow family). The population currently consists of fewer than 60 breeding pairs and is of major conservation concern. These choughs have recently been affected by "lethal blindness" in nestlings—that is, nestlings are born blind or visually impaired and do not survive past fledging. Scientists have shown that a recessive deleterious allele causes blindness; only when an individual inherits a deleterious allele from both mother and father do they become blind. Individuals who inherit only one copy of the allele are known as carriers.

Measuring Genetic Diversity

Let's consider a population of a hundred diploid individuals for which we have genotype information at one locus with two alleles: 'A' and 'a'. This means that individuals can have one of three possible genotypes at this locus: AA, Aa, or aa. Suppose the numbers of individuals of the three types in a sample are:

AA	Aa	aa
32	16	52

The **observed heterozygosity** is simply the proportion of heterozygotes at the locus, which is:

$$H_0 = \frac{n_{Aa}}{n_{AA} + n_{Aa} + n_{aa}} = \frac{16}{32 + 16 + 52} = 0.16.$$

To calculate the **expected heterozygosity**, we must first determine the frequencies of the two alleles in the sample. The frequency of allele 'A' is simply:

$$p_A = \frac{n_{AA} + n_{Aa}/2}{n_{AA} + n_{Aa} + n_{aa}} = \frac{32 + (16/2)}{32 + 16 + 52} = 0.4.$$

Similarly, the frequency of the alternative allele 'a' is $p_a = 1 - p_A = 0.6$. The expected heterozygosity is then

$$H_E = 2 p_A p_a = 2 \times 0.4 \times 0.6 = 0.48,$$

which is simply the probability that an individual either inherits the 'A' allele from the mother and the 'a' allele from the father or vice versa. The expected heterozygosity has another interpretation: it is the probability that two alleles sampled at random from the population are different. Thus, H_E is sometimes also called the **gene diversity**.

We can also calculate the **inbreeding coefficient** at the locus as:

$$F = 1 - \frac{H_0}{H_E},$$

which in this example is equal to

$$1 - (0.16/0.48) = 0.67.$$

Strong inbreeding results in higher values of F. In general, we calculate F from genome-wide data by taking the ratio of average H_0 and average H_E at many loci spread across the whole genome.

Depending on the context, additional measures of genetic diversity may be used. For instance, when characterising the genetic diversity at a specific gene, we may be interested in the number of distinct alleles or variants of the gene that are present in the population. In the example above, there are only two possible alleles—'A' and 'a' at a locus. However, for example, human blood types are determined by a locus with three possible alleles: I_0, I_A, and I_B. And certain genes—such as those involved in mate recognition in flowering plants or immune response in animals—can have > 10 distinct alleles.

8.2 Threats to Genetic Diversity

Now that we know how diversity is measured, let's look at the most common threats that reduce it.

The Danger of Declining Population Sizes

Smaller populations are more vulnerable to unpredictable demographic, environmental, and genetic effects. For example, imagine two populations, one consisting of 10,000 individuals and the other just 100. If a random environmental event, for instance, a natural disaster such as a hurricane, fire, or drought, takes place and kills 50 individuals in each population, that would mean the former loses just 0.5% of its population, whereas the latter loses 50%!

Let us examine the genetic effects of population reduction over several generations. Picture a small population where individuals can carry blue, yellow, pink, and green alleles (see Fig. 8.2), with each colour appearing in different numbers. Now, suppose that the population size decreases suddenly and drastically: by chance, only five blue, one green, one yellow, and three pink alleles remain. The remaining individuals reproduce (Generation 2), and another population reduction occurs, this time leaving individuals with seven pink, two green, and three blue alleles. No yellow allele remains—it is lost from this population. The next (third) generation could then have very different allele frequencies compared to Generation 1.

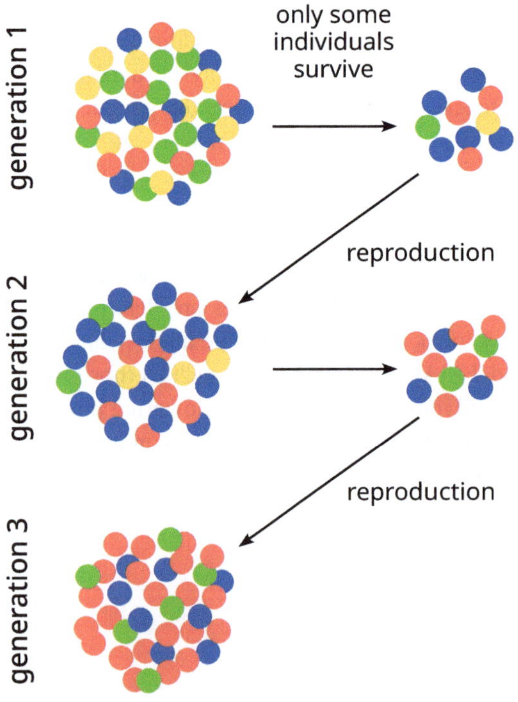

Fig. 8.2 Bottleneck effect. Drastic reduction in size can lead to random changes in the population's genetic composition

As we can see, the population lost some of its genetic diversity (no more yellow alleles) within just a few generations, and some alleles became much more abundant than others. Such a rapid population size reduction is called a **population bottleneck**. Here, reduced genetic variation may mean that the population is unable to adapt to new selection pressures, such as climatic change, a shift in available resources, or a new disease. This is because the current genetic variation is not the product of many generations of selection but a product of random events that change the overall genetic composition, decreasing diversity.

Interesting Fact

An example of a population that went through a bottleneck is the little spotted kiwi, a flightless bird endemic to New Zealand that often appears as its national symbol. Due to habitat destruction and the introduction of mammalian predators into its natural habitat, the species has experienced an extreme decline in abundance and geographical range. Consequently, the last five remaining individuals were put under protection on the predator-free Kapiti Island of New Zealand. As their numbers increased, several other predator-free islands were also populated with individuals taken from Kapiti. However, despite the overall number of individuals rising to over 1000 in around 70 years, a study showed that the genetic consequences of the bottleneck have only become more and more pronounced over the years! When the genetic diversity of the kiwis in Kapiti (the source population) and the other islands (reintroduced populations) was compared, it was shown that the reintroduced populations lost 12% of allele polymorphism, and the level of heterozygosity decreased by 4%. Conservation efforts should seek not only to increase population numbers but also to help maintain the remaining genetic diversity.

As seen in the example of kiwis, though a population may get larger, when individuals come from a very small number of ancestors, the lost genetic diversity does not return, and the population may still suffer from reduced **evolutionary potential**. This often happens when populations are drastically reduced or a small number of individuals migrate to a new habitat, founding new populations. The loss of genetic diversity in populations with a limited number of founders is then called the **founder effect**. In addition to the loss of genetic diversity, there is a high risk of deleterious alleles becoming homozygous and more frequent in the following generations, decreasing the fitness of the population. In particular, in a large (pre-bottleneck) population, the chance of two individuals heterozygous for the same deleterious allele mating with each other is much lower than if these same individuals were in a small (post-bottleneck) population.

Inbreeding Depression Reduces the Fitness of Natural Populations

Breeding between closely related individuals is known as **inbreeding**. In small and isolated populations, the chance of breeding between closely related individuals increases. Some species living in their natural environments have mechanisms to avoid inbreeding. For example, long-term behavioural studies indicate that lions, African wild dogs, and spotted hyenas rarely show breeding between parents and offspring or siblings. Dispersal by young animals is thought to be the principal means of avoiding mating between close relatives. Another mechanism to prevent inbreeding is extra-pair mating, that is, an individual in a usually monogamous pair mates outside of the pair. This tactic is used, for instance, by female purple-crowned fairy-wrens. This species is known to be primarily monogamous (they tend to have just one mate); however, females that are in pairs with closely related males achieve high rates of extra-pair fertilisation by synchronising their breeding attempts with their neighbours. However, when a population becomes small and isolated from the others, the effects of inbreeding become unavoidable. See Box: Inbreeding and Homozygosity for more details.

Inbreeding and Homozygosity

Imagine a male and female, one of whom is a carrier of a recessive deleterious allele, while the other one is homozygous for the (dominant) wild-type allele. According to Mendelian genetics, the proportion of offspring inheriting this recessive allele would be 50%, but there would be no homozygous recessive (sick) individuals. If these offspring mated among themselves (inbred), some of these matings would occur between individuals that have both inherited the deleterious allele (carriers). Then, some of their offspring would be heterozygous (carriers), while others would be homozygous for the deleterious allele (and thus express the disease). Overall, the proportion of the sick (homozygous recessive) individuals would increase, as would the proportion of the carriers. In this condition, we observe decreased levels of overall heterozygosity, and we expect to see the deleterious effects of this allele on the fitness of the individual. If the first generation of offspring had bred with unrelated individuals (outbreeding) who did not carry this deleterious allele, none of these effects would have been observed in the following generation.

Inbreeding depression is the difference in fitness between offspring produced via inbreeding compared to those produced by outbreeding. In almost a hundred of species of mammals, birds, reptiles, and amphibians kept in captivity in zoos, inbreeding depression was reported in terms of a decrease in juvenile survival. In a wild Scottish population of red deer, it has been shown that inbreeding depression is responsible for the low annual breeding success in both males and females and decreased offspring-rearing success in females.

Phenomena such as bottleneck and founder effects are also common in human populations. For example, a genetic disorder, Ellis-van Creveld Syndrome (ECS), that causes polydactyly (extra fingers and/or toes), short limbs, and, occasionally, heart defects is commonly observed in the Amish population living in eastern Pennsylvania, in the United States. While there were just 50 cases of ECS worldwide between 1968 and 2007, the frequency of the allele responsible for ECS was 6.6% in this population! The Amish people living in eastern Pennsylvania are the descendants of a small group of founders who came to the United States from Western Germany during the 18th century. Since then, few new members have joined, and current members tend to marry within the community. The deleterious alleles currently segregating in this population trace back to a single couple from among the initial colonisers. Because of the founder effect and endogamy (predominantly within-group mating), the allele variants became much more common in the subsequent generations.

8.3 Saving Genetic Diversity

The ultimate goal of all species conservation projects is to have self-sustaining populations in their natural habitats. As discussed earlier, genetic diversity provides the raw material for populations to evolve and adapt to environmental change, and this is an essential requirement for populations to be self-sustaining. As it is difficult to build up genetic diversity in a small population, the best way to conserve genetic diversity and prevent extinction is by preventing population decline in the first place, for instance, by combating habitat destruction and pollution.

Complex Conservation Decisions

In theory, decreasing extinction pressure on biodiversity might be straightforward. In practice, however, it is extremely complex, given the increases in human population, land use, and urbanisation. The case of the Mojave Desert tortoise is an interesting example of this. Desert tortoises are native to the Mojave and Sonoran Deserts of the south-western United States. Historically, they have been threatened by changes in land use that destroy their habitat, such as urbanisation, road construction, and off-road vehicle trails. These activities have led to the extreme fragmentation of the Mojave Desert tortoise habitat. Recently, this species has been facing an ever more difficult road to recovery due to the adverse effects of climate change on food availability and the physiology of these animals. One of the ways to combat climate change is to develop renewable energy sources, such as solar power, as a replacement for fossil fuels. However, desert tortoises live in some of the most important areas for solar energy development. If already-fragmented desert tortoise populations become further isolated from each other due to big solar panels, then some populations would definitely go into an **extinction vortex**. To address this, scientists have examined the movements of the remaining populations and tried to create a distribution map of the tortoises. This map also shows important corridors for tortoise movement that facilitate the genetic connection of different populations, a phenomenon known as gene flow. Mapping migration corridors allowed scientists to suggest alternative building sites for the solar panels.

Genetic Rescue

Some endangered species now have populations so small that they have lost genetic diversity and seem to be fixed for multiple deleterious genetic variants. To prevent extinction due to genetic decline, introducing individuals from other populations or subspecies—known as **genetic restoration** or **genetic rescue**—can be beneficial. One of the most famous examples of genetic rescue is the case of the Florida panthers. In the mid-1990s, there were only 22 panthers left in Florida, and this small population showed signs of reduced fitness due to inbreeding, including undescended testicles and heart defects. To "dilute" these inbreeding effects (i.e. eliminate the deleterious variants and return populations to normal levels of genetic variation), eight panthers of the closest subspecies from Texas were introduced into the Florida population. Eventually, molecular genetic variation increased in the previously highly inbred population of Florida. The population tripled in size, and the morphological signs of inbreeding declined after the translocation effort. Although the genetic rescue of the Florida panthers has been an overwhelming success, habitat loss, caused by the development in the north of the state and sea level rise in the south, will remain major concerns in the near future.

Off-Site Conservation

There are some situations in which conserving a species in its natural habitat is not feasible due to various threatening factors. **Off-site conservation** means that individuals of a species are moved to a secure place, reared there, and then released into a new habitat. One example is kakapo conservation in New Zealand. Kakapos are flightless parrots native to New Zealand. The natural populations shrank due to predation of their eggs by non-native species introduced to the country. Recent conservation efforts include finding fertilised eggs in the wild, transferring these eggs to secure artificial incubators, and, once they hatch, introducing healthy individuals to predator-free habitats. Recently, the conservation program was augmented with a genetic aspect where females were artificially inseminated with the semen of genetically "valuable" males (i.e. males that showed high genetic variation within their genomes).

Seed Banks and Frozen Zoos

The long-term off-site conservation strategy for plants is the establishment of seed banks, and for most animals, it is the cryo-preservation (preservation of samples at very low temperatures) of DNA, tissue samples, sperm, egg cells, stem cells, etc. The largest seed bank for plants is located in the Royal Botanical Gardens in Kew, UK. The seed bank has one billion seeds from 130 partner countries and aims to house almost 25% of the world's bankable seeds. The first "frozen zoo", which stores DNA, tissue, and gamete samples from critically endangered animals, was established in the San Diego Zoo. Another project, named "Frozen Ark", was initiated in the United Kingdom and seeks to preserve genetic material from thousands of endangered animals.

All these long-term preservation strategies are, of course, considered to be insurance strategies. If the ongoing on-site and off-site conservation efforts all fail for a species, cryo-preservation may be the very last chance for the revival of a critically endangered or even extinct species. Recently, scientists, in collaboration with the San Diego Zoo, made a breakthrough in the genetic rescue of a critically endangered horse species. Przewalski's horse, known to be the only wild horse species left on Earth, is a distant cousin of domesticated horses. Previously, they were native to the steppes of central Asia. However, around 30 years ago, they were listed as extinct in the wild. Following this, a rigorous conservation program focusing on breeding the 12 remaining individuals was put in place. The offspring that were sired from these 12 individuals were then reintroduced in several locations in China and Mongolia. The main problem was, again, the erosion of genetic diversity within this offspring because they were all full siblings, half-siblings, or, at best, distant cousins. Fortunately, scientists had collected tissue samples when these animals were (relatively) more abundant in the wild, that is, when the genetic diversity of the wild populations was higher. The samples had then been cryopreserved in 1980 in the San Diego Zoo. The recent breakthrough involved cloning one of these cryopreserved cells and implanting it into a surrogate mother. In August 2020, a cloned male Przewalski's horse named Kurt was born in the zoo, and this male will be used in further breeding plans for the horses. In 2023, his genetic twin, Ollie, was cloned from the same cryopreserved cell line. Kurt's and Ollie's offspring will bring some variation to the much-depleted genetic variation of the current zoo population, as well as in the wild, where they will eventually be reintroduced.

The Intergovernmental Science-Policy Platform on Biodiversity and Ecosystem Services recently reported that the unprecedented rates of species extinction have only accelerated. Currently, around one million species are threatened with extinction, and given the projected impacts of increasing land use, exploitation of organisms, and climate change, these negative trends will likely continue through 2050 and beyond unless transformative change is included in policy scenarios. Hopefully, with the help of technology, our imaginations, and inclusive and just policies, we may be able to combat biodiversity decline to a certain extent. However, we must reduce the current human pressure on the natural systems before it is too late!

8.4 Case Study: Solving Complex Species Conservation Problems

Conservation genetics is one of the subfields of conservation biology, as well as a scientific discipline that aims to use genetic approaches to conserve biodiversity. But what do conservation geneticists actually do? My journey as a conservation geneticist started with my PhD studies in Singapore. My topic was the genetic management of a critically endangered freshwater turtle species native to Myanmar, the Burmese roofed turtle (Fig. 8.3).

The Burmese roofed turtle was once abundant in the major river systems in Myanmar. However, in the late twentieth century, their numbers declined due to several human-related pressures, including habitat degradation, overexploitation of adults, overcollection of their eggs, and pollution. By the early twenty-first century, they were considered extinct in the wild.

Later, however, biologists found one adult female and a male in a small pond of a Buddhist temple. These individuals were immediately transferred to a zoo for breeding, thus initiating their off-site conservation. Then, one breeding population was found in the wilderness on the north-western side of Myanmar, on a small stretch of the Chindwin River. The number of wild breeding females and males was unclear, but the beaches that females visit every nesting season were quickly identified, and their on-site conservation began. Those beaches were designated as protected areas, meaning that no fishing was allowed. After the females had laid their eggs, conservationists collected the eggs to incubate them in a secure location. This continued for over a decade, resulting in more than 500 Burmese roofed turtles in captivity.

The ultimate goal of the conservation project was to establish self-sustaining turtle populations in their natural habitats, so the turtles reared in captivity were meant to be released into the wild one day. But, how exactly? If we had more than 500 individuals of one of the rarest turtles on Earth, should we release all of them to the wild? Should all of them be released in the same place? If not, how exactly should we manage the release? Also, after the release, how could we monitor the released individuals? Attaching a transmitter to the released individuals would help track their movement, but how could we understand which of the released females or males started breeding in the wild, contributing to the genetic composition of the future populations?

The answer to the last question largely depends on estimating how many current breeders there were in the wild. Unfortunately, there was very little information about the total number of mothers and fathers who had sired the collected eggs. There were rough estimates of the number of potential mothers based on counting the number of nests laid every season, but this number was unreliable, as females may lay more than one nest in a season. In any case, the number of breeding females was thought to be about ten—only ten mothers for an entire species! My PhD studies focused on these questions, and I worked to come up with solutions based on genetic methods.

First, I did parentage analysis for almost all of the captive turtles (remember that they come

Fig. 8.3 Burmese roofed turtle. From left to right: Juvenile leaving its egg, young turtle, and an adult individual. (Photo credit: Myo Min Win- Myanmar Conservation Fund)

from the eggs collected from the nests of the only remaining wild population), just like the paternity tests done for humans. The only difference in the turtles' case was that neither the mother nor the father was known. The tests are made possible by increasing the number of genetic markers used in the analysis. These markers help scientists examine specific portions of individuals' genomes and build genealogical trees that identify close relatives. My studies revealed that the number of wild female breeders was less than five—not even ten! The situation for the wild population was thus even more dire than conservationists had thought, highlighting the importance of increasing the numbers of wild individuals. Reintroductions were needed as soon as possible! Then, I did additional analyses to assess the genetic diversity remaining in the turtle population. Understanding this helped the conservationists decide which individuals should be returned to the wild and which individuals should be kept in captivity to preserve the genetic stock for the entire species. The government of Myanmar and the two turtle conservation groups, the Turtle Survival Alliance and the Wildlife Conservation Society, were the main partners responsible for the conservation of this species in Myanmar. These organisations decided to keep 25 individuals as an assurance colony (the future genetic stock for the worst-case scenario) and release two groups of 25 individuals into the wild, close to the sites that females visit to lay their eggs. To come up with the groups of individuals, I used the same genetic markers to identify turtles that were genetically least related to each other. If the individuals were not genetically related, inbreeding would be avoided, and genetic diversity would be maximised. Eventually, 25 individuals were sent to the Singapore Zoo as the assurance colony, and the other two groups were also ready for reintroduction.

But another problem still loomed. This species was on the brink of extinction, and the main reason behind this situation was human behaviour. How could we be sure that the released individuals would not be touched? What would local people think about the turtle conservation projects that have been run in the area, and how likely were they to support them? To find answers, I collaborated with social scientists, and we designed social surveys to assess and understand local people's attitudes towards turtle conservation.

To carry out the survey, the turtle conservation team in Myanmar and I visited tens of villages in north-western Myanmar located close to the potential reintroduction sites. However, I was the only person on the team who was not from Myanmar, and this made me feel disconnected from the survey participants. To overcome this, I began learning Burmese, and within a couple of months, I was able to initiate everyday conversations in the language. Eventually, I could ask the volunteers from the villages questions related to turtle conservation (Fig. 8.4).

After the interviews, we conducted outreach programs in the villages and screened turtle conservation movies with the goal of increasing

Fig. 8.4 Working with the locals. The author of the article is talking to the villagers, surveying their attitudes towards turtle conservation in the area

awareness and gaining local support for the ongoing conservation efforts. The hundreds of surveys and interviews conducted showed that there is no major conflict between the local villagers and turtle conservation, which may indicate that gaining local support will not be hard. Most of the local people were actually eager to see more turtles on the river because they believed that turtles are the main indicators of "healthy" rivers. In addition, we received no reports of Burmese roofed turtle hunting or egg collection (mainly because these animals are so rare that most people had only heard of them from their grandparents). On the other hand, many locals could easily identify turtle nests and were curious to taste the eggs of the roofed turtles. Based on our results, we suggested that the policymakers collaborate with local people who could easily identify turtle nests and train these people to identify Burmese roofed turtles in particular. This was primarily based on the idea that as the number of wild turtles increases, the more likely local people are to observe these animals and their nests. With the help of trained local villagers, it would be much easier to monitor the released turtles. Eventually, we concluded that local people have positive attitudes towards the ongoing nearby turtle conservation project and that recruiting additional local collaborators is likely to improve the outcome of the conservation measures.

Finally, two groups of Burmese roofed turtles were released into a river near the interviewed villages. There was also a turtle release ceremony (Fig. 8.5), attended by Buddhist monks, local villagers, local conservationists, and government officers, that aimed to increase public engagement with the project. Turtles selected for release were also fitted with a transmitter that would provide scientists with movement information and ecology data. Finally, the local villagers themselves released the animals into the river.

The conservation story of the Burmese roofed turtle does not end here. As scientists, conservationists, and policymakers, we put every effort into maximising the success of these reintroductions. However, strict ecological and genetic monitoring of these individuals will likely continue for many more years, as the total number of individuals in the wild is still low despite reintroductions. Remember, there are currently only about 100 Burmese-roofed turtles in the wild!

The main take-home message from the Burmese roofed turtle conservation story is that conservation is a complex issue, and scientists from different backgrounds, local people, conservationists, and policymakers must come together to find feasible and sound solutions to address these complex, real-world problems. As a conservation geneticist, I enjoy working in the laboratory, but I am also well aware that theoretical or science-based solutions for such problems may not be directly applicable in the real world. Instead, my work is part of a larger picture: I consider myself a member of a large community consisting of many people with

Fig. 8.5 Turtle releasing ceremony. Locals participating in the Burmese roofed turtle reintroduction event in Northwest Myanmar. (Photo credit: Steven G. Platt—Wildlife Conservation Society)

diverse backgrounds united in their commitment to conserving the astonishing biological diversity on Earth.

DEFINITIONS

Genetic drift: A process in which chance events cause random changes in allele frequencies from one generation to the next. The effects of genetic drift are most pronounced in small populations.

Conservation: The integrated study of ecology, evolutionary biology, and genetics to sustain, protect, and restore biological diversity at all levels.

Inbreeding: Mating between individuals whose recent ancestors were closely related.

Outbreeding: Mating between unrelated individuals.

Inbreeding depression: Decrease in biological fitness of offspring as a result of inbreeding.

Extinction vortex: A scenario in which several factors, e.g. environmental, genetic, and demographic, interact over time to push a population towards extinction.

Heterozygosity: The presence of different alleles at one or more loci on a chromosome.

Polymorphism: The occurrence of two or more clearly distinct forms (morphs or alleles) within the same population of a species.

Genetic rescue: The introduction of new genetic variation into an inbred population that improves its health and long-term persistence.

SUGGESTED READING

- Novak BJ et al. Endangered Przewalski's Horse, Equus przewalskii, Cloned from Historically Cryopreserved Cells. Animals (Basel). 2025 Feb 20;15(5):613
- Çilingir FG et al. Conservation genomics of the endangered Burmese roofed turtle. Conserv Biol. 2017 Dec;31(6):1469-1476
- Land, E. D., and R. C. Lacy. 2000. Introgression level achieved through Florida panther genetic restoration Endangered Species Update 17:100–100

8.5 Questions

1. **Which of the following is the main purpose of building wildlife corridors, e.g. bridges above highways?**
   (A) Protect people from encounters with wild animals
   (B) Slowly enable a species to move to a new habitat
   (C) Connect populations in habitat fragments that are separated by a highway or other human construction and that would otherwise be isolated from each other
   (D) Slow down the introduction of non-native or invasive species
   (E) Create a more variable habitat for local animal populations
2. **Which of the following statements are accurate?**
   (A) Many animal and plant species are so well adapted that they cannot go extinct from natural causes.
   (B) The rate of extinction is constant over time and roughly the same in different clades.
   (C) Based on fossil evidence, the rate of extinction now is similar to the rate in the past.
   (D) Humans are the only species that can drive another species to extinction.
   (E) One of the main extinction events in Earth's history was caused by the accumulation of oxygen produced by newly evolved photosynthesising organisms.
3. **Which of these problems can occur in a population created by cloning a few individuals?**
   (A) Increased probability of recessive disorders.
   (B) All individuals have identical mating preferences, resulting in a single breeding female.
   (C) All individuals have the same sex, so they cannot reproduce.
   (D) Decreased ability to adapt to new environmental conditions.
   (E) Decreased range of immune responses.
4. **Genetic analysis of a population shows that there are 64 individuals homozygous for one allele and 32 heterozygous, but no individuals homozygous for the other allele. What could explain this pattern?**
   (A) One of the alleles is recessive and lethal.
   (B) One of the alleles is dominant and lethal.
   (C) Only heterozygous individuals can reproduce.
   (D) There is a strong selection against heterozygous individuals.
   (E) Individuals reproduce by self-fertilisation.
5. **Figure 8.6 shows the population numbers, mean age, and estimated heterozygosity for the Florida panther (*Puma concolor coryi*) population over more than 20 years of research and conservation efforts. In 1995, eight female pumas of the same species—but different subspecies (*P. c. stanleyana*)—from Texas were released, which can be regarded as a successful conservation intervention. Based on this information, which of the following was true about the population before the intervention?**
   (A) The population was living in a very small area, and the conservation effort managed to establish a much larger protected area for the Florida panther population.
   (B) The mean age and heterozygosity of the population were decreasing rapidly before the introduction.
   (C) The population managed to survive with very low numbers for about a decade.
   (D) Between 1988 and 1996, no young individuals were detected.
   (E) In 1992, the population consisted of 36 closely related individuals.
6. **Based on the graphs above, which of the following is an observation or a logical interpretation of what happened in the population after the introduction of the eight individuals from Texas?**
   (A) The average number of surviving offspring per individual increased.
   (B) The total population numbers increased rapidly.

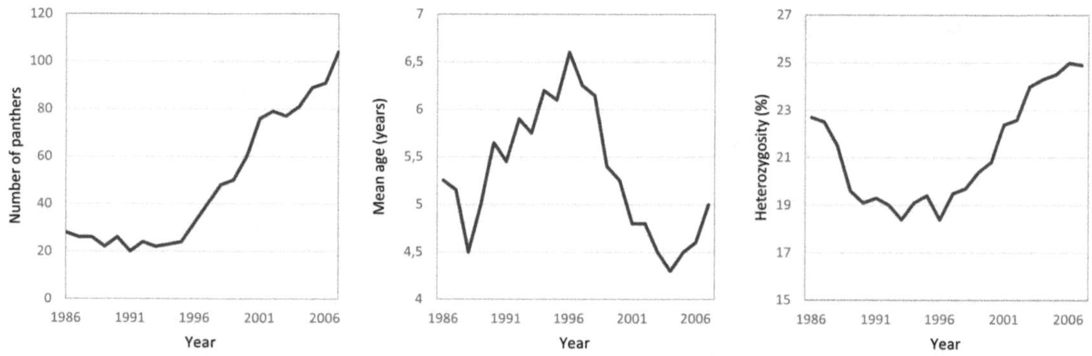

Fig. 8.6 Population numbers, mean age, and estimated heterozygosity for the Florida panther. (Figures based on Johnson WE, Onorato DP, Roelke ME, et al. Genetic restoration of the Florida panther. Science. 2010;329(5999):1641–1645. doi:10.1126/science.1192891)

(C) The new individuals started to breed with the original population.

(D) The estimated heterozygosity increased.

(E) The immigrating individuals were very young, so it took them several years to start breeding.

8.6 Project: Studying the Genetic Diversity of a Wild Population

Biology of a Weird Fluffy Small Mammal

In a small fragment of Carpathian forest, isolated from the rest of the Carpathian mountains, only 20 individuals of a weird, fluffy, small mammal species are left. These animals are very peculiar—they live only 1 year and give birth in the winter. In autumn, mature animals burrow in deep nests, and no one knows what happens next. But in the spring, a new generation of these animals leaves their nests. The older generation is never seen again.

Due to their weird breeding and nesting habits, only a few populations of this animal species are left in the world. In order to save the Carpathian population from extinction, geneticists decided to genotype all of the individuals in it. They will use the data to determine how best to protect them. They looked at five genes (loci). The results of the genotypic analysis are shown on the following page.

In this project, you will follow the steps of these scientists and try to save these wonderful, weird, fluffy small mammals from extinction!

Year 1: Investigate the Wild Population

Look at the population in Fig. 8.7. Each individual is represented by two rectangles: two chromosomes, each with five genes.

1. Count the number of different genotypes for the first two genes, and fill in Table 8.1.
2. For both genes, calculate:
   (a) The **observed heterozygosity** H_0
   (b) The **expected heterozygosity** H_E
3. Considering these two genes, calculate the average H_0 and average H_E.
4. Considering these two genes, calculate the **inbreeding coefficient** F.
5. As an additional measure of genetic diversity, consider the remaining three genes. Which of them has more than two distinct alleles? How many alleles are there for each?

The scientists have all their numbers, but spent months arguing with policymakers about what to do next. Ideas were all over the board: Some wanted to freeze the entire population and keep it for later advancements in science, and others wanted to fence the entire area immediately and declare it a nature reserve. Some wanted to go all natural and do nothing, while one hunter just wanted to catch them all and stuff them for his personal collection! A year passed, and no one did anything except for the animals—they bred!

Year 2 Spring: Reproduction Time!

1. All individuals reproduced. Cut the diploid individuals into haploid gametes, shuffle them, and pair them randomly. In this way, you get the individuals of the second generation. Keep the two paper strips making up an individual together, for instance, with a paper clip.
2. Take a picture of the genotypes of the new generation or write the genotypes down.
3. Count the number of different genotypes for the first two genes, and fill in Table 8.2.
4. For both genes, calculate:
   (a) The **observed heterozygosity** H_0
   (b) The **expected heterozygosity** H_E
5. Considering these two genes, calculate the average H_0 and average H_E.
6. Considering these two genes, calculate the **inbreeding coefficient** F.
7. Answer the following questions:

A1	A1	A1	A1	A1	A1	A1	A1	A1	A1
B1	B2	B1	B2	B1	B2	B1	B2	B1	B1
C1	C1	C1	C1	C1	C1	C1	C1	C1	C1
D1	D1	D1	D3	D1	D2	D2	D2	D2	D3
E1	E1	E2	E2	E2	E2	E2	E3	E3	E3

A1	A2	A1	A2	A1	A2	A1	A2	A1	A2
B1	B1	B2	B2	B1	B2	B1	B2	B1	B2
C1	C1	C1	C1	C1	C1	C1	C1	C1	C1
D3	D3	D2	D2	D1	D2	D1	D3	D1	D1
E1	E2	E2	E3	E2	E3	E2	E2	E2	E2

A1	A2	A1	A2	A1	A2	A1	A2	A1	A2
B1	B2	B1	B2	B1	B2	B2	B2	B2	B2
C1	C1	C1	C1	C1	C1	C1	C1	C1	C1
D1	D2	D2	D2	D2	D3	D3	D3	D3	D3
E2	E3	E2	E3	E3	E3	E3	E3	E3	E3

A2	A2	A2	A2	A2	A2	A2	A2	A2	A2
B2	B2	B1	B2	B1	B2	B1	B2	B1	B2
C1	C1	C1	C1	C1	C1	C1	C1	C1	C1
D2	D3	D2	D2	D1	D2	D1	D3	D1	D1
E3	E3	E2	E3	E2	E2	E2	E2	E1	E3

Fig. 8.7 The first generation of the weird fluffy mammal

Table 8.1 Number of individuals of each genotype for the first two genes

Genotypes	AA	Aa	aa	Genotypes	BB	Bb	bb
Number of individuals				Number of individuals			

Table 8.2 Number of individuals of each genotype for the first two genes

Genotypes	AA	Aa	aa	Genotypes	BB	Bb	bb
Number of individuals				Number of individuals			

(a) Did the proportion of genotypes for the first two genes change within the population? How?
(b) Did the inbreeding coefficient change? How?

Again, there was a lot of arguing in the group. Scientists could not persuade the local authorities to act. Unfortunately, there were frequent storms in late summer, resulting in a massive landslide in the area occupied by the small mammals. Only

Table 8.3 Number of individuals of each genotype for the first two genes

Genotypes	AA	Aa	aa	Genotypes	BB	Bb	bb
Number of individuals				Number of individuals			

four individuals were found alive: two males and two females!

Year 2 Autumn: The Great Bottleneck

1. Shuffle all the individuals and choose four survivors randomly.
2. Count the number of different genotypes for the first two genes, and fill in Table 8.3.
3. As an additional measure of genetic diversity in these four individuals, consider the remaining three genes. Which of them has more than two distinct alleles? How many alleles are there for each?
4. Were any alleles lost as the population passed through the bottleneck?

Finally, people were forced to act. They took the remaining individuals, released them in a fenced and protected area, provided food, let the animals dig holes and enter them, and spent the winter hoping. Luckily, a population of 20 individuals emerged from the burrows in spring! Local authorities were pleased, patting each other on their shoulders and smiling in front of the cameras. The geneticists, however, had other work to do.

Year 3: Breeding in Captivity

1. Decide which two surviving individuals represent males and which represent females.
2. Disconnect the two pieces of paper representing diploid individuals. Keep male and female gametes in separate piles.
3. For each gamete, make four additional identical ones using the empty templates provided in Fig. 8.8. You should have five copies of each gamete in two piles, one representing males and one representing female gametes.
4. Shuffle each pile individually.
5. Take one gamete from each pile and make 20 new pairs of gametes. These pairs represent the new generation of 20 individuals.
6. Take a picture of the genotypes of the new generation or write the genotypes down.
7. Count the number of different genotypes for the first two genes, and fill in Table 8.4.
8. For both genes, calculate:
   (a) The **observed heterozygosity** H_0
   (b) The **expected heterozygosity** H_E
9. Considering these two genes, calculate the average H_0 and average H_E.
10. Considering these two genes, calculate the **inbreeding coefficient** F.
11. As an additional measure of genetic diversity in these four individuals, consider the remaining three genes. Which of them has more than two distinct alleles? How many alleles are there for each?

Analysis

1. Did the proportion of genotypes change from generation to generation for the first two genes? How much does the third year's population differ from the original one?
2. Did the observed heterozygosity for the first two genes change from generation to generation?
3. Compare the inbreeding coefficient for the first and the last generation. Explain your observations.
4. Imagine that you still have the samples from the first year and the resources to clone one or two animals. Look at the individuals in the first year.
   (a) Which of them would you clone? Why?
   (b) What would happen if you introduced the cloned individual(s) to the current (year 3) population?
5. Imagine that gene D plays a role in the process of storing fat reserves. The allele D1 codes

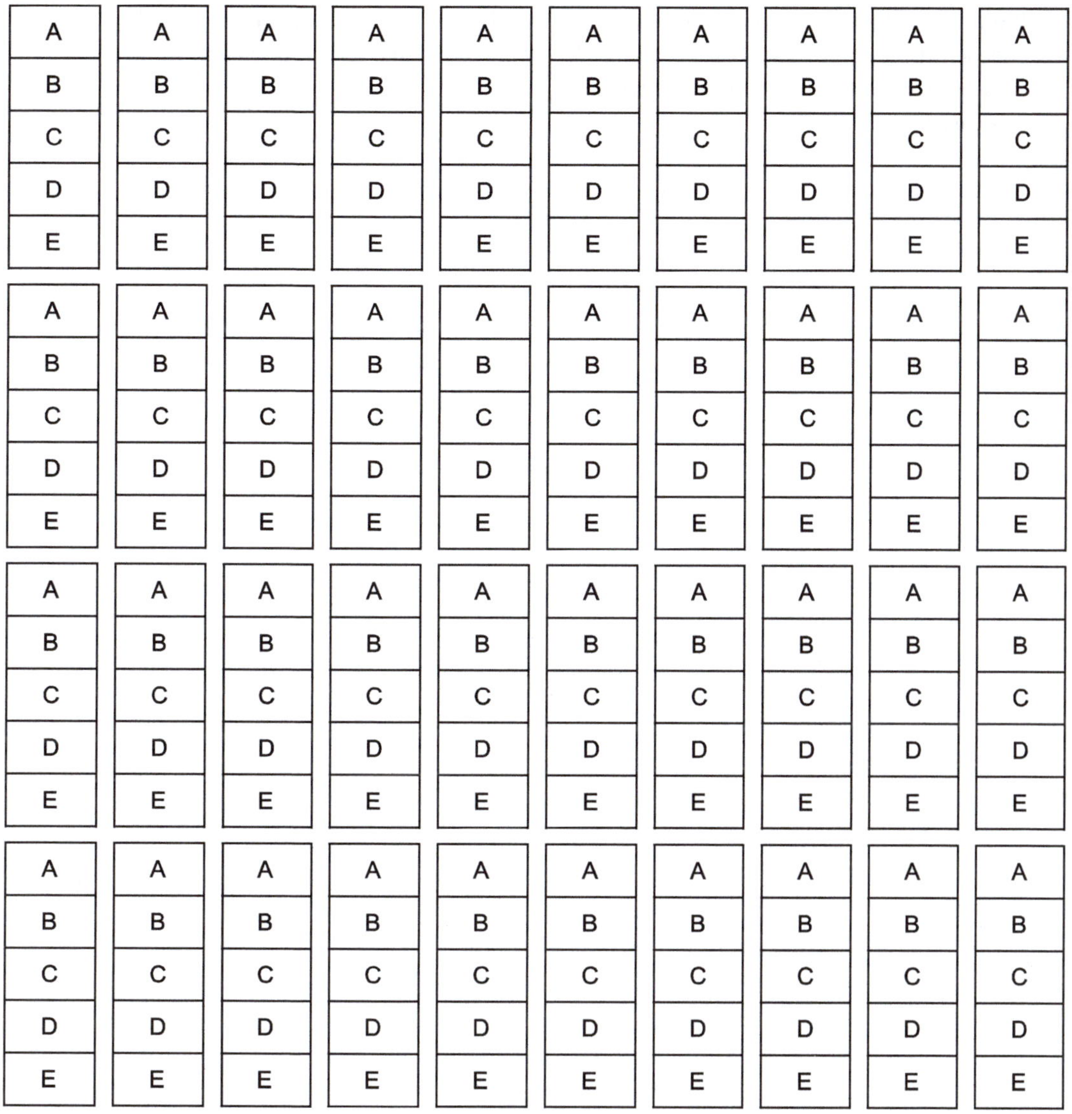

Fig. 8.8 **Genotype templates**

Table 8.4 Number of individuals of each genotype for the first two genes

Genotypes	AA	Aa	aa	Genotypes	BB	Bb	bb
Number of individuals				Number of individuals			

for a defective receptor protein on fat cells, resulting in lower fat accumulation. If an individual is homozygous for D1, it is very likely that its fat cells will not accumulate enough fat to enable its winter reproduction. Look at the changes in your population.

(a) Would the genetic composition of the population have changed during the observed time period if we assume that D1D1 homozygotes do not reproduce? Explain.
(b) Would the individuals who survived the bottleneck event have survived in this case?

(c) Would this additional information make you choose different individuals for cloning? If so, which individual(s) would you choose in this case?

This project was developed by Barbora Trubenová and Kristína Hudáková.

From Early Hominins to *Homo Sapiens* 9

Kristína Hudáková

▶ All humans currently living on planet Earth are members of the species *Homo sapiens*. We are the most numerous and widespread species of mammals inhabiting the Earth. How did we become so successful? Where do we come from? Who were our ancestors, and what do we know about them?

What we know about human evolution changes rapidly. Every year, multiple articles announcing new discoveries are published, forcing scientists to re-evaluate their previous understanding.

In addition to newly discovered fossil evidence of our past, advances in molecular techniques have enabled scientists to not only analyse large quantities of DNA in a short time but also to extract valuable information from very old samples.

In this chapter, we will explore how all this new information allows us to trace the evolutionary history of our ancestors and their close relatives, starting with their split from chimpanzees until the emergence of our own species, *Homo sapiens*.

The primate family Hominidae (also known as great apes or hominids) consists of eight living species: three species of orangutans (genus Pongo[1]), two gorilla species (genus Gorilla), two species of chimpanzees (genus Pan), and one human species (genus Homo). The closest living relatives of humans are chimpanzees, followed by gorillas and then orangutans (Fig. 9.1). All of these species are the only surviving members of their evolutionary lineages. Similarly, we are the only surviving species of our lineage (Fig. 9.2).

The last common ancestor of humans and chimpanzees is estimated to have lived between 6.5 and 7.5 million years ago. Since then, the lineages have diverged, and our ancestors (and their relatives) accumulated changes that we now recognise as typically human features. They began living on more open ground rather than in the tree canopy of dense forests; they also started to walk upright, use tools, share food, live in large groups and developed languages. Most importantly, the brains of our ancestors enlarged, enabling humans to develop complex societies, use symbols, and change the world in ways no other animal ever could.

K. Hudáková (✉)
eduB, Bratislava, Slovakia
e-mail: kristina.hudakova@edub.sk

[1]The Tapanuli orangutan (*Pongo tapanuliensis*) was described as a distinct species only in 2017. Nater A, Mattle-Greminger MP, Nurcahyo A, et al. Morphometric, Behavioral, and Genomic Evidence for a New Orangutan Species. Curr Biol. 2017;27(22):3487-3498.e10. doi:10.1016/j.cub.2017.09.047.

B. Trubenová et al. (eds.), *Everyday Evolution: Unveiling Life's Hidden Patterns*,
https://doi.org/10.1007/978-3-032-07050-0_9

Fig. 9.1 Humans and their closest living relatives. The closest living relatives of humans are the chimpanzees, followed by gorillas and orangutans

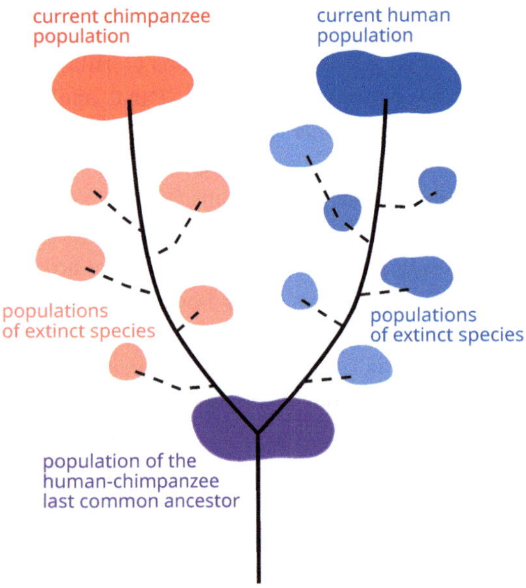

Fig. 9.2 Human and chimpanzee evolutionary lineages. The evolutionary relationships of the extinct species are not resolved

9.1 Fossils

The most classic line of evidence for evolution comes from fossils—fossilised remains of organisms that can be dated based on their position in geological strata, radiocarbon dating, or other techniques. Before DNA analyses became available, fossils were the only source of information about the evolution of human ancestors.

The hominin fossil record (like the fossil record in general) is rather fragmented and incomplete, with fossil evidence from a particular site and time period often containing only a few teeth and pieces of bone. More complete skeletons, such as those of Ardi or Lucy, are rather an exception. However, scientists can gather quite a lot of information even from this sparse material.

When a new hominin fossil is discovered, scientists ask two types of questions:

1. What kind of animal was it? What can we learn from this fossil about its way of life?
2. How does this fossil fit into the hominin phylogeny? Where is its place in the human evolutionary tree?

As hominins evolved from their arboreal quadruped ancestors, the skeletal features associated with bipedalism emerged. These changes evolved over millions of years—a time during which our ancestors walked the land but also took refuge in trees. Thus, fossils of different extinct species show different degrees of adaptation to bipedal locomotion.

Bipedalism gave several advantages to our ancestors and relatives: Walking upright on open grass- and shrub-covered land enabled our ancestors to spot dangerous predators. They could also use their free hands to carry food and, eventually, to make and use tools.

Hominin skeletons are adapted to bipedal locomotion in several ways (Fig. 9.3):

- The **opening in the skull** through which nerves leave the brain for the spinal cord (the **foramen magnum**) is located at the bottom of the skull instead of the back, as is the case in quadrupeds.
- The **skull** itself is rather round, with a flat face, instead of elongated as in quadrupeds. Thus, the shape and position of the skull put most of its weight right above the spine.
- The **curvature of the spine** helps balance the heavy skull in an upright position.
- The **pelvis** in bipeds is broader and more horizontal than in quadrupeds, with hip joints positioned closer to the backbone, which aids in stability. The broader pelvis and a femur

Comparing the DNA of Humans and Chimpanzees

It has been widely stated that humans and chimpanzees are more than 98% genetically identical. However, this figure, primarily based on initial comparisons of single-nucleotide differences in highly alignable regions, has been refined by more complete genomic sequencing efforts in recent years, revealing a much greater overall divergence.

What does this mean? And how can we be so different if our DNA sequences are so similar?

First, we should make clear what is being compared: Scientists may look at how DNA is organised into chromosomes and where the individual genes are located within these chromosomes. The human genome is organised into 46 chromosomes, while chimpanzees have 48. Further, human chromosome 2 originated from a fusion of two shorter chromosomes that are separate in chimpanzees. In addition, several sections in other chromosomes are inverted—these sections are present in both species in the corresponding chromosomes, but their positions are reversed. Some of these inversions are thought to have occurred in the ancestors of humans, others in the ancestors of chimpanzees.

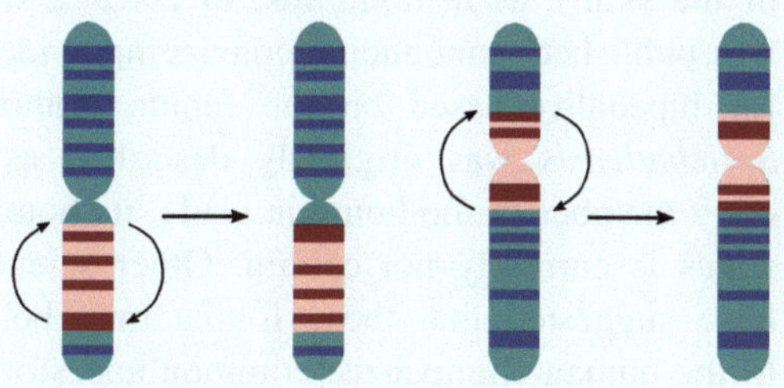

Chromosomal inversions

Scientists also compare the nucleotide sequence—the sequence of As, Gs, Cs and Ts—of chimpanzees and humans to identify differences. Initial comparisons of human and chimpanzee DNA sequences included only protein-coding sequences, resulting in 99% identity between these. However, later analysis, including non-coding regions, reported lower similarity.

Some examples of single-nucleotide differences in protein-coding sequences include olfactory (smell) receptor genes, immunity and cell-recognition genes, language-related genes (FOXP2), and genes associated with brain development (MCPH1 and ASPM). The genes connected with language and brain development are estimated to have been under strong positive selection along the human lineage.

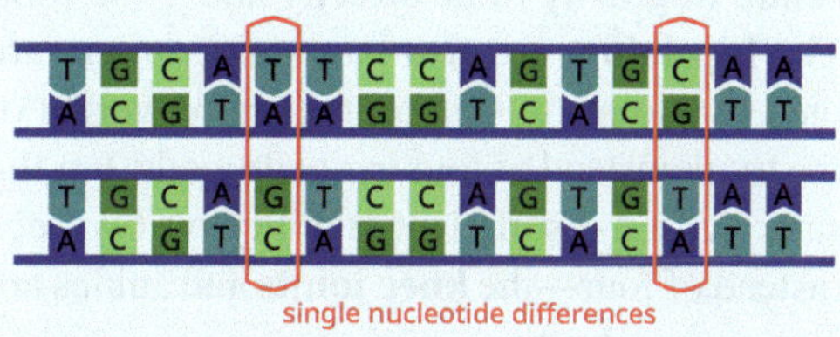

Single nucleotide differences

Recent analysis of ape genomes reported that a large proportion (12.5–27.3%) failed to align when compared across species and with the human genome. This indicates that a large part of the genomes are highly divergent and cannot be directly compared base by base in traditional alignments. Furthermore, this research found that structurally divergent regions constitute an average 10% of sequence per lineage (including humans).

These structural variations contribute significantly to the overall genetic differences. Considering all types of genetic variations, including structural differences and unalignable regions, the total genetic difference between humans and chimpanzees is considerably higher than previously reported.

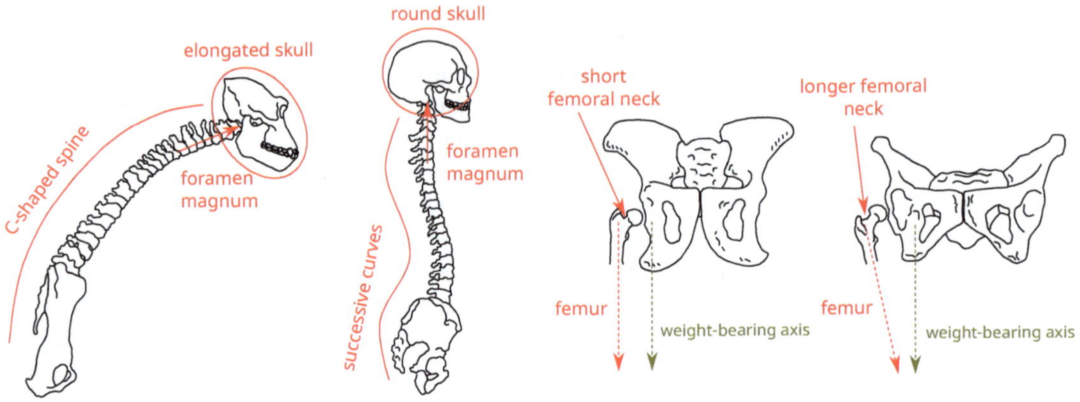

Fig. 9.3 Skeletal features associated with upright posture and bipedalism. (Drawings based on Figure 1 in Janković, Ivor. (2015). Certain medical problems resulting from evolutionary processes: Bipedalism as an example. Periodicum Biologorum. 117)

angled inwards towards the knee cause the centre of gravity to lie directly above the foot.

- The **hip joints** are also larger since there are only two joints to support the entire weight of the trunk instead of four in quadrupeds. For the same reason—sufficient stability on two legs instead of four—the **knee joints** and ankles are also more robust.
- Finally, the **feet** have an arched structure to support the standing body.

On the other hand, features that suggest a quadrupedal arboreal way of life are:

- Arms that are rather long—comparable to or longer than legs.
- Prehensile toes on both arms and legs, allowing for grasping and hanging from branches using all four appendages.
- Curved phalanges, also associated with grasping branches.

9.2 The Earliest Hominins

Several fossils have been discovered from around the estimated time of the human-chimpanzee divergence. While some of these fossils have been attributed to one of the two lineages, there are still ongoing debates about others. Let's have a look at the most important of the early hominin fossils.

Sahelanthropus tchadensis was described based on a cranium and several fragmentary lower jaws found in 2001 and 2005 in northern Chad. It is estimated to have lived between seven and six million years ago, around the time the chimpanzee and human lineages diverged. Later, a femur of the same age found in the same location was also attributed to this species. The skeletal remains show a combination of features: the skull is elongated with prominent brow ridges and a small brain case, as in chimpanzees, but has a short midface, small canine teeth, and a foramen magnum located more forward (on the underside of the skull), as in hominins. In 2022, a study was published, announcing convincing evidence for bipedality based on the femur. Although *S. tchadensis* was originally described as an early member of the hominin clade, its hominin status is currently not certain. Other scientists have suggested that these fossils may belong to the human-chimpanzee common ancestor, an early member of the chimpanzee clade, an early member of the gorilla clade, or a closely related extinct clade.

The fossils of *Orrorin tugenensis* (femora, humerus, phalanges, part of a jaw, and several teeth) were found in 2000 in central Kenya and are estimated to be about six million years old. "Orrorin" means "original man" in the local language. Its unique combination of ape-like and human-like features indicates that it is one of the

Is It a Bone or a Stone?

The word fossil comes from a Latin term that means "obtained by digging". We use it to describe preserved remains, impressions, or traces of once-living organisms, such as bones, shells, or imprints, that are estimated to be over 10,000 years old. Fossils usually form when the dead body of an animal or a plant is quickly covered by sediment, such as sand, mud, or volcanic ash. Otherwise, scavengers, decomposers, and environmental factors would completely remove all traces of it within a few months or years (depending on the organism's size and circumstances).

Within the sediment, soft tissues usually decompose, and only the hard bones, teeth, or shells remain. In most cases, even these will be completely dissolved within a few years by the acids naturally present in many fertile soils.

However, in rare cases, when the environment is favourable—salty or mildly alkaline, rich in minerals and poor in oxygen—these hard parts may undergo fossilisation. The empty spaces within (filled with liquid or gas during life) fill up with water, which contains minerals taken up from the surrounding sediment (e.g. iron compounds or calcium carbonate). As the water penetrates the bones, some of these minerals precipitate and crystallise within the empty spaces, eventually forming rocks in the shapes of the body part. This is the most common mode of fossil preservation, and examples include teeth, bones, shells, and wood. If we were to compare a very old bone with a fossilised bone, the fossil would be much heavier, thanks to the accumulated minerals.

earliest hominins. The shape of the upper part of the femur indicates that this species was bipedal when on the ground, while its phalanges suggest that it could grasp branches while climbing trees. The canines are ape-like but smaller. Also, the molars are smaller and covered with thick enamel.

Ardipithecus ramidus is a species estimated to have lived about 4.4 million years ago. It is represented by a much more abundant fossil record than the previously mentioned species. Since 1992, scientists have discovered a number of its fossils in Ethiopia, enabling them to put together almost 45% of a complete female skeleton (nicknamed "Ardi"). In addition, fossil remains of about 100 other individuals of this species have been discovered. The ability to walk upright is suggested by the femur and the shape of the pelvis, which is broader and shorter than that of arboreal primates. The combination of grasping thumbs and big toes with a rigid foot indicates that this species could most probably walk on two legs but also use all four limbs to grasp branches when moving through trees. The brains of these primates were still small, even smaller than those of today's chimpanzees (just about 20% of the modern human brain). As with the other species mentioned above, the exact place of *A. ramidus* in the hominin phylogeny is unclear.

9.3 Australopithecines

The genus[2] Australopithecus is a collection of species with similar characteristics, most probably closely related to one another, whose fossil remains come from several sites across East and South Africa. The first fossils were discovered in 1924, and since then, the remains of several hundred individuals have been found, including

[2]Australopithecus is most probably not a true genus (it is not a clade); the relationships between the species classified under this genus are not certain, nor are their relationships to the genus Homo clear. Some scientists suggest that not all species classified as Australopithecus should remain in this genus. On the other hand, other scientists suggest that the related genera Paranthropus and Kenyanthropus should also be included in the arbitrary genus Australopithecus.

What Teeth Can Tell Us

Teeth are the most enduring part of the skeleton and are thus the predominant material in the hominin fossil record. This is because the enamel that covers teeth is a very durable material (97% mineral), much stronger than bone. Thus, teeth are more likely to resist the chemical and physical forces acting on them for millions of years. Fortunately for us, teeth also evolve relatively quickly, as they are under direct selective pressure due to changes in food availability and/or sexual selection in males. For example, canines are important in male-male competition in many primate species.

Because teeth evolve rapidly, scientists can use teeth not only to classify species but also to infer other aspects of the species' biology (e.g. the diet and social system). Even if only part of a jaw is available, scientists can infer the individual's possible diet, social behaviour, environment, and age at the time of death.

In all known modern and fossil apes, the large and protruding upper canine tooth in males is continuously sharpened by wear against the lower third premolar tooth, together forming the canine premolar honing complex. In these species, the large and protruding upper canine tooth is used as a weapon in social conflicts. In contrast, modern humans (*Homo sapiens*) have small canines which function more like incisors.

The reduced size of canine teeth is a key characteristic of hominins, present to varying extents in all members of the hominin clade. Several explanations have been suggested, ranging from changes in the diet and social behaviour of our ancestors to dental crowding due to the shortening of the front part of the skull. Over the course of human evolution, several selective forces acted differently on canines in different species, resulting in considerable variation in canine size and shape among the early hominins.

all parts of the skeleton. It was a highly successful genus, inhabiting the Earth for more than two million years. Several species differing in characteristics, such as brain size, likely coexisted during certain time periods.

Australopithecines are considered to be the closest known relatives of the genus Homo. If we look at their skeletons, we can clearly see traits typical of hominins.

The pelvis, femur, and bones of the foot clearly indicate bipedal walking, the canine teeth are small and lack sexual dimorphism, and the molars are larger with thick enamel. The skull, though encasing a relatively small brain, is quite human-like: it has a short facial part and small teeth, especially the canines, which are human-like instead of ape-like. Perhaps most importantly, the foramen magnum is located underneath the skull, indicating that Australopithecines were upright creatures that walked on two feet. On the other hand, the body and brain sizes of most Australopithecines are relatively small, similar to those of chimpanzees, and their long arms still have fingers capable of grasping branches. This combination of traits suggests that bipedalism evolved first in our ancestors, and brain size increased later. The famous "Lucy", known for its extraordinarily complete fossil, is also an Australopithecine, more specifically a member of the species *Australopithecus afarensis*.

9.4 The Genus Homo

The earliest members of the genus Homo were the first to be designated as humans. The oldest known fossil of this genus—a lower jaw with teeth—was discovered in Ethiopia and is estimated to be 2.75 to 2.8 million years old.

Laetoli Footprints

An unusual discovery was made in 1978 when fossil hominin footprints were excavated. These fossil tracks are estimated to be 3.66 million years old. They were most likely made by members of the species *Australopithecus afarensis*, as fossil remains of this species were found in the same sediment layer in this locality. The fossil track is about 27 m long, contains about 70 footprints, and runs from south to north. It was made by a small group of individuals with feet of different sizes. Two individuals walked side by side. A smaller individual followed behind, stepping into the footprints of one of the larger individuals. In 2015, another trackway was discovered, very close to the first one (only about 150 m away), and going in the same direction. These footprints were made by two individuals, one of whom was much taller and heavier. These two sets of footprints might have belonged to members of a single social group, with the largest individual possibly being a male who was walking with females and children.

How was the trackway made?

A lucky combination of events enabled the preservation of the footprints. First, an eruption of a nearby volcano, Sadiman, covered the landscape with a layer of fine ash. Then, light rain fell onto the ash, turning it into something like wet cement. Many animals, among them members of the hominin species *Australopithecus afarensis*, wandered across the surface before it hardened, leaving behind footprints in the wet ash. Further volcanic eruptions covered the footprints with more ash. The layers of volcanic ash fossilised over millions of years. Eventually, erosion uncovered the footprints.

The Laetoli footprints are the oldest known hominin footprints. They provide evidence of habitual, full bipedalism long before the evolution of the modern human brain and long before the earliest known stone tools.

A part of the Laetoli fossil trackway. Fidelis T Masao and colleagues 2016. Test-pit L8 at Laetoli Site S. Fidelis T Masao and colleagues, CC BY 4.0 via Wikimedia Commons; eLife 5:e19568. https://doi.org/10.7554/eLife.19568

Members of the genus Homo share several distinct characteristics:

- rounded skull with a relatively large braincase
- flatter face with smaller jaws and teeth
- distinct body adaptations to bipedalism
- larger bodies with longer legs and shorter arms
- lower degree of sexual dimorphism
- use of stone tools

However, it is not possible to evaluate all these features in a fossil, particularly for the earliest members of the genus, for which the fossil record is highly fragmented. Furthermore, early Homo fossils exhibit significant variability in these traits, suggesting a "patchy" evolutionary pattern, with different changes occurring across different species, regions, and time periods.

During the period when the genus Homo appeared, the local environment in Africa was undergoing rapid changes. Grasslands expanded and contracted multiple times, resulting in changes in plant and animal communities. The unstable climate and changing landscape likely exerted strong selective pressures, favouring the evolution of larger brains and prosocial behaviours in Homo species. Larger brains may have enabled behavioural flexibility, driving innovations such as increased meat consumption, cooperative hunting, and the exploration of new environments beyond Africa. The combination of larger brains and smaller teeth in early Homo reflects a shift in adaptive strategy—one that involved using tools to exploit new food sources, distinguishing them from earlier hominins.

Several species have been classified within the genus Homo, but the relationships among them remain unclear. Several of them left Africa and managed to populate diverse environments; the most successful of these was our own species—*Homo sapiens*.

Homo habilis and *Homo erectus*

Among the earliest members of the genus Homo, *Homo habilis* and *Homo erectus* represent key milestones in human evolution: the emergence of tool-making behaviour, the transition to a more modern body form, larger brains, and a range that extended beyond Africa.[3]

Homo habilis, traditionally regarded as the earliest known member of the genus Homo, was long considered the first toolmaker—earning it the nickname "handyman"—due to the discovery of stone tools and cut-marked animal bones near its fossil remains. While later discoveries suggest that earlier hominins, possibly *Australopithecus afarensis*, may have also used tools, archaeological sites from the time of H. habilis nonetheless show a significant increase in the quantity of stone tools, signalling a growing reliance on these technologies. This species exhibited several key traits that distinguished it from earlier hominins, including smaller molar teeth and a larger brain. Notably, its hand is anatomically similar to that of modern humans, enabling a "precision grip"—an important adaptation linked to making and using tools. *H. habilis* lived in parts of sub-Saharan Africa approximately 2.4–1.4 million years ago. Fossils have been found primarily in East and South Africa in regions that included open grasslands, wooded areas, and lakeshores—environments rich in resources for scavenging, tool-making, and survival. The fossil record of *H. habilis* shows considerable variation, suggesting it may not have been a single, uniform species but rather a diverse group with a range of anatomical traits.

Homo erectus is the longest-surviving species in the genus Homo, enduring for nearly two million years, from around 1.9 million years ago to as recently as 108,000–117,000 years ago. Its long survival and broad ecogeographic range resulted in considerable variation between populations. *H. erectus* was also the first human species to venture out of Africa, successfully populating diverse environments across Eurasia and Indonesia for over a million years, thriving in both tropical and tem-

[3] The classification of *Homo habilis* is debated, with some scientists proposing a split into *H. habilis* and *H. rudolfensis* due to variability in anatomy. Similar debates surround *Homo erectus*, suggesting differentiation between African *H. ergaster* and Asian *H. erectus* specimens. Additionally, early European fossils are sometimes classified as a separate species, *H. antecessor*.

perate regions. This wide geographic spread highlights its exceptional adaptability, far exceeding that of any earlier hominin. Over time, its brain size increased to around 900 cubic centimetres, while tooth size decreased, reflecting dietary and behavioural adaptations. The species coexisted with *H. habilis* and possibly other Homo species in Africa for some time but was the last to remain, persisting for a considerable period after the extinction of the other species. *H. erectus* shows clear evidence of being fully committed to life on the ground, with long, powerfully muscled legs, a narrow waist, and short forearms—all of which made it highly suited for endurance activities like walking and running. This enabled the species to cover long distances, follow prey, and access scattered resources, which played a key role in its survival and expansion into diverse environments. *H. erectus* exhibits an increased reliance on meat, as evidenced by archaeological findings such as cut-marked bones and the use of advanced tools. Additionally, there is evidence suggesting that *H. erectus* was using fire as early as 1.7–2.0 million years ago, making it one of the first hominins to do so.

Archaic Homo Species

After *Homo erectus*, several human species emerged that are often referred to as archaic Homo species. These species, predating *Homo sapiens*, exhibited a mosaic of primitive and more advanced traits—more developed than earlier hominins, yet not fully modern. For some of these species, we don't just have fossils—we also have DNA, which has provided some unexpected and fascinating insights.

Living roughly between 700,000 and 200,000 years ago, ***Homo heidelbergensis*** was the earliest known Homo species to have a brain size comparable to that of modern humans, with estimates ranging from 1100 to 1400 cm^3. The species also had short, stocky bodies, likely an important adaptation for conserving heat in colder environments. *Homo heidelbergensis* fossils have been discovered across Europe, Africa, and possibly Asia, indicating the species' extensive range and adaptability. They regularly used fire, hunted large animals, and utilised a variety of tools, including both wooden and stone implements. *H. heidelbergensis* is also considered the first Homo species for which evidence exists of constructing shelters made of rock and wood. The 430,000-year-old hominin fossils from the Sima de los Huesos site in Spain represent the oldest known instance where both mitochondrial DNA (mtDNA) and nuclear DNA have been sampled. Intriguingly, while nuclear DNA analysis indicates a close relationship with Neanderthals, the mtDNA reveals a closer connection to Denisovan hominins, suggesting a complex evolutionary history for this population.

A recently discovered Homo species, found on the Indonesian island of Flores, has attracted much attention due to its unusually small stature—earning it the nickname "Hobbit". Initially, some researchers proposed that the fossils might represent diseased individuals of a known species, such as *Homo erectus*. However, it is now widely accepted that these remains belong to a distinct species, *Homo floresiensis*, which was about 1 meter tall and weighed around 25–30 kilograms. Living on Flores between 100,000 and 50,000 years ago, their small size is largely attributed to island dwarfism, a process where isolation on an island can lead to a reduction in body size in large animals. Despite their small brain size—around 380–400 cm^3—these hominins were capable toolmakers. They utilised relatively simple stone tools and likely processed animals and cooked food, as suggested by the discovery of bones with cut marks in their dwellings. Their anatomy includes several unique traits, such as unusual wrist bones and shoulder structures, which distinguish them from other Homo species. Two main hypotheses exist regarding their origins: one posits an early divergence from the Homo lineage, potentially close to *Homo habilis*, while the other proposes that they evolved from *Homo erectus*. The debate about their exact evolutionary history is still ongoing.

The Denisovans are a population of hominins identified as a distinct species based on genomic data. Indeed, DNA had been for 15 years the pri-

Why Humans Lack Fur

Humans appear nearly hairless compared to our closest relatives—chimpanzees, gorillas, and orangutans. Yet this isn't entirely accurate. We possess roughly the same number of hair follicles as chimpanzees, but our body hair is so fine that it offers little insulation. Only certain areas—such as the scalp, armpits, and pubic regions—retain thicker hair. The reasons for our loss of dense body hair remain a subject of scientific debate, with several theories attempting to explain this evolutionary shift.

The most widely accepted theory is thermoregulation. As early humans moved from wooded areas to open savannahs, they needed efficient cooling. Reduced body hair, combined with an increase in sweat glands, allowed them to dissipate heat more effectively during prolonged activities such as endurance hunting. Over time, the development of fire and clothing provided warmth at night, enabling humans to maintain this hairless condition even in cooler environments. Simultaneously, darker skin pigmentation likely evolved to protect the mostly hairless body from harmful UV radiation. However, this theory has its critics. Some argue that early hominins often lived at high altitudes with cooler temperatures, making heat loss at night a risk. Others question whether they were really active during the hottest parts of the day, noting that many primates tend to be more active in the morning and late afternoon.

Another theory proposes that humans lost their dense body hair to reduce infestations by parasites like lice and biting flies. Bare skin would have been less attractive to these pests, improving health and making hairless individuals more appealing as mates. As humans began wearing clothing which could be removed and washed, this adaptation gained further benefit.

An unusual hypothesis suggests that hair loss in humans may be linked to enhanced social communication. Humans possess a unique type of colour receptor in their eyes, making them highly sensitive to subtle changes in skin colour. This sensitivity enables the detection of nonverbal cues—such as blushing, paleness, or flushing—that signal emotions or underlying health issues. These cues are only visible on exposed skin, which may explain why humans retained minimal facial and body hair, ensuring that these social signals are easily seen.

While no single theory fully explains why humans lost their dense body hair, it is likely that a combination of factors—thermal regulation, parasite reduction, UV protection, social communication, and sexual selection—all played a role.

mary source of information about this species due to the scarcity of fossil evidence—including a finger bone, a few teeth, long bone fragments, a partial jawbone, and a skull fragment—leaving much of their anatomy unknown. The fossils identified as Denisovan originate primarily from Denisova Cave in Siberia, Russia, and are dated to between approximately 285,000 and 25,000 years ago. One specimen was also found on the Tibetan plateau. In 2025, a nearly complete skull originally discovered in north-eastern China was identified as Denisovan using advanced protein and mitochondrial DNA analysis from dental calculus (hardened tartar). This is the first definitively classified Denisovan skull, providing new insights into the morphology of this species. Bone fragments from Denisova Cave have yielded both mtDNA and nuclear DNA sequences. This genetic data confirms that Denisovans were a distinct species from both Neanderthals and modern humans. Genetic data further suggests that Denisovans diverged from a common ancestor shared with Neanderthals approximately 700,000 years ago.

9.5 DNA in the Study of Human Evolution

To study human evolution, scientists analyse nuclear DNA and mitochondrial DNA (mtDNA) to reconstruct ancestry, migration patterns, and genetic adaptations. Nuclear DNA is housed within the nucleus of a cell and contains the vast majority of genetic material. Because nuclear DNA is inherited from both parents, it provides a detailed and complex picture of ancestry. Mitochondrial DNA is found in the mitochondria and is inherited exclusively from the mother, making it a powerful tool for tracing direct maternal ancestry. Despite being a very tiny fraction of the human genome, its short sequence length makes it notably easy to isolate, sequence, and align.

Both contemporary and **ancient DNA** can be used to study human evolution, and each approach has its advantages and limitations. Contemporary DNA analysis involves studying the genetic material of present-day humans and our closest relatives, the great apes. Analysing genetic variation within and between human populations provides clues about migration patterns, population bottlenecks, and adaptations to local environments. Studies of global mtDNA and Y-chromosome diversity, for example, have traced the dispersal of modern humans from Africa to the rest of the world (see Chap. 5). Genomic data can help us identify genes that have been under selective pressure, such as those related to brain size, skin pigmentation, or adaptation to high altitudes. For example, genes related to lactase persistence show evidence of positive selection in populations with a long history of dairy farming.

Ancient DNA (aDNA)

Ancient DNA (aDNA) refers to DNA extracted from specimens ranging from several hundred to hundreds of thousands of years old. In recent years, it has led to unprecedented insight into the lives and migrations of past populations, thereby significantly enriching our understanding of human history.

When an organism dies, enzymes (both within the cells and those from decomposing microorganisms) begin to break down the bonds between the nucleotides that form DNA. Water exposure and temperature fluctuations further accelerate this process. Important changes during DNA degradation include the fragmentation of DNA molecules into ultra-short fragments, the conversion of the four nucleotides into various derivatives, and the cross-linking of DNA to other molecules. Degradation also involves the chemical modification of nucleotides, such as cytosine deamination, which converts cytosine into uracil and can lead to errors in aDNA sequencing.

The rate of DNA degradation depends on local conditions, most notably temperature, humidity, and the presence of microorganisms. The best environments for preserving DNA are dry and cold places, such as permafrost, high mountains, and caves. Even in ideal conditions, DNA would degrade over time. Researchers analysing DNA degradation across a large number of ancient bones estimated the half-life of DNA to be 521 years. This means that after 521 years, half of the bonds between nucleotides in the backbone of a sample would have broken. After each 521-year period, the remaining DNA halves again, making it increasingly fragmented over time. This limits the recovery of genetic information from ancient specimens.

The haploid human genome (which consists of a single set of chromosomes) is approximately 3.2 billion base pairs (3.2 Gbp) long, while the diploid genome is about 6.4 Gbp.[4] Considering the estimated DNA half-life, even under ideal conditions, practically every bond would be destroyed after approximately 6.8 million years. However, scientists suggest that DNA would become too degraded for meaningful analysis much earlier, likely after 1.5 to two million years, when the remaining fragments would be too short to reconstruct genetic information. For the study of human evolution, this implies that scientists can recover and analyse DNA from species within the

[4]The male genome is slightly shorter than the female genome, due to the shorter Y chromosome.

genus Homo, such as Neanderthals and Denisovans, but cannot obtain genetic data from older hominin genera, including Australopithecus and Ardipithecus.

9.6 The Neanderthals and Us

Homo Neanderthalensis

In 1856, quarry workers in Germany's Neander Valley uncovered a partial skeleton that was clearly human-like but distinctly different from any known remains. This find became the first ancient human species ever identified. Officially classified as *Homo neanderthalensis* in 1864, it marked a turning point in our understanding of human evolution and helped lay the foundation for the field of paleoanthropology.

Neanderthals inhabited Europe and parts of western and central Asia from approximately 400,000–40,000 years ago. Their remarkable adaptability enabled them to inhabit a vast range, stretching from Portugal and Wales in the west to the Altai Mountains of Siberia in the east, thriving in diverse environments, from cold steppe regions like Ice Age England and Siberia to warmer, temperate woodlands in places such as Spain and Italy.

Anatomical features reveal that Neanderthals were highly adapted to their Ice Age environments, both physically and behaviourally. Their bulky trunks, broad shoulders, wide hips, and short limbs helped minimise heat loss by reducing surface area—an important adaptation for surviving glacial climates. Adults typically stood between 1.50 and 1.75 m tall and weighed around 64–82 kg. Brain size in Neanderthals averaged around 1,520 cubic centimetres—slightly larger than that of modern humans, although proportional to their larger body size. Neanderthal skulls were long and low, with strong brow ridges. Their faces projected forward, dominated by a large, wide nose—possibly an adaptation to cold, dry air, helping to warm and moisten it before reaching the lungs.

Contrary to other ancient humans, thanks to the abundance of their fossil record, we can reconstruct not only the appearance of Neanderthals but also aspects of their way of life. Excavations at sites where multiple individuals, including children, were buried suggest that Neanderthals lived in small, cohesive, possibly family-based social groups.

Neanderthals were skilled hunters and toolmakers. Their sophisticated tools, such as spears and flint axes, suggest that they regularly hunted large animals like mammoths, bison, and reindeer. Neanderthals also learned to control fire at least 200,000 years ago, an ability essential for surviving in the cold environments they often inhabited.

While the typical image of Neanderthals is of highly carnivorous hunters and scavengers who ate large mammals, they had a more varied diet than previously believed. Remains preserved in dental calculus reveal they also consumed a variety of plants and fungi. Neanderthals coexisted with modern humans for an estimated 2,600–5,400 years in various regions of Europe and Asia and went extinct between 35,000 and 50,000 years ago. Although their disappearance is well documented, the exact causes remain a subject of debate. It is clear, however, that their small, isolated populations, competition with *Homo sapiens*, and the impacts of climate change all played a significant role in their eventual disappearance.

Homo sapiens

All humans alive today belong to the species *Homo sapiens*, the scientific name meaning "wise man" in Latin, first introduced by Carl Linnaeus in 1758.

Homo sapiens is characterised by a relatively light skeleton and an upright posture. Our distinctive skull features a rounded and high braincase, a vertical forehead, and a small, retracted face with reduced jaw and teeth. Modern humans also possess a projecting chin, a feature unique among hominins, which may have evolved in connection with speech and to reduce stress on the jaw. Our species has a large brain, averaging around 1,300

Compassionate Neanderthals

One of the most remarkable aspects of Neanderthal life was their apparent compassion and social care. Neanderthals didn't abandon their sick or elderly; instead, they likely lived in supportive communities that provided long-term care for vulnerable members. Numerous skeletons show evidence of healed injuries and conditions that would have required ongoing assistance from others. For instance, remains from Shanidar Cave in Iraq include an individual with serious injuries who survived for years, and an elderly man from France was found to have lived well beyond the age when he could care for himself, despite suffering from severe arthritis and tooth loss. These discoveries challenge stereotypes, revealing a species capable of deep social bonds and a profound sense of social responsibility for disabled community members.

Archaeological discoveries now strongly support the idea that Neanderthals practised intentional and possibly symbolic burial, indicating that death held meaning for them. At Shanidar Cave, for example, a burial site was found surrounded by pollen from wildflowers, suggesting ritual or symbolic practices. Other burials further reinforce this, with the presence of grave goods, such as animal horns, and the deliberate positioning of bodies, suggesting their funerals carried emotional or symbolic significance. This offers a poignant reminder that these ancient humans may have honoured their dead in ways not so different from our own.

cm^3, although there is variation in brain size between individuals and populations.

Homo sapiens is defined not only by anatomical traits but by exceptional cognitive and behavioural innovations. Their ability to cooperate and adapt enabled them to thrive in diverse environments and eventually spread over the globe.

Our species evolved in Africa around 300,000 years ago amidst significant climate change. While genetic and fossil evidence consistently point to an African origin, recent discoveries challenge the long-held view that modern humans emerged from a single ancestral population. Instead, findings based on extensive analysis of modern human DNA[5] indicate a more complex story.

The study suggests that modern humans emerged from two ancestral populations that diverged around 1.5 million years ago and reconnected approximately 300,000 years ago, shaping the genetic diversity of *Homo sapiens*. The first population, which eventually contributed around 80% of the genetic makeup of modern humans and was also ancestral to Neanderthals and Denisovans, experienced a severe bottleneck before expanding. The second population remained distinct until about 300,000 years ago, when interbreeding between the two groups led to it contributing roughly 20% to modern human DNA. Importantly, this ancient mixing of the two populations happened before modern humans began to spread out of Africa across the globe. This makes the way these two groups interbred an important event that shaped the genetic makeup of all human lineages before they dispersed worldwide.

Over generations, *Homo sapiens* migrated out of Africa in several waves. The most successful dispersal occurred between 50,000 and 70,000 years ago, eventually leading to their spread across the entire globe (see Chap. 5). During this expansion, they encountered and interbred with other Homo species, such as Neanderthals and Denisovans. Today, traces of this interbreeding remain in the DNA of modern populations—individuals of non-African descent typically carry

[5]The data used in the study is from the 1000 Genomes Project, a global initiative that sequenced DNA from populations across Africa, Asia, Europe, and the Americas.

a small percentage (1–2%) of Neanderthal DNA, while populations in Oceania and Southeast Asia show genetic traces (4–6%) of Denisovan ancestry.

Scientists continue to explore how uniquely human traits—such as language—emerged and identify the underlying genes and evolutionary forces acting on them. For example, a recent study identified a single amino-acid difference in the NOVA1 protein, which is found only in Homo sapiens and absent in Neanderthals and Denisovans. Experimental introduction of this exclusively human variant into mice altered their vocalisations. Scientists believe that this change in the NOVA1 protein may have contributed to the development of the neural system involved in more complex vocal communication.

From the earliest bipedal steps in Africa to our global spread and complex societies, the human story reflects continuous adaptation and innovation. As new evidence fills gaps in the fossil record and genomic data, this evolutionary journey continues to unfold, with every new discovery deepening our understanding of what it means to be human.

DEFINITIONS

Hominids: the group consisting of all modern and extinct Great Apes: modern humans, chimpanzees, gorillas, and orangutans, together with all their ancestors

Hominins: the group consisting of modern humans, extinct human species, and all our ancestors and relatives, including members of the genera Homo, Australopithecus, Paranthropus, and Ardipithecus, defined in part by the shared ability of its members to walk bipedally

Clade: a group of organisms which contains a common ancestor and all descendants of this common ancestor, living and extinct

Fossils: preserved remains, impressions, or traces of once-living organisms, such as bones, shells, or imprints, typically over 10,000 years old

Fossil record: all the fossils found till now

Last common ancestor (LCA) of two species: the most recent population from which both of them evolved

SUGGESTED READING

- Cousins T, Scally A, Durbin R. A structured coalescent model reveals deep ancestral structure shared by all modern humans. Nat Genet. 2025 Apr;57(4):856-864
- Ghasidian E et al. Modelling Neanderthals' dispersal routes from Caucasus towards east. PLoS One. 2023 Feb 23;18(2):e0281978
- Tajima Y et al. A humanized NOVA1 splicing factor alters mouse vocal communications. Nat Commun. 2025 Feb 18;16(1):1542
- Fu, Qiaomei et al. Denisovan mitochondrial DNA from dental calculus of the >146,000-year-old Harbin cranium. Cell 2025

9.7 Questions

1. **The position of the foramen magnum within the skull is important for our understanding of hominin evolution because:**
   (A) Larger and more forward-located foramen magnum enabled hominin brains to grow larger.
   (B) The spinal cord passing through the foramen magnum is better protected from injury in hominins than in other primates.
   (C) Backwards location and vertical orientation of the foramen magnum enabled the upright position of the enlarged hominin skull.
   (D) The foramen magnum is located more forward and downward in species capable of upright walking compared with arboreal quadrupeds.
   (E) The foramen magnum is found only in bipedal primates.
2. **Which of the following statements is/are correct?**
   (A) Some of the distant ancestors of humans were identical to extant chimpanzees.
   (B) The upright posture of humans followed the evolution of the enlarged brain.
   (C) Humans, chimpanzees, and gorillas evolved from a common ancestor.
   (D) *Homo sapiens* is the most advanced species produced by evolution and is no longer affected by natural selection.
   (E) Humans evolved from chimpanzees.
3. **Which of the following traits appeared first, as hominins diverged from other primates?**
   (A) Using fire and making stone tools.
   (B) Upright posture and bipedal locomotion.
   (C) Reduced jaws and canine teeth.
   (D) Enlarged skulls and brains.
   (E) Opposable thumbs on hands and feet.
4. **When scientists compared the DNA of humans and chimpanzees, they found out which of the following?**
   (A) Humans and chimpanzees have the same number of chromosomes.
   (B) The common ancestor of chimpanzees and humans underwent whole genome duplication.
   (C) Several genes were under strong positive selection in the ancestors of humans.
   (D) Chromosomal inversions have a greater impact on the differences between these two species than single-nucleotide mutations.
   (E) Mutations in non-coding DNA regions have a large impact on the biology of these species.
5. **Which of the following statements about the use of DNA in human evolution studies is correct?**
   (A) Ancient DNA (aDNA) can be reliably recovered from hominin fossils up to 5 million years old.
   (B) Mitochondrial DNA (mtDNA) is a tiny fraction of the human genome but is easier to isolate and sequence than nuclear DNA.
   (C) Comparing aDNA with the DNA of existing humans can give us information about selected genes.
   (D) Only aDNA of hominin samples can be reliably sequenced and analysed.
   (E) Half of all aDNA samples can no longer be sequenced after a period equal to one half-life.
6. **Based on the combined fossil and recent genetic evidence discussed in the chapter, which of the following statements accurately reflect the complex evolutionary relationships among members of the genus Homo?**
   (A) Homo sapiens and Neanderthals evolved as distinct lineages without any genetic exchange after their initial divergence.
   (B) All modern human populations have 2–4% Neanderthal ancestry.
   (C) Only 4% of the genetic material of Neanderthals has been incorporated into the genome of modern human populations.
   (D) An ancestral population common to Homo sapiens, Neanderthals, and Denisovans originated in Africa and represents a shared root for these distinct hominin groups.
   (E) *Homo sapiens* were the only Homo species to have migrated out of Africa.

7. **Figure 9.4 shows estimates of the predicted stature (height) of fossil hominins over time for the interval between 4 and 1 million years ago. Which of the following statements is/are consistent with the graph?**
   (A) The Laetoli footprints indicate a large variation in height in the species that left them.
   (B) For *Paranthropus boisei*, the stature estimates based on incomplete femora suggest taller individuals than the stature estimates based only on actual femora lengths.
   (C) The stature estimates suggest an increase in average hominin height over time of their existence.
   (D) Stature estimates for members of the genus Homo show greater variation than the stature estimates for members of the genus Australopithecus.
   (E) No complete femora were included in the analysis for *Paranthropus robustus*.

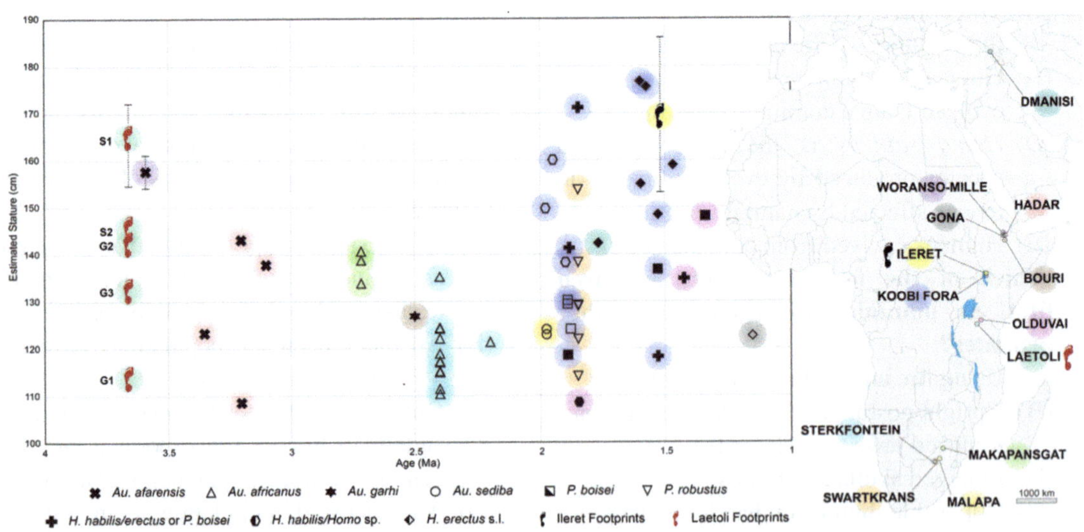

Fig. 9.4 Estimates of the predicted stature (height) of fossil hominins over time. Solid symbols (or crosses in bold) refer to stature estimates based on actual femur length; open symbols refer to stature estimates based on estimated femur length, in turn based on femur head diameter. For Laetoli and Ileret, stature estimates are based on footprint length. For Laetoli, Ileret, and Woranso-Mille, the average value and range of predicted stature are shown. Colours are associated with the geographical location of each fossil/footprint site on the map. (Source: Fidelis T Masao, Elgidius B Ichumbaki, Marco Cherin, Angelo Barili, Giovanni Boschian, Dawid A Iurino, Sofia Menconero, Jacopo Moggi-Cecchi, Giorgio Manzi (2016) New footprints from Laetoli (Tanzania) provide evidence for marked body size variation in early hominins eLife 5:e19568. https://doi.org/10.7554/eLife.19568)

9.8 Project: Further Chapters of the Story

The chapter on human evolution feels somewhat unfinished. After all, there is so much more to say about our own species—*Homo sapiens*. Why and how did our species become so widespread, numerous, and successful? What do we currently know about the evolution of our brain, communication, and social behaviour? What adaptations are specific to different environments—from equatorial rainforests, through high plateaus, to polar regions? And based on what research can we claim to know this?

In this project, you will act as a contributing author to an advanced high school or introductory college biology textbook. Your task is to expand upon our knowledge of *Homo sapiens* by delving into a specific aspect of our evolution.

1. Choose one specific question or topic about the evolution of *Homo sapiens* that genuinely interests you (e.g. the evolutionary drivers of increased brain size, the origins of language, the genetic evidence for human migration, or adaptations to extreme climates).
2. Conduct thorough research using a variety of credible sources (academic articles, reputable scientific web sites, textbooks). You may utilise artificial intelligence (AI) tools to assist with brainstorming, summarising information (which you must verify), or refining your writing, but the synthesis and critical thinking must be your own.
3. **Write a concise, scientifically accurate text (approximately 500 words or one page) that could serve as a new section continuing the discussion of human evolution in this chapter.**
   Your text should:
   - Clearly address your chosen question or topic.
   - Integrate findings supported by at least two distinct types of scientific evidence (e.g. fossil, genetic, neurological). Briefly explain how this evidence contributes to our understanding.
   - If relevant to your topic, briefly acknowledge a current scientific debate or an unanswered question within the field.
4. Attach a short accompanying reflection (approximately 150 words) in which you:
   - List the primary sources you consulted (e.g. specific book titles, article names/DOIs, Web site URLs)
   - Describe how you used any AI tools, detailing specific tasks (e.g. "used AI to generate keywords for search", "used AI to summarise a research article for initial understanding", "used AI for grammar checking")
   - Critically evaluate AI's contribution: provide one example where AI was particularly helpful and one example where you had to correct or disregard its output due to inaccuracy or irrelevance
5. Propose one relevant visual element (e.g. a diagram, a map, a graph) that would enhance understanding of your text if it were published. Briefly explain the purpose of this visual (1–2 sentences).

This project was developed by Kristína Hudáková.

Part III

Broad-Scale Patterns and Processes

Evolution of Social Behaviours 10

Barbora Trubenová and Kristína Hudáková

▶ Everywhere around us, organisms engage in complex social behaviours. Cells interact to form multicellular organisms, individuals interact within groups and populations, and species interact with each other in complex networks of ecosystems. It is hard to believe that all these behaviours are also encoded by genes and are the result of millions of years of evolution!

As we have seen, natural selection favours traits that help individuals to survive and reproduce. It is easy to see why a behavioural trait that benefits the actor—the individual who displays the behaviour—but harms everyone else should be selected for. Such a selfish trait increases the fitness of its bearer, giving it a competitive advantage and allowing it to spread across the population. It would seem that all bacteria, plants, and animals would behave selfishly.

However, many organisms help others at their own expense. Buffaloes risk their lives to defend the young of others against lions. Meerkats make warning calls when they spot danger, giving away their position. Termites even sacrifice their lives to protect the nest. Even cells in multicellular organisms are helping each other to survive. One might wonder what would happen if every one of our cells decided to act independently.

The evolution of these interactions appears "unnatural" at first glance and has puzzled scientists for decades. Even Darwin recognised such behaviours as a challenge to his theory. In this chapter, we will explore how these behaviours can evolve under the conditions of natural selection.

10.1 Very Important Definitions

Let's start with a few crucial definitions. The **direct fitness** of an individual is given by its ability to survive and reproduce—this is basically the fitness we talked about in previous chapters. We say *direct* here because later in this chapter, we will introduce a different type of fitness: **indirect fitness**.

We call a particular behaviour *costly* if it decreases the direct fitness of an individual in the *long term* (during its lifetime) or *beneficial* if it increases its direct fitness in the long term. It is very important to talk about lifelong benefits and costs because a single act or interaction can be costly in the short term but still improve long-

B. Trubenová (✉)
D-USYS, ETH Zürich, Zürich, Switzerland

Eawag, Dübendorf, Switzerland

eduB, Bratislava, Slovakia
e-mail: barbora.trubenova@env.ethz.ch

K. Hudáková
eduB, Bratislava, Slovakia
e-mail: kristina.hudakova@edub.sk

B. Trubenová et al. (eds.), *Everyday Evolution: Unveiling Life's Hidden Patterns*,
https://doi.org/10.1007/978-3-032-07050-0_10

term fitness. For instance, a squirrel storing nuts for winter is paying a cost in autumn, but it is definitely worth it when the snow comes. Overall, this behaviour is beneficial.

From an evolutionary perspective, any behaviour (or interaction) is considered to be **social** if it influences not only the fitness of the **actor** (the individual performing the behaviour) but also the fitness of **recipients** (its social partners who are affected by its actions). A bear (an actor) defending its territory from other bears or wolves (recipients) is also displaying social behaviour.

Depending on their lifelong effect on both the actor and the recipients, we can distinguish four main categories of social behaviours: **cooperation**, where both the actor and recipient benefit; **altruism**, where the recipient benefits at a cost to the actor; **selfishness**, where the actor benefits at a cost to the recipient; and **spite**, where both actor and recipient suffer a fitness loss.

		Effect on recipient	
		+	-
Effect on actor	+	Cooperation	Selfishness
	-	Altruism	Spite

Note the very important difference between cooperation—a symmetric behaviour where everyone profits in the long term—and altruism—an asymmetric behaviour where only the recipients benefit while the actor suffers decreased fitness in the long term. For instance, wolves hunting in a pack are cooperating—they are more successful when hunting together, and thus they all profit from the behaviour. On the other hand, an ant soldier sacrificing her life is paying the ultimate cost to protect the queen, who benefits from this behaviour—this is, therefore, an example of altruism.

Interestingly, these are not the universally accepted definitions of cooperation and altruism—in fact, there are no universally accepted definitions. Researchers use different definitions, leading to confusion and even heated arguments! One might encounter many contradictory statements, especially regarding the evolution of altruism, making it important to check all definitions carefully.

10.2 Evolution of Cooperation

It is easiest to explain cooperation if there is no cost and everyone benefits immediately from each interaction. For instance, many prey species, such as gazelles, deer, and capybaras, as well as many species of birds, feed together in large herds or flocks. There are numerous benefits to living in a bigger group: a smaller chance of being eaten by a predator, a higher chance of finding a mating partner, and less time invested in being vigilant, meaning more time for feeding. Every member of the group benefits from the presence of the others.

Such a cost-free mechanism for establishing cooperation is also common between species: think about sea anemones and clownfish, spider crabs and algae, and humans and our gut bacteria. These behaviours are immediately beneficial for all participants, and there is no cost. Such relationships between species are often called **mutualism**.

However, sometimes cooperative behaviour consists of many single acts that are costly for the actor in the short run—for instance, donating food or protecting someone. Even then, such behaviours can be beneficial in the long term if all individuals in a group help each other and good deeds are rewarded later.

But what happens if a mutation occurs, causing an individual to behave selfishly in a group of helping individuals? Such an individual—a cheater—would get all the help (benefits) from the others, but would not pay any cost! For instance, if a lazy wolf, which avoided hunting and thus saved its energy (no cost), ate with the rest of the pack, it would have higher fitness than other members of the pack. Over the next generations, such selfish wolves would replace the cooperators.

Even though cooperative behaviour increases the fitness of all participating individuals, selfish, non-helping cheaters always benefit more than helpers. Thus, it would seem that alleles encoding cooperation should never spread in a population of non-helpers, while cheaters should always be able to invade (and replace) a population of cooperators.

Nevertheless, cooperation has been documented in many species, suggesting that genes for helping behaviour must be able to spread and persist within the population. Some mechanisms must exist that ensure that cheaters cannot exploit the cooperators in the long term. Let's look at a few of these.

Reciprocity

One way to ensure that helping behaviour is beneficial overall, even if any single act is costly, is through reciprocity (Fig. 10.1). Reciprocity is based on the expectation that a good deed is repaid by another good deed. It requires that the benefit the recipient receives is greater than the cost the actor has to pay. Then, even though the act of helping is costly in the short term, the behaviour pays off because the benefit that is received later is well worth it.

However, for reciprocity to work, individuals must be able to discriminate between cooperators and cheaters and behave accordingly. In a population where everyone offers help without conditions, individuals who cheat can easily take advantage, as they incur no costs while still reaping the benefit.

The simplest way to stop cheaters from gaining benefits is if individuals assist only those who have previously helped them. It pays to help another individual because the same individual is likely to repay the help later. Cheating individuals can cheat maybe once, but next time, they will receive no help and, in the end, will have lower fitness than cooperators. This mechanism is called **direct reciprocity** and requires relatively high intelligence and memory. It can be captured by the principle: *"You scratch my back, and I'll scratch yours."*

Direct reciprocity may explain cooperation in vampire bats (*Desmodus rotundus*). These bats

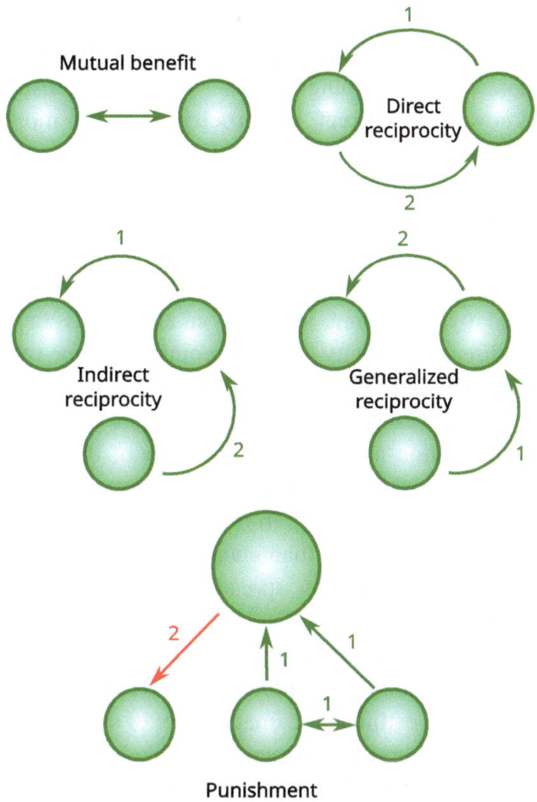

Fig. 10.1 Mechanisms of cooperation. Different mechanisms can ensure that cooperative behaviour is beneficial in the long term. Green arrows represent help, while the red arrow represents punishment. Numbers represent the order of the interactions. In mutual benefits and direct reciprocity, two individuals help each other, while in indirect and generalised reciprocity, multiple individuals and interactions are involved. Punishing a non-cooperative individual (bottom), a costly behaviour for both the actor and the recipient, is rewarded by the help of others

leave their cave every night to feed on the blood of the livestock. If an unlucky bat does not find a cow to feed on, it returns home hungry and may die after a day or two. Luckily, it can be saved by its more fortunate (and sated) bat friend, which regurgitates part of its own food to feed the hungry bat. Scientists have observed that bats are more likely to save those who have helped them previously. Even though this behaviour is costly in the short term (giving away food), it pays off in the long term, as the donor can be saved from death later.

Another way to protect cooperators from cheaters is for individuals to help only those known for helping others. This mechanism, known as **indirect reciprocity**, is a bit more complicated. It is captured by the principle *"I scratch your back, and someone else will scratch mine."*

Indirect reciprocity is based on *reputation*, or *image score*, which is increased by helping others and by spreading information about this fact. Indirect reciprocity does not require the same two individuals to ever meet again, which is advantageous. Cooperation is rewarding as it fosters a reputation as a valuable member of the community, and those with such a reputation tend to receive more help from others. Cooperators have higher fitness than cheaters, and alleles for such behaviours can spread and persist in the population.

However, cooperation based on indirect reciprocity requires individuals to have memory and be able to estimate the image score of their social partner. Therefore, cooperation is likely to be limited to fairly intelligent animals such as primates, especially humans. The spread of information and communication is also important for this mechanism to work. In fact, several scientists think that indirect reciprocity encouraged the evolution of human language, intelligence, and even morality!

Both direct and indirect reciprocity require individuals to know how likely their social partners are to cooperate. On the other hand, several scientists suggested that another, more primitive mechanism could work: individuals help anyone if they have received help previously. Such a mechanism is called **generalised reciprocity** and requires far fewer brain cells. Rather, it is based on changes in the individuals' physiological or neurological state and so can work in simpler organisms. Generalised reciprocity can be captured by the principle *"You scratch my back, and I will scratch someone else's."*

However, theoretical studies are a bit skeptical about this explanation—mathematical models show that it is very unlikely that this mechanism is robust enough to work in large populations. It is hard to imagine how it could resist cheaters. Yet, in one experiment, scientists trained rats to pull a stick in order to release food for another rat. They observed that rats pulled more often to help an unknown partner after they had been helped than if they had not received help before. Therefore, it might be worth investigating this phenomenon further.

Interesting Fact

The sense of fairness seems like a clearly human trait. However, scientists have discovered that some animals, such as Capuchin monkeys and dogs, also appear to recognise when they are treated unfairly. In one very cool experiment, two monkeys were given a piece of cucumber as a reward for doing a simple task. As long as both received the same reward, both were happy. But when one of them received a much better prize—a grape—the other one became very angry and even refused to do the task!

Punishments and Rewards

Another way in which cooperative behaviour may be favoured by natural selection is if individuals who cooperate are rewarded while those who do not are punished (Fig. 10.1). This mechanism is similar to indirect reciprocity—only well-behaved, cooperative individuals are rewarded. However, unlike in the case of indirect reciprocity, rewards and punishments assume some kind of hierarchy or asymmetry in the relationship, such as an alpha male punishing subordinates.

An example of a punishment is the case of the pied kingfisher—*Ceryle rudis*, a type of cooperatively breeding bird— where unrelated males help take care of the young. If they do not, they are likely to be evicted from the flock. Living alone decreases their fitness by making it harder to survive and find a mate—so it is worth helping! The cost to the group is a loss of one individual, but the cost to the evicted member is much higher—most likely his life. An example of rewards ensuring cooperation is the case of bonobo females, which often offer to have sex in exchange for food. Everyone is better off—the female gets her banana, and the male has an opportunity to sire offspring.

Sometimes, it is difficult to distinguish between the effects of punishments and rewards. Punishment can be interpreted as a lack of rewards and vice versa. However, the only important thing is that it should pay to help, either by receiving a reward or by avoiding punishment.

Cooperation is an extremely important phenomenon underlying major transitions in evolution, allowing the emergence of complex systems (see Chap. 15). Symbioses between prokaryotes led to the evolution of eukaryotes, the cooperation of cells led to the evolution of multicellular organisms, and the cooperation of solitary individuals led to the evolution of complex societies, such as bee hives, termite and ant colonies, or even human society. Without cooperation, we would not be here.

10.3 Evolution of Altruism

Much more difficult to explain than cooperation are behaviours that are *not* beneficial to the actor, even in the long term—for instance, sacrificing one's own life. Why do ants sacrifice themselves for the nest if they will not live to enjoy any benefits later on? Alleles for such suicidal behaviour certainly should not spread in the population, right?

Imagine an animal with a gene that predisposes it to protect its den, even if it means sacrificing its own life. Such an individual is less likely to leave offspring than those that sit snug in the den while an attack takes place. This is definitely altruistic behaviour, as it lowers the direct lifelong fitness of the actor. Thus, there should be fewer individuals carrying the "altruistic allele" in the next generation and even fewer in the next one. As a result, we would expect any alleles for altruistic behaviour to disappear within a few generations.

Contrary to this expectation, we observe exactly this behaviour in many species, including ants, bees, and termites. Each of their colonies consists of thousands of workers who spend all their lives building the nest, protecting it, and taking care of the queen. But the queen is the only one who reproduces, so only her genes are passed to future generations! Workers do not reproduce—so their direct fitness is zero. How is it then possible that altruistic behaviour still exists and was not erased by natural selection?

Kin Selection

> *I would gladly lay down my life [die] for two brothers or eight cousins.*
>
> JBS Haldane (c.1952)

This famous remark, jokingly stated by the British scientist JBS Haldane, captures the essence of kin selection—the most accepted explanation for the evolution of altruism.

So far, we have only considered one way of passing on one's genetic material to future generations: by reproducing oneself. This defines an individual's direct fitness (Fig. 10.2). However, there is another option: one can pass on one's genetic material by helping one's kin (relatives), who carry identical copies of one's genes, to pass on their DNA to the following generations. This indirect transmission of an individual's genetic material (via their relatives; see Fig. 10.2) contributes to its indirect fitness. The sum of direct and indirect fitness is called **inclusive fitness**—and this is actually what matters the most because this is the fitness that natural selection maximises! Imagine the following situation: if an individual helps their sibling, even if it is not good for them in the long term, the allele for this kind of altruistic behaviour is probably also present in the sibling and may be passed on to the next generation.

However, there are two important conditions for kin selection to work: first, individuals must be able to direct their help only towards their relatives, and second, the benefit must be high enough to pay off. Let's look at these two conditions in detail.

Fig. 10.2 Direct and indirect fitness. Direct fitness of individual A is given by its two offspring, which share 50% of A's genes. However, the offspring of B (who is a full sibling of A) also share 25% of A's genes—that is what defines the indirect fitness of A. From A's perspective, two nieces or nephews carry as much genetic material as one son or daughter. Total (inclusive) fitness is the sum of both direct and indirect fitness

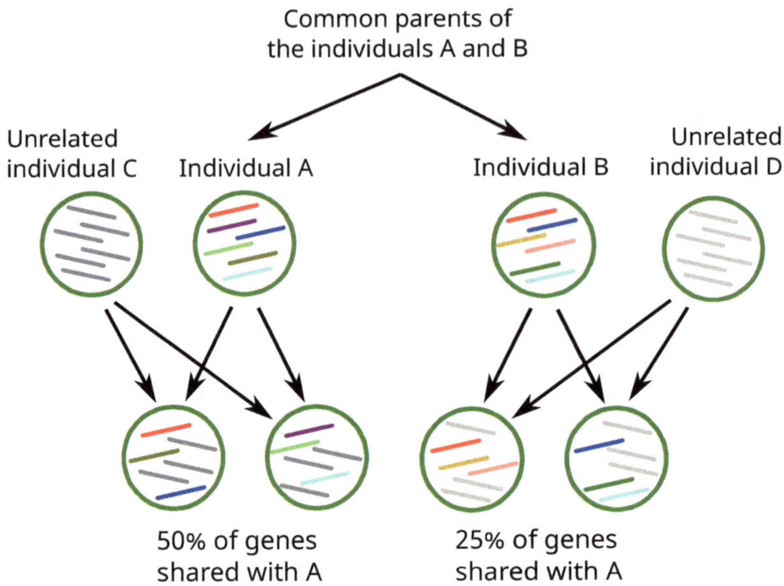

Recognising One's Kin

To direct the helping behaviour only towards relatives, it is useful to know who they actually are. However, most organisms do not store information about marriages and births as we do. So, how can they recognise their kin and know whom they should help?

Many animals rely on phenotypic similarity, e.g. smell or appearance, to assess who is their kin and who is not. Others rely on shared history—a common nest, den, or pack.

Many birds, such as long-tailed tits (*Aegithalos caudatus*), can discriminate kin and non-kin based on their calls. Siblings have similar calls because they learn them from their parents while still in the nest. Mammals often rely on smell. For instance, related Belding's ground squirrels produce similar odours, allowing them to recognise each other. House mice recognise their kin by the smell of their urine.

Another way to direct helping behaviour only towards close relatives is by not moving far away from them. In this case, everyone around is a close relative, and an individual can simply help everyone around. Scientists call this mechanism **limited dispersal**. This mechanism does not require high intelligence or memory and is likely important in a wide range of taxa, from bacteria to humans. It is especially a common mechanism in microorganisms, where a single cell can multiply, creating large colonies of clones. Thus, all nearby cells are closely related.

Hamilton's Rule

The other important condition is that the benefit should be high enough to pay off. But what does "high enough" actually mean?

Helping relatives survive and reproduce is a less efficient way of spreading your genetic material than reproducing yourself. In sexually reproducing organisms, with the exception of identical twins, no two individuals share all of their genes. Closer relatives share more genes: on average, siblings share approximately 50% of their genes, while first cousins share only about 12.5% (see Box: Relatedness for details).

From the perspective of a single gene, there is, therefore, about a 50% probability that a sibling will share an identical copy of this gene inherited from the same parent. Thus, this gene copy will benefit overall only if the benefit that sibling X receives is at least twice as large as the cost that sibling Y has to pay. Alternatively, at least two siblings should each receive a benefit at least as large as the cost. This is what Haldane meant—from the perspective of a gene, an individual's life is worth the lives of two of his or her siblings or eight of his or her cousins.

Mathematically, this condition for the evolution of altruism is formulated as **Hamilton's rule**:

$$rB > C$$

where B is benefit that the recipient(s) receives, r is relatedness between the donor and the recipient, and C is the cost that the actor has to pay. We will look at Hamilton's rule more closely in the project.

One cute example of kin selection comes from American red squirrels (*Tamiasciurus hudsonicus*). Occasionally, a female squirrel adopts an orphaned young that is her close relative. Even if this adoption comes at a cost to her (she has to care for more offspring), the benefit (saving the life of a close relative) outweighs the cost, and her inclusive fitness is increased. Scientists observed that adoptions always occurred only between close kin and that orphaned baby squirrels without nearby kin were never adopted, which supports Hamilton's rule. Interestingly, adoption of close kin has been reported in multiple mammalian species. A study by Riedman in 1982 documented that over 120 different kinds of mammals and 150 different kinds of birds have been seen taking care of young that aren't their own and even adopting them.

Other common examples of kin selection include the previously mentioned societies of bees, ants, and termites. All these organisms belong to the order Hymenoptera, which have haplodiploid sex-determination systems. This means that males are haploid and pass on their entire genetic material to their (female) offspring, while females are diploid and pass on half their genetic material to their offspring. Thus, sisters are highly related—sharing 3/4 of their genome. This unusually high degree of relatedness between workers means that helping their sisters (and the mother to reproduce) actually makes more sense than reproducing themselves.

10.4 Strangeness of Spite

So, we can explain the evolution of selfishness, cooperation, and even altruism. But what about spite? Why should organisms perform behaviours that are costly not only for their competitors but also for themselves?

Spiteful behaviour—a behaviour that lowers the fitness of both the actor and the recipient—can be selected if it reduces the fitness of the unrelated recipient more than the fitness of the actor or its close relatives. This way, it reduces the competition, and individuals who carry the spiteful gene variants are better off. This mechanism is very similar to kin selection but with vicious protagonists.

But...what about dolphins saving people? A leopard saving a baby baboon? Female dogs adopting abandoned kittens? YouTube is full of cute, sometimes moving, sometimes downright sticky-sweet videos of animals helping each other. In many of these examples, neither kin selection nor reciprocity can explain such behaviour. So, how do we explain it?

Well, actually, we do not. These examples are extremely rare! Most leopards would just eat the baby baboon, and most dogs are certainly not very nice to cats. So these weird behaviours may be the result of confusion (or a very, very rare mutation). They are not evolutionarily stable—gene variants for them would not spread through populations. That is, unless we, humans, start to select for these traits, making them pay off. This is where human influence comes into play. Through domestication and selective breeding, humans have repeatedly altered the evolutionary trajectory of other species. Consider different dog breeds, all descended from wolves, yet exhibiting an astonishing range of temperaments and physical traits. These characteristics, many of which would be detrimental in the wild, have been amplified and fixed in populations precisely because humans selected for them. You can read more about artificial selection in Chap. 1.

Altruistic Green-Beards

One fairly weird and interesting way in which genes for altruism can spread and survive in the population is called the "green-beard effect". Imagine the following situation: an allele that causes its bearer to behave altruistically is located very close on the genome to the allele that causes its bearer to have a distinctive characteristic—for instance, a green beard.[e] All and only altruistic individuals have green beards. If these altruistic individuals only help each other, even if some of them sacrifice their own lives in the process, the gene for altruism (and the green beard) will be associated with increased fitness overall and will spread in the population. This is a special case of kin selection: those who have green beards (and the altruistic gene) are more related to each other than those who do not have it.

However, what happens if a mutant arises that has a green beard but is not altruistic? This individual—a cheater—will have the highest fitness of all because it does not have to pay any cost but receives the benefit. The cheater's offspring will be fitter than helping green-beards; the cheater's genes will eventually spread, and the gene for altruism will be lost. Thus, for this mechanism to work, it is essential that such a mutation does not occur—both altruism and the green beard must sit so close to each other that recombination does not separate them, or they must be encoded by the same gene.

[e]Based on an example introduced by Richard Dawkins in his book *The Selfish Gene*.

DEFINITIONS

Social behaviour: an interaction between two or more individuals, typically (but not necessarily) of the same species, that influences each other's fitness

Cooperation: a mutually beneficial interaction where both individuals gain from each other's actions

Selfishness: behaviour that benefits an individual at a cost to others

Spite: a rare behaviour that harms both the actor and the recipient

Altruism: a behaviour that benefits another individual at a cost to the actor's own fitness

Kin selection: a mechanism ensuring that altruistic behaviour is favoured by natural selection if it increases the reproductive success of genetically related individuals, even at a cost to the actor

Inclusive fitness: an individual's total genetic success, including their own reproductive success (direct fitness) and the reproductive success of their relatives, weighted by their degree of genetic relatedness (indirect fitness)

Hamilton's rule: a mathematical inequality ($rB > C$) that explains when [altruistic] behaviour can evolve: when benefit (B) to the recipient, multiplied by their genetic relatedness (r) to the [altruistic] actor, exceeds the cost (C) experienced by this actor

SUGGESTED READING

- Nowak MA. Five rules for the evolution of cooperation. Science. 2006 Dec 8;314(5805):1560-3
- Gardner A, West SA. Spite and the scale of competition. J Evol Biol. 2004 Nov;17(6):1195-203
- Riedman ML. The Evolution of Alloparental Care and Adoption in Mammals and Birds. The Quarterly Review of Biology. 1982; 57(4):405-435
- Clutton-Brock TH, Parker GA. Punishment in animal societies. Nature. 1995 Jan 19;373(6511):209-16

Relatedness

As we have seen in previous chapters, in most sexually reproducing organisms, individuals inherit half of their genetic material from their father and half from their mother (well, ignoring those pesky mitochondria for now). So, we share 50% of our DNA with each of our parents. But what about our siblings? In theory, we may (with a very, very small probability) actually share all of our DNA, or none. However, most of the time, this number is also very close to 50%, sometimes a bit more, or a bit less.

Let's look at a pair of homologous genes across two siblings. Suppose one parent carries A and B alleles of this gene and the other parent C and D. One offspring inherits A from the mother and C from the father. What about the second offspring? She may have inherited any combination: AC, AD, BC, or BD with equal probability. In the first case, the two offspring have an identical pair of alleles; in the second and third cases, they share one allele out of two; in the last case, they share no alleles. Thus, on average, the two siblings share $(1 + 1/2 + 1/2 + 0)/4 = 1/2$ of their alleles, so that the coefficient of relatedness is $r = 0.5$.

Now, if we average across all the genes spread across all $23 \times 2 = 46$ chromosomes, it is even more likely that any two siblings share about half of their genetic material, maybe a little less, maybe a little more. Averaging across all siblings in the world, the average relatedness one would expect between them is 0.5.

Ok, but it has been widely published that we share more than 90% of our DNA with chimpanzees! Does it mean that the relatedness between a human and a chimp is 0.90? No, not really. When talking about relatedness between individuals of the same species, we ignore parts of the genome that are identical across all individuals, and only consider those parts that are variable. Think about it: we, humans, all share genes that code for two eyes, one nose, red blood cells, and 33 vertebrae. However, when talking about relatedness between individuals of the same species, we ignore parts of the genome that are identical across all individuals and only consider those parts that are variable, such as those coding for eye or skin colour, height, or blood type. It is the combination of these tiny features that makes us unique. So actually, about 99.9% of your DNA is identical between you and your sibling, parent, or other people. But the remaining 0.1 % of DNA can differ between humans. And out of that 0.1%, half of it you share with your closest relatives.

This chapter is based on the PhD thesis of Barbora Trubenová entitled "Indirect Genetic Effects and the Evolution of Cooperation", 2014. The University of Manchester.

10.5 Questions

Before answering the questions, look at the project section, and learn how to determine the coefficient of relatedness!

1. **Scientists observed grazing mammals in groups of varying sizes (Fig. 10.3). Graph A shows the mean number of times an individual in a given group raised its head per hour to scan for predators (individual alert rate). Graph B shows the total number of raised heads per hour for a given group (group alert rate). What can you conclude *from these graphs*?**
   (A) The bigger the group is, the less time each individual spends scanning the environment.
   (B) Bigger individuals can spend more time grazing and less time scanning for predators.
   (C) Overall, the total number of scans for the whole group was smaller in larger groups.
   (D) Animals grazing in the centre of the herd were less vigilant compared to those on the margins.
   (E) Larger groups are at lower risk of being attacked by predators.

2. **Some species of ants "farm" aphids: they protect aphids and guide them to the plants the aphids eat, and then the ants eat the honeydew the aphids release. Which of the following mechanisms could explain the evolution of this relationship between aphids and ants?**
   (A) Mutual benefits
   (B) Direct reciprocity
   (C) Indirect reciprocity
   (D) Generalised reciprocity
   (E) Kin selection

3. **Altruistic behaviour:**
   (A) Can evolve only if individuals interact repeatedly
   (B) Never occurs in natural populations
   (C) Can evolve only if it increases the long-term inclusive fitness of the altruists
   (D) Is expected to evolve when there are plentiful resources to share
   (E) Can be only explained if it occurs between relatives

4. **Ester and John had two sons: Peter and Paul. Peter married Jane and had two sons, Richard and Emil. Paul married Anna, and they had two identical twin daughters, Isabella and Rachael. Which of the following statements are true?**
   (A) The coefficient of relatedness between Peter and Ester is the same as between Paul and Rachael.
   (B) The coefficient of relatedness between Peter and Paul is the same as between Isabella and Rachael.
   (C) The coefficient of relatedness is higher between Isabella and Richard than between Rachael and Emil.

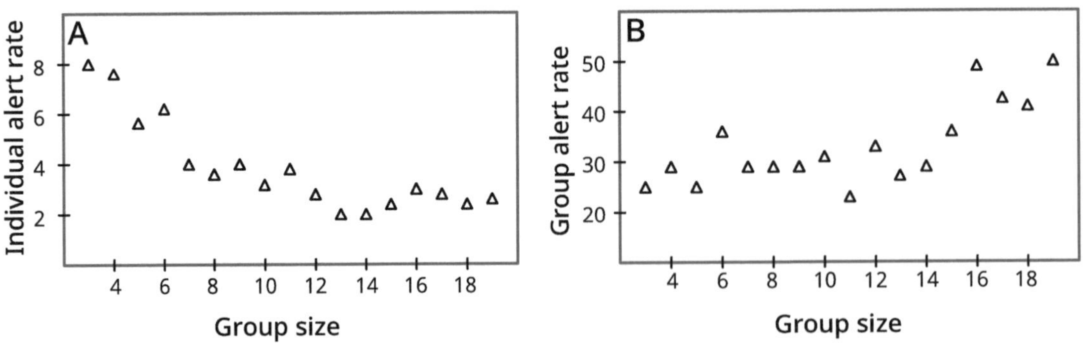

Fig. 10.3 Individual and group alert rate as a function of group size. (The graphs were created based on: Herrera, E.A. (2013). Capybara Social Behavior and Use of Space: Patterns and Processes. In: Moreira, J., Ferraz, K., Herrera, E., Macdonald, D. (eds) Capybara. Springer, New York, NY. https://doi.org/10.1007/978-1-4614-4000-0_11)

(D) The coefficient of relatedness between Anna and Ester is the same as between Jane and John.
(E) Coefficient of relatedness is the same between Jane and Isabella as between Paul and Emil.

5. **Imagine that Jane and Anna are identical twins. Which statements would be correct in such a case?**
(A) The coefficient of relatedness between Peter and Ester is the same as between Paul and Rachael.
(B) The coefficient of relatedness between Peter and Paul is the same as between Isabella and Rachael.
(C) The coefficient of relatedness is higher between Isabella and Richard than between Rachael and Emil.
(D) Coefficient of relatedness is the same between Anna and Ester as between Jane and John.
(E) Coefficient of relatedness is the same between Jane and Isabella as between Paul and Emil.

6. **You have a rare allele for violet eye colour. In which individual or group of individuals would you have the highest probability of finding this allele?**
(A) Your brother
(B) Your mother
(C) Four brothers of your mother
(D) All four of your grandparents
(E) Your aunt's family: your aunt, her husband, and their four children

7. **A squirrel adopting and saving a litter of her sister will:**
(A) Increase her indirect fitness
(B) Always increase the likelihood that her genes will be passed into the next generation
(C) Pay a cost in the form of possibly reducing the chances of survival of her own offspring
(D) Increase her inclusive fitness, if her own as well as adopted offspring survive
(E) Pay a cost in the form of energy expenditure in order to feed more offspring

8. **Suppose an altruistic behaviour brings 50 fitness units to the recipient, and the "cost" to the actor of this behaviour is 15 units. This altruistic behaviour can evolve if it is directed towards:**
(A) Only full siblings
(B) Nephews or nieces
(C) First cousins
(D) Full siblings or half-siblings
(E) Anyone, because the benefit is much higher than the cost

10.6 Project: One Sister Is Always the Best

We said in the chapter that while you share *exactly* half of your genes with your mother and half with your father, you only share *approximately* half of your genes with your sister or brother. Let's look at the situation with siblings in detail.

How to Determine Relatedness

If we have information about individuals' genomes, we can easily compare them to see how similar they are. We mostly look at many different sites of DNA (nucleotides) called polymorphisms, in which we, humans, differ from each other. However, if we compare close relatives, the differences between them are caused mainly because they inherited different chromosomes. So if John inherited two maternal chromosomes A^- and B^* and two paternal A^- and B^+ and his sister Maria inherited A^- and B^x from her mother and A' a B^+ from her father, John and Maria have two chromosomes from four identical (A^- and B^+), and thus the coefficient of relatedness between them is 2/4 = 0.5. Use this method in the project below.

If we do not know anything about the genomes involved, but we know familial relationships (e.g. when looking at pedigrees), we simply assume that the coefficient among full siblings is 0.5, just like the coefficient between parents and their children. When we deduce a coefficient of relatedness among more distant relatives, we do so by finding a path between them through siblings and parents. For each link, we introduce a coefficient of 0.5, multiplying them to find the final relatedness.

For example, there are three links between you and your cousin (you, your mother, her sister, her child (your cousin)), so the coefficient of relatedness between the two of you is $0.5 \times 0.5 \times 0.5 = 0.125$, as we mentioned in the chapter. Therefore, we can say that the probability that a specific gene is shared between you two is 0.125 or that you very likely share approximately 12.5% of your genes (Fig. 10.4). Use this method to answer questions in the "Question section".

Material

You will only need a pen, paper, and Internet access. Go to https://en.edub.sk/everyday-evolution, and look for Random genome generator.

Relatedness with Simple Genomes

1. Imagine two mating individuals. Their genotypes are shown in Table 10.1. The genotype of this species contains three pairs of chromosomes, A, B, and C. Assume that chromosomes do not recombine.
2. Each offspring randomly inherits one chromosome from each pair from its mother (red) and one from its father (blue), e.g. $\frac{A^*}{A^\wedge}$ $\frac{B'}{B^\dagger}$ $\frac{C^*}{C^\dagger}$
3. Use the random generator on our website to simulate matings and generate six offspring. Fill their genomes in Table 10.2.

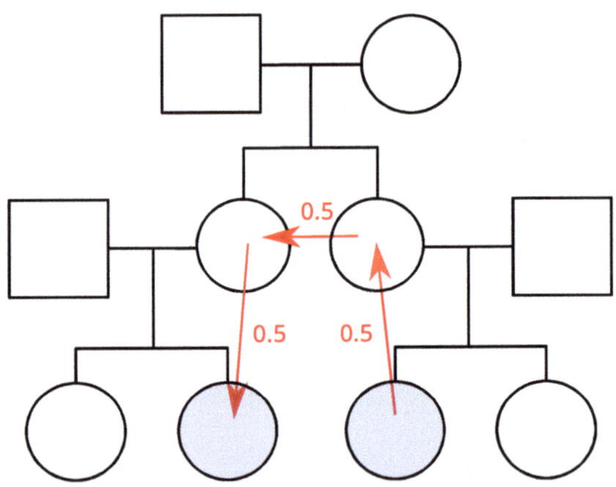

Fig. 10.4 Coefficient of relatedness. Coefficient of relatedness between two first cousins is $0.5 \times 0.5 \times 0.5 = 0.125$, because they are separated by three links with relatedness 0.5

Table 10.1 Parental genotypes

Individual	Mother	Father
Genotype	$\frac{A'}{A^*}$ $\frac{B'}{B^*}$ $\frac{C'}{C^*}$	$\frac{A^\wedge}{A^\dagger}$ $\frac{B^\wedge}{B^\dagger}$ $\frac{C^\wedge}{C^\dagger}$

Table 10.2 Offspring genotypes

Individual	Offspring 1	Offspring 2	Offspring 3
Genotype	— — —	— — —	— — —
Individual	Offspring 4	Offspring 5	Offspring 6
Genotype	— — —	— — —	— — —

Table 10.3 Coefficients of relatedness between siblings with three pairs of chromosomes

Offspring	1	2	3	4	5	6
1						
2						
3						
4						
5						
6						

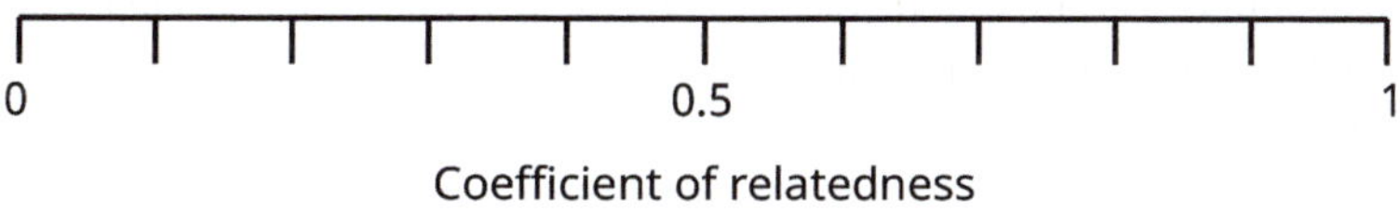

Fig. 10.5 **Coefficients of relatedness.** Mark all coefficients of relatedness between all pairs of siblings on the line with a coloured pen

Table 10.4 Parental genotypes

Individual	Mother						Father					
Genotype	$\frac{A^*}{A'}$	$\frac{B^*}{B'}$	$\frac{C^*}{C'}$	$\frac{D^*}{D'}$	$\frac{E^*}{E'}$	$\frac{F^*}{F'}$	$\frac{A^\wedge}{A^\dagger}$	$\frac{B^\wedge}{B^\dagger}$	$\frac{C^\wedge}{C^\dagger}$	$\frac{D^\wedge}{D^\dagger}$	$\frac{E^\wedge}{E^\dagger}$	$\frac{F^\wedge}{F^\dagger}$

Table 10.5 Offspring genotypes

Individual	Offspring 1	Offspring 2	Offspring 3
Genotype	— — — — — —	— — — — — —	— — — — — —
Individual	Offspring 4	Offspring 5	Offspring 6
Genotype	— — — — — —	— — — — — —	— — — — — —

4. Calculate the coefficient of relatedness between each offspring pair (fraction of shared chromosomes), and fill in Table 10.3.
5. Mark each calculated relatedness on the line in Fig. 10.5.
6. Calculate the average relatedness across all pairs of offspring.

Table 10.6 Coefficients of relatedness between siblings with six pairs of chromosomes

Offspring	1	2	3	4	5	6
1						
2						
3						
4						
5						
6						

Relatedness with Complex Genomes

1. Now, assume that there are six pairs of chromosomes. Genotypes of parents are shown in Table 10.4
2. Use the random generator to again create six offspring, just like you did in the previous section, and fill in Tables 10.5 and 10.6.
3. Mark each calculated relatedness on the line in Fig. 10.6.

Analysis

1. Compare Figs. 10.5 and 10.6. Look at the lines with marked relatedness. What do you observe? Where is the density of marks the highest?
2. Humans have 23 pairs of chromosomes. What do you think the mark density on the relatedness line would look like in that case?

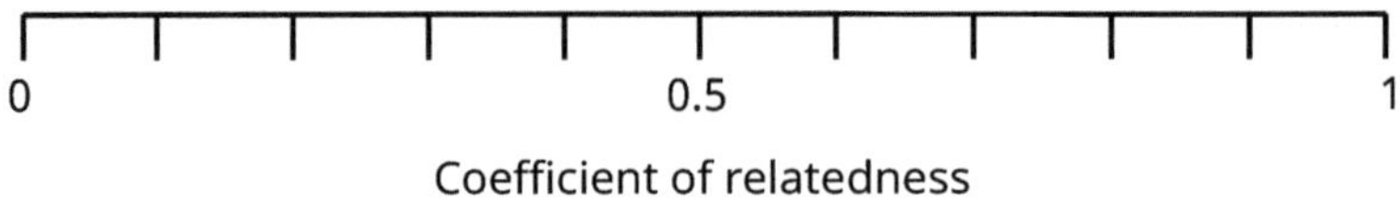

Fig. 10.6 **Coefficients of relatedness.** Mark all coefficients of relatedness between all pairs of siblings on the line with a coloured pen

3. What do you think will happen with the density of marks if you take into account that recombination occurs and results in the reshuffling of thousands of genes?

This project was developed by Barbora Trubenová and Kristína Hudáková.

Branches in the Tree of Life: How New Species Evolve

11

Himani Sachdeva

▶ The amazing variety of life forms around us can often inspire awe and wonder. Scientists estimate that the number of species on our planet may be anywhere from a few million to a whopping one trillion. The huge uncertainty in this estimate is largely due to the difficulties in enumerating bacterial species. But what exactly are species, and how do they evolve?

When we talk about distinct species, we often imagine organisms or populations that are visibly different: We might represent these using a species tree or phylogeny. But what if we zoom into any one part of this tree, let's say the family of penguins: Do all penguins belong to one species, or are there several different species of penguins? One might argue that the emperor penguin and the chinstrap penguin have rather different morphologies and are perhaps different species (Fig. 11.1), but what about small versus large emperor penguins? These have distinct body types; does this mean they are distinct species?

After nearly a century of heated debate and as many as 30 different definitions of species, most scientists have come to agree that the most useful definition of species is provided by the so-called **Biological Species Concept**. This definition basically says that two populations can be considered distinct species if they do not interbreed successfully, i.e. either if individuals from different populations never mate or if mating between them produces inviable or infertile offspring.

This means that if two populations are "complete" biological species, then novel mutations that arise and fix in one population are unlikely to spread to the other population (as populations cannot exchange alleles if they do not mate). Thus, two populations that have stopped producing successful hybrids become more and more genetically dissimilar and evolve very different phenotypes over time. As more time passes, one or both of these populations may branch again into populations that do not interbreed, eventually giving rise to a tree of the kind we see in Fig. 11.2.

Interesting Fact

Scientists typically represent the relationships among different life forms as an evolutionary tree, in which each node corresponds to a distinct species (living or extinct), and species that are more closely related share a more recent ancestral node. In fact, we can represent the relationships among all eukaryotic life forms as one big *Tree of Life* (Fig. 11.2). Note, however, that this kind of tree is, at best, an approximation. The evolutionary history of

H. Sachdeva (✉)
Faculty of Life Sciences and Faculty of Mathematics, University of Vienna, Vienna, Austria
e-mail: himani.sachdeva@univie.ac.at

B. Trubenová et al. (eds.), *Everyday Evolution: Unveiling Life's Hidden Patterns*, https://doi.org/10.1007/978-3-032-07050-0_11

Fig. 11.1 Closely related species of penguins. The cladogram shows the evolutionary relationships of various penguin species. Photos show chinstrap (left) and emperor (right) penguins. The penguin cladogram: Cole, T.L., Zhou, C., Fang, M. et al. Genomic insights into the secondary aquatic transition of penguins. Nat Commun 13, 3912 (2022)

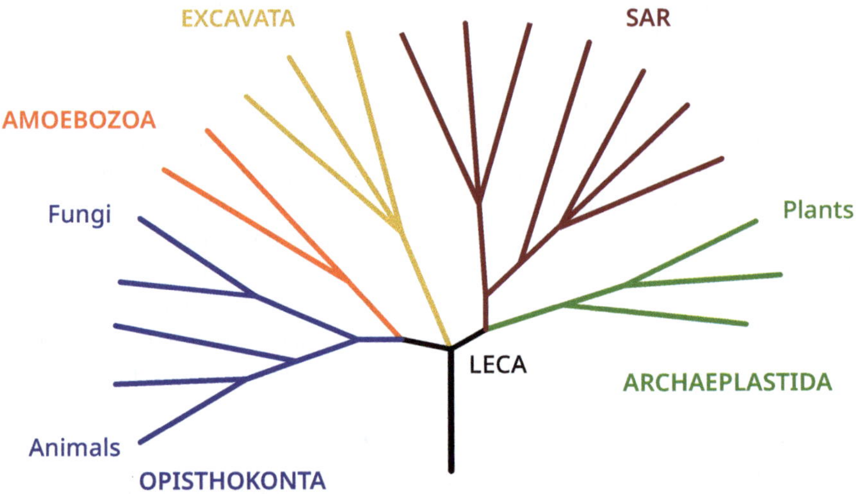

Fig. 11.2 Eukaryotic Tree of Life. A phylogeny showing how the major eukaryotic clades are related and the last eukaryotic common ancestor (LECA). (Adapted from Iwamoto, Masaaki, Hiraoka, Yasushi, Haraguchi, Tokuko. (2016). Uniquely designed nuclear structures of lower eukaryotes. Current Opinion in Cell Biology. 40. 66–73. 10.1016/j.ceb.2016.02.019. CC-BY-4.0 http://creativecommons.org/licenses/by/4.0/)

organisms may also involve processes such as **horizontal gene transfer**, which involves the transfer of genetic material between different types of bacteria, bacteria and their eukaryotic hosts or even different species of eukaryotes. Thus, at least in some contexts, it may be more appropriate to represent evolutionary relationships as a network rather than a tree. See also Chap. 15.

Some Challenges in Defining Species

The Biological Species Concept (BSC) not only provides a way of classifying species, but it also explains why species become more and more different over time. However, in practice, it is difficult to apply the BSC to many natural populations. For instance, if two populations are completely separated and have never met, then it is generally impossible to tell whether or not they can interbreed successfully! Another serious limitation of the BSC is that it doesn't apply to *asexual* organisms, e.g. most bacteria and archaea and some plants. Over the past century, scientists have debated many alternative definitions. For instance, the **Morphological Species Concept** defines a species as a group of individuals that are much more similar to each other (e.g. in their visible characteristics) than to any other group of individuals. This definition of species has been widely used in taxonomy. However, it has its own problems: Many populations that don't interbreed (and can thus diversify even further over time) simply don't look different from each other. Such "cryptic" (i.e. difficult to distinguish) species are especially common in insects and reptiles.

Thus, to understand how the Tree of Life emerges, we must start by focusing on a single such branching or **speciation** event and ask: How does a single population split into two subgroups whose members cannot successfully mate with each other? In other words, how do **reproductive barriers**, i.e. factors that block successful mating between two subgroups, evolve in a population in which no such barriers exist in the first place?

Crucially, for these barriers to remain intact (i.e. for two species to remain distinct) over many thousands or millions of years, it's not enough if one population just "decides" to not mate with the other population or if all hybrids happen to die in one generation. Instead, the preference for mating partners or the mechanisms which cause hybrids to fail must be passed on reliably from one generation to the next. This means that the failure to interbreed must be a result of *genetic* changes in the two populations. Such genetic changes accumulate gradually over many thousands or even millions of generations, meaning that speciation is typically a very slow process. Below, we look at a few evolutionary scenarios that could lead to the emergence of reproductive barriers between two subgroups of a population, thus setting the stage for them to become distinct species over time.

Interesting Fact

Speciation is usually, but not always, a slow process. One route to rapid speciation is when diploid (2n) parents sometimes produce unreduced, i.e. diploid (2n) gametes, which fertilise to form polyploid, e.g. tetraploid (4n) zygotes. Because of the mismatch in chromosome number between parents and offspring, these offspring can no longer hybridise with the parental species and, thus, immediately become a new species. This kind of *speciation by polyploidisation* is fairly common in flowering plants and ferns.

11.1 From Geographic to Reproductive Barriers

Imagine a population spanning a large land or oceanic region, as in the case of an aquatic organism. Any one individual, let's say a shrimp in the sea, explores only a limited region in its lifetime. However, if we track its descendants over time, then these are found (typically) farther and farther from their original ancestor as the generations pass. We say that this population has a large geographic range.

Now, imagine that a major geological or climatic change splits this population's range into two. A famous example of this is when the isthmus of Panama (a natural land bridge between North and South America) formed about 3 million years ago, separating various marine populations (including shrimp) into two separate (Pacific and

Atlantic) populations that were cut off from each other. Another way in which a small part of a population can break off from the main population is if a few individuals migrate extremely far and establish a new colony (e.g. if a log of wood drifts off from the mainland to an island, carrying along a few hapless crickets). Since such long-range migration is very rare, another such event may not occur for many hundreds of thousands of generations, effectively isolating the new population from the main population for a very long time.

Now, each of these two subpopulations (e.g. the mainland and island populations or subpopulations on either side of a geographic barrier) evolves *independently*. This basically means that if a new mutation arises in one of the subpopulations, it rarely, if ever, spreads to the other population. Over time, each of the two populations acquires other unique mutations. The longer the time since the geographic split, the greater the number of mutational differences between the two populations.

So now we know how two subpopulations can separate and become genetically distinct. But what happens if the subpopulations meet again (typically, hundreds of thousands or even millions of years) after the original split? This has happened several times in the past, for instance, when an Ice Age ended, melting the glaciers that blocked migration between two subpopulations. This kind of renewed contact between populations is called **secondary contact**. In modern times, several instances of secondary contact have been due to anthropogenic (i.e. human-induced) activity, such as the construction of the Panama Canal that linked the Atlantic and Pacific Oceans, reuniting the long-separated shrimp.

Secondary contact between previously separated populations can have one of several different outcomes. In some cases, individuals from the two populations meet, mate and produce perfectly fit offspring. Over time, subpopulations essentially fuse back into a single population – as if they had never been separated.

In other cases, one of the populations may have evolved to grow faster or compete more aggressively for scarce resources. Thus, when the two populations come into contact, that population will outcompete the other and drive it to extinction. However, in some cases, some genetic material from the population that goes extinct may survive in the descendant population if individuals from the two populations occasionally mate. An interesting example of this is from when *Homo sapiens* came into contact with Neanderthals in Europe approximately 40,000 years ago. While Neanderthals are now extinct, comparison of human and Neanderthal DNA shows that 1-4% of the genome of a typical modern-day ethnically Eurasian person is derived from Neanderthal ancestor(s). This suggests that *Homo sapiens* and Neanderthals occasionally mated during secondary contact.

A third possibility is that both populations manage to co-exist after secondary contact but do not mate or only mate rarely. Alternatively, individuals from the two populations may mate readily but produce hybrid offspring with poor survival and/or fertility. Infertile hybrids have been observed in various instances of interspecies mating. Notable examples include tigons (hybrid offspring of male tigers and female lions) and ligers (hybrid offspring of male lions and female tigers) that emerge when tigers and lions occasionally mate in captivity (e.g. in zoos). Such hybrids are almost always infertile and thus represent a kind of evolutionary dead end.

The third possibility is the most interesting, as it means that some reproductive barriers (that reduce genetic mixing between the two populations) are already in place. Thus, in contrast to the first and second scenarios, in this third scenario, the two populations may already be on their way to becoming distinct species. However, this also raises several questions: What exactly causes hybrids to fare poorly? What happens if the hybrid offspring are only slightly worse off (i.e. have only moderately lower survival rates or numbers of offspring) than their parents? In the long run, will the two populations stop exchanging genes and become two completely separate species anyway? In the following pages, we will review partial answers to these questions that have emerged from nearly a century of speciation research.

11.2 The Many Causes of Hybrid Failure

Let us start by examining some of the reasons why hybrids produced by mating between two genetically distinct subpopulations might find themselves in trouble. In the next section, we will examine why some long-separated subpopulations may not even be able to mate easily after secondary contact.

Genetic Incompatibilities in Hybrids

One of the first hypotheses for why hybrids fail was proposed by William Bateson (1909) and later independently by Theodosius Dobzhansky (1936) and Hermann Muller (1942). They hypothesised that hybrids fail if the alleles they inherit from one parent at a locus are *incompatible* with alleles inherited from the other parent at a second locus. Let us see how this might work.

Suppose we have an ancestral population that is fixed for allele A at one locus and allele B at a second locus so that all individuals have the AABB genotype. Now, the population splits into two subpopulations; let's call these 1 and 2. Over time, an individual in subpopulation 1 undergoes an $A \rightarrow A'$ mutation at the first locus. This mutation may spread through the subpopulation (just by chance or because it is beneficial) so that over time, most individuals in subpopulation 1 have genotype $A'A'BB$. Suppose also that subpopulation 2 undergoes a $B \rightarrow B'$ substitution at the second locus in a similar way, such that long after the split, most individuals in this population are $AAB'B'$. If the two populations now come into secondary contact, hybrids formed by mating between $A'A'BB$ and $AAB'B'$ must have genotypes $AA'BB'$ (Fig. 11.3). Note that this is the *first* time the alleles A' and B' have ended up together in the same genome. What if they are incompatible, i.e. do not function well together? In this case, hybrids would have poor survival or fertility and would be at a disadvantage with respect to the parental populations.

But what exactly causes alleles A' and B' to be incompatible? Let us look at one example involving hybrids of the platyfish *Xiphophorus maculatus* and the swordtail fish *Xiphophorus helleri*. The platyfish has spots on its dorsal fin, but the swordtail is spot-free. Scientists have discovered that this difference is primarily due to two genes: one gene that produces spots and a second gene that regulates the first gene and prevents the spots from becoming too large. One might think of the platyfish as carrying an allele A (that produces spots) and an allele B' (that regulates them). The swordtail, on the other hand, has neither the spot-producing gene nor the gene that regulates it (let's say it carries alleles A' and B). This means that some[1] hybrids formed by mating between the two will inherit the spot-producing gene but not the regulator (imagine these as carrying alleles A' and B' in the same genome): These individuals then have spots on their dorsal fins but no way of regulating their growth. Thus, spots tend to be very large and develop into malignant tumours, reducing survival in hybrids.

Environment-Dependent Hybrid Disadvantage

A second reason for hybrid disadvantage has to do with how well hybrids compete with the two parental types in their natural environments. One of the best-studied examples of this is in stickleback fish in the lakes of British Columbia, Canada. Sticklebacks in these lakes come in two different types or morphs: the *benthic* type that is generally found close to lake shores and the *limnetic* type that lives mostly in open waters (Fig. 11.4). Benthics primarily eat invertebrates, e.g. large insect larvae, and have broad bodies and wide jaws. Limnetics feed primarily on zooplankton and have slender bodies, narrow jaws and gill structures that are quite different from those of the benthic type. Hybrids between the two types have features that are *in between*

[1] The reason why some (but not all) hybrids inherit the spot-producing gene is that this gene is actually located on the X chromosome in platyfish and is thus present only as a single copy in males. Therefore, male (hybrid) offspring of male platyfish and female swordtail will not inherit the spot-producing gene.

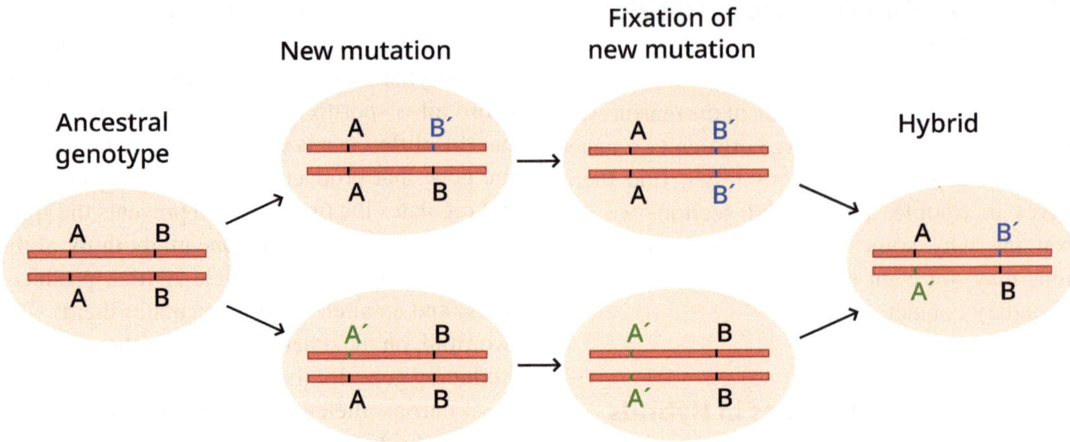

Fig. 11.3 Schematic for evolution of Bateson-Dobzhansky-Muller incompatibilities. The ancestral population splits into two. Mutations A' and B' arise and become fixed in the two populations independently. If A' and B' are incompatible, then hybridisation between the two populations will result in unfit $A'B'$ hybrids. (Based on figure by OrientationEB, CC BY-SA 4.0 https://creativecommons.org/licenses/by-sa/4.0, via Wikimedia Commons)

Fig. 11.4 Two adult stickleback morphs. The letters A and B represent spines that are present in limnetic but not in benthic fish. Rennison, D.J., Rudman, S.M. and Schluter, D. (2019), Genetics of adaptation: Experimental test of a biotic mechanism driving divergence in traits and genes. Evolution Letters, 3: 513-520. doi:10.1002/evl3.135. CC-BY-4.0 http://creativecommons.org/licenses/by/4.0/

those of the two parents. However, they are not very common in the wild.

Do hybrids have lower survival or growth rates, and does this (partially) explain their rarity in the wild? To test this, researchers designed an ingenious experiment. They first caught both types of fish from the lake and then crossed them to create various kinds of hybrids: e.g. those that have one benthic and one limnetic parent or one benthic grandparent and three limnetic grandparents and so on. They then reared these hybrid juvenile fish together in the lab with juveniles who had no mixed ancestry. This experiment showed that growth and survival were *not significantly different* between hybrid and non-hybrid juveniles when they were raised in the *same environment* with the same food sources.

Next, the researchers transferred non-hybrid juveniles (of both the pure limnetic and pure benthic type) and hybrid juveniles to an enclosure along the lake shore to test how they fared in the natural benthic habitat. They then transferred individuals of all three types into an enclosure in the open lake (i.e. a natural limnetic habitat) and repeated the experiment. In both cases, the *hy-*

brids grew significantly slower than the pure type that was common in (and most likely well adapted to) that habitat. Scientists now believe that the two types of fish have evolved different morphologies that help them *specialise* on the different food sources available in the different habitats: small insects on the lake shore versus zooplankton in the open waters. Hybrids with their intermediate jaws, gill structures, etc., simply cannot compete well with these specialists on their home turf!

Unlike in the example of the tumour-prone hybrids, here, hybrid disadvantage is not because of genetic incompatibilities that reduce hybrid survival in *all* environments (including the lab). Instead, hybrids are disadvantaged because their genotype (and phenotype) does not match the available habitats or microenvironments within the lake. This highlights the importance of studying both the genetics and the ecology of species in order to understand how they adapt and diversify.

Interesting Fact

Hybrids are not always inferior to parental populations. In many cases, hybrids actually perform better – a phenomenon known as **hybrid vigour** or **heterosis**. This can occur, for example, when the two parental populations are nearly fixed for different sets of recessive deleterious alleles. Hybrids then inherit only one copy of the recessive allele (from one of the parents) at any gene and thus have higher fitness than either parent. Hybrid vigour has been observed in many plant and animal species, and hybridisation is often used to improve crop yields and their resistance to common pathogens.

11.3 The Many Causes of Mating Failure

So far, we have considered why populations that have been separated for some time might produce unfit hybrids when they come into contact again. However, as mentioned above, it could also be that after a long period of separation, members of the two populations simply cannot mate with each other anymore. Let us explore a few possible reasons for this.

Sometimes, two populations might have evolved to *mate at different times* of the year. This is especially common in flowering plants: When a formerly continuous population splits into a northern and southern population, then the northern population typically evolves a shorter, more intense flowering season that occurs later in the year. As a result, even when geographic barriers disappear and the geographic ranges of northern and southern populations start overlapping again, individuals from the different subpopulations rarely cross-pollinate and produce hybrid flowers.

Of course, populations don't always need to evolve different mating seasons in order to stop interbreeding. They could also evolve to mate at different times of the day! For example, mating between the fruitfly *Drosophila persimilis* and the closely related species *Drosophila pseudoobscura* in Hawaii is extremely rare because the former mates early in the morning, while the latter mates later in the afternoon.

Yet another reason why mating between different populations may fail is if they have evolved *different genitalia* that do not "fit". A curious case is that of the left-coiling versus right-coiling snails of the Japanese snail *Euhadra*, which have shells that coil clockwise versus anticlockwise. These snails are thus mirror images of each other. Mating between the two types is very difficult because of the different locations of the genital organs.

In populations with intense competition for mates, males commonly evolve striking courtship displays, colourful plumage or mating songs, while females evolve preferences for these specific displays. When different populations come into contact, females may respond only to the *specific mating rituals* of males from their own population and reject males from the other population. Something similar was also observed when shrimps from both sides of the isthmus of Panama were brought together in the lab: Males and females from the different populations snapped aggressively at each other and failed to mate.

Evolution of Mating Traits During Speciation

While identifying the specific traits that prevent mating between two closely related populations may be straightforward in many cases, it is generally very difficult to figure out *why* populations have evolved these different mating organs, behaviours or seasons. These differences could have evolved while the populations were separated (as described above) – either by chance or due to some selection pressure in one environment (e.g. longer winters at higher latitudes) that has nothing to do with the other population. But different mating behaviours and timings could also evolve when and *because* two genetically distinct populations come into contact and start producing unfit hybrids. Let us see how this might happen in a hypothetical scenario involving our two stickleback morphs.

Recall that limnetic and benthic sticklebacks have very distinct morphologies (which allow them to utilise different food sources) and produce hybrids that are less competitive than either pure parental type. Suppose we initially have a population in which any stickleback individual is equally likely to mate with any other individual (limnetic or benthic) it encounters. Now imagine that a benthic female acquires a mutation that affects her mating behaviour, making her more likely to choose mates that look similar to herself. This female will mate more often with other benthic males and produce fewer hybrid offspring than females who are not "choosy". As a result, the choosy female has more surviving offspring (since hybrids have lower fitness than pure types). These (mostly) benthic offspring also inherit the choosiness mutation and, in turn, transmit it to their own offspring. Over time, mutations that cause individuals to prefer mating partners of their own type would spread in the population – simply because natural selection favours behaviours that prevent individuals from wasting reproductive effort on offspring that are unlikely to survive. Thus, in this case, the evolution of choosiness "reinforces" and completes speciation – causing two populations that produce less fit (but not completely inviable) hybrids to eventually evolve into completely separate species that do not hybridise at all.

How do we know if this kind of selection for choosy behaviour (i.e. selection against mating with the "wrong type") actually occurs in nature? Indirect evidence comes from studies of closely related *Drosophila* species that can mate with each other (at least occasionally) but typically produce less fit hybrids. Researchers collected individuals from pairs of species with overlapping ranges, as well as pairs of species that are geographically far apart and hence unlikely to have been in recent contact. They then tested how often individuals from different species mated with each other in the lab, as well as how often hybrid offspring from such mating events survived.

This experiment revealed a striking pattern: *Drosophila* species that were found far apart in nature were much *more likely* to mate with each other when brought together *in the lab* than *Drosophila* species that were in close contact in nature, even though hybrids were equally unfit in both cases. This suggested that species whose members frequently encounter each other in the wild had evolved to be choosy, i.e. to reject the other species, presumably because natural selection favoured (choosy) individuals who did not waste their reproductive resources on unfit hybrid offspring. By the same logic, there would be no selection pressure on a species to evolve a bias against another species it had never encountered before. Thus, this pattern is precisely what we would expect if mating preferences evolved primarily to avoid mating with the "wrong type".

11.4 Speciation Without Geographic Separation?

A common feature of all the mechanisms discussed so far is that two populations must evolve to be sufficiently different – either by fixing different sets of mutations that are incompatible with each other, or by evolving specialised traits that improve fitness in a specific habitat, or by evolving different mating traits or via some combination of all three. Such differences between populations can evolve relatively easily if the populations are completely separated. Does this mean that every occurrence of speciation involved the splitting of a population into two by an impassable mountain range or water body, or by continents drifting apart?

As one might imagine, this is highly unlikely! Part of the answer is that *partial* (rather than complete) separation between populations is extremely common and can occur even in the absence of geographic separation. Two fish populations may forage at different depths or at different distances from the shore within a lake so that

African Cichlids and the Geographic Context of Speciation

One of the most extraordinary examples of speciation ever recorded is in the freshwater lakes of the East African Rift Valley. These contain more than 1600 different species of cichlid fish, with some individual lakes, such as Lake Victoria or Lake Malawi, alone supporting 700-800 different species. Species can differ in feeding habits, nesting sites and male colouration (which is one of the criteria by which female cichlids choose mates). A key question is: Did a *single* lake population branch into multiple species, or did *multiple* populations in different lakes first evolve to be very different, and then come into secondary contact (via the rivers and streams that connect some of the lakes)? To add to the complexity, the lakes themselves have undergone major changes in water levels, with the water channels that connect the major lakes appearing and disappearing through the ages. Current evidence suggests that both kinds of branching events (i.e. speciation within a single lake and speciation of geographically separated populations in different lakes) have contributed to this remarkable explosion of diversity. However, piecing together the full evolutionary history of the East African lake cichlids remains an interesting puzzle for evolutionary biologists and can help us to more generally understand the factors that promote speciation in the absence of geographic barriers.

Different cichlid species from (A) Lake Victoria[1] and (B) Lake Malawi[2]. ([1]By Ole Seehausen and [2] by Reade Roberts. Adapted from Maan ME, Sefc KM. Colour variation in cichlid fish: developmental mechanisms, selective pressures and evolutionary consequences. Semin Cell Dev Biol. 2013 Jun-Jul;24(6–7):516–28. CC BY-SA 3.0, https://creativecommons.org/licenses/by-sa/3.0/deed.en)

they rarely encounter each other. Two closely related parasite populations may live in the same geographic area but utilise different hosts. A well-studied example of this is *Rhagoletis pomonella*, the apple maggot fly, which has two types or morphs: an ancestral type that lives on hawthorn trees (native to the midwestern United States) and a newly emerged type that lives on apple trees (which were introduced in North America 200 years ago). The two types of flies are not physically separated: Apple and hawthorn trees are often in close proximity, and flies may sometimes feed on their non-preferred type of fruit. However, flies have a tendency to return to their "own" kind of tree to mate and deposit eggs, which reduces the number of matings between apple flies and the hawthorn flies. The two types are not considered complete species yet, as they do (infrequently) mate and produce hybrids. Scientists speculate that the two types are on the way to becoming different species.

Given that the two types of flies maintain a preference for different fruit trees generation after generation, it is very likely that such differences in preferences and behaviour are not simply learnt from parents but are actually due to underlying genetic differences (which can be passed on *reliably* from parent to offspring). In fact, scientists have managed to identify the specific changes (in genes that influence how flies detect odours associated with different fruits) that may be responsible for the different host preferences in the two types of flies.

The question then is: (How) can populations evolve sufficient genetic differences and eventually become distinct species if they are only *partially* separated? This question has intrigued evolutionary biologists since Darwin and is yet to be fully resolved. For speciation to occur without separation, at the very least, there must be a strong selective advantage to becoming different. This could be the case if there is intense competition for food or breeding sites or host plants, so individuals that evolve to utilise a different food source or habitat efficiently can improve their survival and reproductive output.

However, in order for specialisation, e.g. to a different food source, to evolve, alleles that modify neurobiology (e.g. how individuals perceive different odours) or morphology (e.g. jaw shape in the case of the speciating sticklebacks) must spread and reach high frequency (typically at several different genes) in one subgroup of the population, while the other subgroup maintains different alleles. Can this happen if the two subgroups mate and exchange alleles frequently? This question does not have a yes/no answer and can only be answered (even in theory) if we first ask ourselves: What numbers are we looking at? For example, by what factor does the survival rate of an individual (or its offspring) increase if it utilises a different food source or habitat? How many genes influence specialist traits, and what is the rate of recombination between these genes, i.e. how likely is it that novel genetic combinations (required for specialisation) will be broken apart by recombination during mating? How often does an individual encounter mating partners that are phenotypically and genetically similar or different?

To understand how these (and other) factors influence speciation, evolutionary biologists have made extensive use of mathematical models and simulations. Models of this kind ignore much of the complexity of natural populations but provide a powerful way of asking: Which of many possible outcomes is to be expected when there is a tug-of-war between different evolutionary processes? Models thus help us understand the conditions that promote speciation in the absence of geographic barriers. However, it is far more difficult to determine whether these conditions actually occurred within a natural population or even to what extent populations have been geographically separate in the past.

11.5 Modern Tools to Unravel an Ancient Mystery

Speciation is challenging to study because it occurs over timescales that make direct observation impossible. While in some cases the fossil record can tell us something about the prehistoric (and generally extinct) ancestors of existing species, for the most part, we must rely on our knowledge of how organisms are in the present.

Systematic observations of differences in microhabitats, feeding and mating habits, shape, size and so on, between closely related species in the wild form an important part of speciation research. These observations help us refine our hypotheses and models and can provide important clues about the processes underlying speciation in specific cases.

A second line of evidence comes from experimentally crossing closely related species in the lab. As we have discussed above, such experiments can tell us a great deal about the causes of hybrid failure and the evolution of mate choice in natural populations. However, artificial crossing experiments are feasible only in small model organisms such as *Drosophila*, which makes it difficult to assess whether their findings generalise to other species and taxa.

In the last few decades, speciation research has been revolutionised by advances in DNA sequencing. We have already come across one use of sequence data: for constructing species trees and estimating how long ago species split. However, genomic data can provide a lot more information. For instance, scientists have developed statistical methods to "scan" DNA sequences for regions of the genome that are unusually different between closely related species. Such regions are of interest because they may contain genes that influence traits that are under different selection pressures in diverging species. In the case of sticklebacks, for example, these could be genes that influence body shape or jaw morphology, which would thus typically differ more (between the limnetic and benthic types) than most other genes. These highly differentiated regions of the genome are thus an excellent starting point in the search for "speciation genes" that underlie differences between closely related species.

We have focused on just one aspect of speciation, namely, how a single population branches into two species that gradually stop exchanging genetic material. Each of these may split further, generating a tree of species, sometimes referred to as the *Tree of Life*. Why do some branches in this tree have many sub-branches and others only a few, i.e. what factors influence the *rates* of speciation and extinction, and why are these so different across different parts of the tree? How have rare but life-altering evolutionary transitions (e.g. the emergence of multicellular organisms) influenced the extent of biodiversity? These are some of the big open questions that continue to inspire scientific exploration.

DEFINITIONS

Hybrids: Offspring of individuals from two geographically or genetically distinct populations

Secondary contact: Re-establishment of contact between two populations that were geographically separated in the past

Biological Species Concept: Species definition in which a population is considered a species if its members cannot breed successfully with other populations

Reproductive barriers: Factors that block or reduce successful reproduction between two populations, typically encoded by genes

Speciation: Process by which multiple reproductive barriers between two populations evolve over time

Heterosis: Higher fitness (survival or fertility) of hybrids compared to parental species

SUGGESTED READING

- Schluter D. Ecology and the origin of species. Trends Ecol Evol. 2001 Jul 1;16(7):372-380.
- Hatfield T, Schluter D. Ecological speciation in sticklebacks: environment-dependent hybrid fitness. Evolution. 1999 Jun;53(3):866-873.
- Henning F, Meyer A. The evolutionary genomics of cichlid fishes: explosive speciation and adaptation in the postgenomic era. Annu Rev Genomics Hum Genet. 2014;15:417-41.
- Linn C Jr et al. Fruit odor discrimination and sympatric host race formation in Rhagoletis. Proc Natl Acad Sci U S A. 2003 Sep 30;100(20):11490-3.

- Seehausen O et al. Genomics and the origin of species. Nat Rev Genet. 2014 Mar;15(3):176-92.
- Daniel L. Powell et al., Natural hybridization reveals incompatible alleles that cause melanoma in swordtail fish. Science. 2020; 368, 731–736.
- Matute D. R., Cooper B. S. Comparative studies on speciation: 30 years since Coyne and Orr. Evolution. 2021; 75(4), 764–778.
- Cole TL et al. Genomic insights into the secondary aquatic transition of penguins. Nat Commun. 2022 Jul 19;13(1):3912.

11.6 Questions

1. **Which of the following factors or events are likely to lead to speciation?**
   (A) Fragmentation of a large lake into several small ones.
   (B) Introduction of a new food source in a population with intense competition for food.
   (C) The emergence of a new mutant individual that reproduces faster than and outcompetes wildtype individuals in a given habitat.
   (D) Migration of a small subpopulation across a large desert.
   (E) Rise in temperature leading to reduced average fertility of an insect population.
2. **Consider the experiment involving the benthic and limnetic types of stickleback fish, as discussed in the chapter. Which of the following conclusions can we draw from the results of the experiment?**
   (A) Individuals of the two types have different mating habits and rarely produce hybrids.
   (B) Hybrids inherit incompatible alleles from the two parents and are not viable.
   (C) Hybrids have metabolic deficits and are unable to grow as fast as the "pure" types.
   (D) Hybrids have intermediate morphological traits that make them less competitive than the pure types in both lakeshore and deep-water environments.
   (E) Any pure type can outcompete the hybrids in all possible natural environments.
3. **An ancestral population with genotype $AABB$ is separated into two subpopulations, with no possibility of contact between the two. The first subpopulation undergoes an $A \rightarrow A'$ substitution and evolves the genotype $A'A'BB$. The second subpopulation undergoes a $B \rightarrow B'$ substitution and evolves the genotype $AAB'B'$. Scientists now cross individuals from the two subpopulations to produce F1 hybrids. These have genotypes $AA'BB'$ and are fully fertile. They now cross the F1s with each other to produce F2 hybrids. These can have nine possible genotypes. Five of these genotypes have never been generated in nature or in the lab before. These five novel genotypes are $AA'BB$, $AA'B'B'$, $AABB'$, $A'A'BB'$ and $A'A'B'B'$. Scientists discover that F2s with the genotypes $A'A'B'B'$ and $AA'B'B'$ are infertile, while all other F2s are fully fertile. Which of the following facts could explain these observations?**
   (A) The allele B' is recessive and causes infertility when expressed in an individual.
   (B) The allele B' is dominant; the allele A' is dominant; A' and B' are incompatible with each other, i.e. cause infertility when expressed in an individual.
   (C) The allele B' is recessive; the allele A' is dominant; A' and B' are incompatible with each other, i.e. cause infertility when expressed in an individual.
   (D) The allele B' is dominant; the allele A' is recessive; A' and B' are incompatible with each other, i.e. cause infertility when expressed in an individual.
   (E) The allele B' is recessive; the allele A' is recessive; A' and B' are incompatible with each other, i.e. cause infertility when expressed in an individual.
4. **The researchers repeat this experiment for a different pair of species. This time, they find that the F1 hybrid and all F2 hybrids are fully fertile, *except* the ones with genotype $A'A'B'B'$. Which of the following facts could explain these observations?**
   (A) The allele A' is dominant and causes infertility when expressed in an individual.
   (B) The allele B' is dominant; the allele A' is dominant; A' and B' are incompatible with each other, i.e. cause infertility when expressed in an individual.
   (C) The allele B' is recessive; the allele A' is dominant; A' and B' are incompatible with each other, i.e. cause infertility when expressed in an individual.
   (D) The allele B' is dominant; the allele A' is recessive; A' and B' are incompatible with each other, i.e. cause infertility when expressed in an individual.
   (E) The allele B' is recessive; the allele A' is recessive; A' and B' are incompatible

with each other, i.e. cause infertility when expressed in an individual.

5. **Scientists discover two new populations of mice with very different physical characteristics. They hypothesise that these populations are already two distinct biological species (as defined by the Biological Species Concept). Which of the following observations or experiments would provide *strong* support for their hypothesis?**
   (A) The two populations have nonoverlapping geographic ranges.
   (B) The two populations have overlapping ranges, but no hybrid mice with intermediate physical characteristics have ever been observed.
   (C) The mice avoid mating with each other in the lab.
   (D) The two populations have very different feeding habits.
   (E) DNA sequencing reveals that the two populations have many genetic differences across the genome.

6. **The island of Arcadia is home to two different populations of birds. One population feeds on the large nuts of the hickory tree and has a wide and strong beak. The other population feeds on the small seeds of apple trees and has a long and narrow beak. Hybrids between the two have intermediate beaks and can eat both nuts and seeds, but not as easily as the specialist birds. Which of the following conditions is likely to favour the evolution of mating preferences for one's own kind?**
   (A) An increase in the size of hickory nuts over time due to natural selection on hickory trees.
   (B) A sudden outbreak of a fungal disease that kills 90% of both types of birds.
   (C) The introduction of a large predatory bird that feeds on both types of birds as they fly from one tree to another.
   (D) Both types of birds have very distinct songs that can be recognised from afar.
   (E) A reduction of the number of both fruits due to drought.

11.7 Project: Discovering Species

Part 1: On Lizards

Imagine that you discover two populations of lizards, one living in the mountains of Eastern Turkey and the other on one of the small islands in the Aegean Sea. Members of these two populations look very similar; the only significant difference is the relatively larger heads of individuals in the island population.

1. Suggest an experimental procedure which would enable you to decide whether lizards living in these two populations belong to the same species. Describe the procedure as a series of steps you would take. Specify the type of data you will gather and explain how this data will help you to decide whether members of the two populations belong to the same species.
2. If you decide that these two populations belong to the same species, suggest a plausible explanation for members of the same species to live in populations so separated from one another. What might have happened there?
3. If you decide that members of these two populations do not belong to the same species, suggest an explanation for members of two distinct species being so similar to one another.

Part 2: On Unicorns

In Table 11.1, you can see short DNA sequences of ten individuals belonging to the family *Unicornidae*. Imagine a speciation event as a simple splitting of one original population into two separate populations, each eventually evolving into a separate species. Reconstruct the evolutionary history of the *Unicornidae* family – the sequence of speciation events between the closely related species.

Detailed instructions below can aid you in this process. You will probably need pencils of different colours. The word "locus" in the text below means the position of the letter (nucleotide). For instance, sequences AAT and ACT differ in the second locus.

Questions and Instructions

1. Study the sequences in Table 11.1 and highlight the mutated (diverse) loci (positions at which you can see different variants in sequences).
2. Try to group the sequences according to their similarity. Individuals of the same species may differ a little, but individuals of different species are likely to differ much more. How many different species do these individuals belong to?
3. Identify which mutations are on the individual level (differ among individuals of the same species). Indicate these mutations in Table 11.1.
4. For all individuals you identify as members of the same species, suggest the most probable sequence of their most recent common ancestor. If you are not sure about some parts of the sequence, indicate that part and explain why you are not sure about it.
5. Identify which loci (nucleotides) differ between species but are shared by all the individuals of the same species. Indicate these in Table 11.1.
6. If two (or more) species share the same mutation at a specific locus, but not with the other species, these species are likely to be more closely related to each other – they share a recent ancestor. Group the identified species into pairs, if possible, based on these similarities, and record the pairs.
7. If you identified any such pairs, suggest the most probable DNA sequence for their most recent common ancestor if you can. If you are sure about some part of the sequence, but not about all the letters, indicate these in the sequence.
8. Can you determine the sequence of the most recent common ancestor of all the species shown? Are there any loci at which you cannot be sure about the original nucleotide? If yes, how many? Which are those?
9. Draw a speciation tree within the family *Unicornidae* by drawing the branching pattern (the sequence of splitting events), including all species, leading to the observed species

Table 11.1 DNA sequences of ten individuals from family *Unicornidae*

Sample	DNA sequence
Individual 1	A T A T G G C G A T G C C A A G C A C T C G T G A C
Individual 2	A T A T G C C C T T G C C C A G C T C T G G T A A C
Individual 3	A C A T A G C A T T G C C A G T C A C A G G T A A C
Individual 4	A T A T G G C G A T C C C A A G C A C T C G T A A C
Individual 5	A T A T G C C C T T G C C A A G C T C T G G G A A C
Individual 6	A C A A G G C G T C G C C A G T C A C A G C T A G C
Individual 7	A C G T A G C G T T G C C A G T C A C A G G T A A C
Individual 8	A T A T G G C G A T G A C A A G C A C T C G T A A C
Individual 9	C C A T A G C G T T G C C A G T C A C A G G T A A C
Individual 10	A C A A G G C G T T G C C A G T C A C A G C T A G A

diversity. Write which individuals belong to each species.

10. We know that these species have been collected from an area, across which a geographic barrier lies. Based on the DNA data, could you divide these species into two groups, each lying on one side of the barrier? Explain your reasoning.
11. We also know that two of the species share the same habitat. Can you provide at least one plausible explanation why these two evolved to be separate species?

This project was developed by Kristína Hudáková and Barbora Trubenová.

Let's Talk About Sex!

12

Réka Kelemen

> No, in this chapter, we will not discuss what you are (probably) thinking about. Well, not quite. We will think about sex from an evolutionary perspective and will be guided by a central question: Why even bother having sex?

Let us back up for a second. What even is sex? Is it the act of reproducing oneself? No, if we put a single bacterial cell onto a plate with bacterial food, it will reproduce very happily without meeting any other bacterial cells. Sexual reproduction somehow implies that one is not alone and that there are multiple parties involved in the act of reproduction. But why do we need someone else? Why do male frogs or birds call to attract females when this also draws the attention of predators? Why do females risk their lives looking for mates instead of reproducing in safety, like a bacterial cell on a plate? Why even have males when they cannot bear young themselves, meaning that only half of the individuals in a population are devoted to producing the next generation?

The central significance of sex appears to lie in its ability to combine and reshuffle genetic material from different cells, a process broadly defined as the act of combining genetic material from more than one source. Sexual processes are prevalent across organisms, which suggests that they confer more benefits to these organisms than harm. But what are these benefits? This has been one of the most puzzling questions in evolutionary biology, and for over a century, many scientists have devoted thought and effort to answering it. In this chapter, we will explore some of the ideas and evidence that are currently out there.

12.1 The Twofold Cost of Sex

Let us start with a simple illustration of this evolutionary puzzle. Consider a population of asexual single-celled organisms. Each individual in this population consumes food throughout its lifetime and then clones itself to produce two daughter cells. Each member of the population contributes its fair share to the hard work of producing offspring; therefore, the population grows as quickly as possible.

Let us now consider a sexual population consisting of males and females. In this case, everybody eats, but only females devote their physical resources to producing offspring. Males are only there to fertilise females; they typically do not contribute to the hard and resource-intensive task of reproduction. If half of the population is male, which is most often the case in sexual species, then the population will produce only half as many offspring as it would have with all asexually

R. Kelemen (✉)
Institute of Science and Technology Austria, Klosterneuburg, Austria

B. Trubenová et al. (eds.), *Everyday Evolution: Unveiling Life's Hidden Patterns*, https://doi.org/10.1007/978-3-032-07050-0_12

Unusual Forms of Sex

Did you know that not all sexual species have male and female individuals? Some snail and earthworm species consist only of hermaphrodites, which means that each individual has both male and female reproductive organs and produces two kinds of gametes: a large one, called the egg, and a tiny one, called the sperm. Many plants are also hermaphrodites — one plant can produce both male and female flowers or flowers that have both male and female parts. Do hermaphrodites then escape the cost of sex? Not quite. They must still invest their body's resources in developing both male and female organs to produce two kinds of gametes.

Hermaphrodites can reproduce in a variety of ways. In hermaphroditic snail species, two snails meet, mate and fertilise each other's eggs with sperm. Other hermaphrodites, such as the mangrove killifish, don't need others to reproduce. Instead, each individual unites its own eggs and sperm inside its body to make offspring. The egg and sperm are still produced by randomly reshuffling the individual's two sets of chromosomes. Thus, even in this mode of sexual reproduction, the offspring can have genetic combinations that differ from the parental ones.

If you thought that was wild, wait for this. Many fish are sequential hermaphrodites, which means the same individual can be male for some time and female for some time, but not both at the same time. The colourful clownfish lives in groups where the largest fish is a female. This female reproduces with the second-largest fish in the group, a male, while all the other fish are males that do not reproduce. When the sole female dies or is removed from the group, the reproductive male turns into a female, and the next largest male becomes the new reproductive male!

Sexual reproduction can also work without eggs and sperm. In many microorganisms, the two haploid cells that fuse during sex are of the same size, and so there are no males or females at all. These organisms thus avoid the twofold cost of males while still enjoying the benefits of genetic reshuffling.

From left to right: snail, earthworm and clownfish

reproducing individuals. Clearly, this is a large cost to the population, which is often also referred to as "the cost of males".

Since sex is such an inefficient way of reproducing, we would expect asexuals to easily outcompete their sexually reproducing relatives and take over any population. However, this is not what happens — in fact, in nature, most animals, plants and fungi reproduce sexually!

12.2 The Need for Speed in Adaptation

So why is sex widespread across most life forms? Important hints come from tiny animals, such as rotifers or water fleas, that can switch between being sexual and asexual. Such organisms typically reproduce asexually, resulting in offspring that are almost identical to the parent. However, when

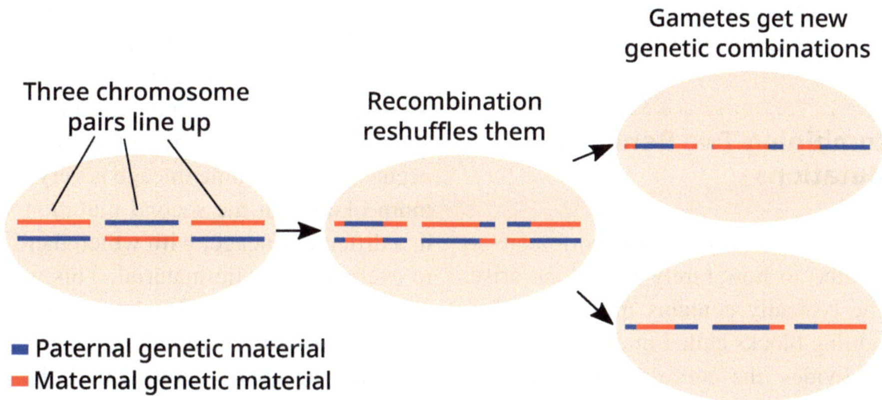

Fig. 12.1 The principle of recombination. This cartoon shows how genetic material from the maternal (red) and paternal (blue) sets of chromosomes is shuffled during sexual reproduction. A cell aligns its two sets of chromosomes. Then, the chromosomes in each pair recombine or exchange pieces, creating mosaics (or combinations) of the maternal and paternal copies. One of the recombined chromosomes from each pair then ends up at random in the gamete. Note: This is a simplified cartoon of sexual cell division, which focuses only on one chromatid of each meiotic chromosome and does not show the other chromatid

environmental conditions worsen, they instead look for a mate and produce the next generation sexually. What happens in this round of reproduction? Each animal produces one type of gamete — a sperm or egg cell. As you might recall, gametes are *haploid*, i.e. they contain only one set of chromosomes, while all other cells of the organism are *diploid* (with two sets of chromosomes). Gametes are formed when a diploid cell in the reproductive organs undergoes a special kind of cell division known as meiosis. During meiosis, the two chromosomes within each pair recombine, i.e. exchange pieces. This results in a new pair of chromosomes with shuffled genes (Fig. 12.1). The gamete then randomly inherits *one* of these reshuffled chromosomes from each pair. Then, during sexual reproduction, gametes fuse to form a zygote with two sets of chromosomes, which can then develop into a new diploid individual.

Obviously, there is a large amount of change involved in this round of reproduction. The offspring are not identical to each other or to the parents; rather, they each inherit a different random combination of their parents' genetic material. Some of these offspring may, in fact, have new combinations that give them a competitive advantage over their parents and other siblings, allowing them to thrive and reproduce in the new environment. Therefore, it appears that one strategy to adapt to stressful environmental changes is to create genetically diverse offspring — then the likelihood that *at least some* of these offspring will survive is higher than if all offspring were identical.

One example of a rapidly changing environmental factor that can put a lot of pressure on populations is the threat of pathogens. Pathogens, such as bacteria or viruses, infect their hosts by targeting specific host proteins. One can imagine this as the pathogen having a "key" that it tries to fit into the host cell's "lock". If the key fits, it can access the organism's resources and use them for itself. The pathogen changes its key constantly (either via random mutation or genetic reshuffling) so that it can open the lock more easily, and the only way the host can survive is by constantly altering the lock to avoid infection. Such a situation is often referred to as an **evolutionary arms race**. How does the host achieve rapid changes in its lock? This can again happen either via mutation or by reshuffling genetic material to create new versions of the lock. Experiments on the worm *Caenorhabditis elegans* have shown that populations that reshuffle their genetic material through sexual reproduction are better and faster at combating bacterial pathogens than genetically identical populations that only rely on mutations as a source of variation. But how exactly does recombination speed up adaptation?

12.3 Combining Two Beneficial Mutations

One of the reasons sex and recombination are so useful is related to how rarely mutations arise. The genome typically contains millions or billions of building blocks called nucleotides. Each time a cell divides, the entire genome is replicated. Sometimes, errors occur during replication, resulting in an incorrectly copied nucleotide. These copying errors, or mutations, in turn, result in changes in nucleotides at certain sites in the genome. Mutations are generally quite rare, and in every generation, only a small minority of the sites in an individual's genome are affected by a mutation. The human genome, for example, contains 3 billion base pairs,[1] and only about 100 of these mutate in one generation. Of these 100 mutations, only one to two are expected to land in a gene since genes constitute only a small part of the whole genome (most of the genome is noncoding DNA). From previous issues, you might recall that genes provide the instructions for building proteins that do most of the work in the cell — this means that such mutations can change how cells and organisms function, for better or for worse.

Now, imagine that there are two specific sites in the genome that, when mutated, become more beneficial for the organism. Moreover, an individual with both mutations receives the benefits of both. For instance, as a hypothetical example, suppose one mutation results in stronger muscles and the other improves vision. The question is: How common are individuals that carry both mutations in an asexually reproducing population versus in a sexually reproducing one?

In the asexual scenario, this can only happen if an individual gets a beneficial mutation that it passes on to its descendants, and then, one of these descendants gets the second beneficial mutation. Crucially, both beneficial mutations must occur in the *same* lineage (line of descendants). Since very few sites mutate in one generation in any individual, the chance that both mutations occur within the same lineage is very low. It is far more likely that the second mutation will occur in a different lineage, with which there is no way to exchange genetic material. This will result in one individual and its descendants having strong muscles and a *different* individual and its descendants having good vision. If both these families have individuals with similar fitness, then they will compete with each other for a long time, but not be able to combine their good mutations. The population will be doomed to the competition of suboptimal families until a lucky individual in one of the families gets the missing mutation, finally creating the fittest combination of traits.

Now, consider a sexual population. A mutation hits any specific site rarely, and as we saw above, it is far more likely that the two desired mutations will occur in unrelated individuals in the population. However, this time, there is an easy way to create an individual who has both mutations on the same stretch of DNA. If an individual with strong muscles mates with a sharp-sighted individual, their offspring might get both mutations, one from the maternal chromosome and the other from the paternal chromosome. When this individual grows up and produces sperm or egg cells, recombination will reshuffle the two chromosomes and might create the desired combination: the strong-muscle mutation and the good-vision mutation on one superior chromosome. All descendants of this individual who inherit this super-combination will be fitter and thus typically have more surviving offspring than other individuals in the population (who carry at most one of the mutations). Some of these offspring will again inherit the super-combination and pass it on to some of their offspring and so on, causing it to spread through the population (see Fig. 12.2).

Thus, in this example, evolution is much faster and more efficient in the sexual case, where individuals exchange genetic material with each other. Although many sexual populations suffer the twofold cost of sex, meaning they produce

[1] Base pairs is a standard expression used when we describe the length of a DNA molecule. The base is one part of a nucleotide — the one in which nucleotides differ.

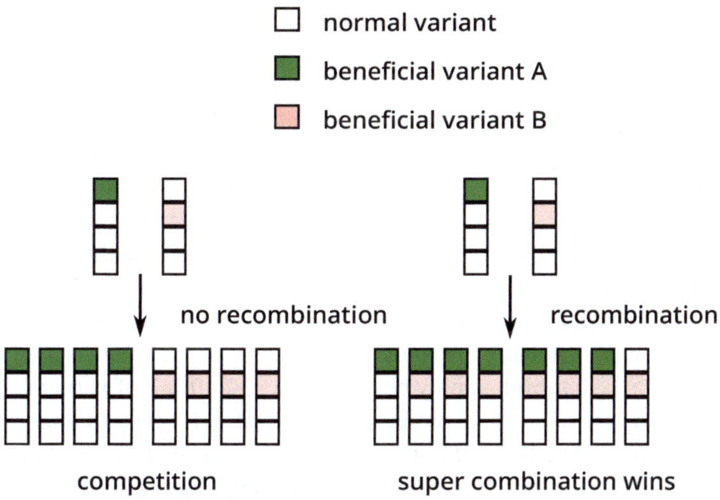

Fig. 12.2 The advantage of recombination. The first scenario represents an asexual population where recombination cannot reshuffle variants between genomes. In this case, the similarly fit genomes with either beneficial mutation will compete in the population. The second scenario shows that in a sexual population, recombination can reshuffle genomes and create a superior genome with both beneficial mutations that spreads rapidly through the population

fewer offspring than asexual populations in each generation, they can evolve towards the fittest trait combinations faster than asexual populations. This is especially important if the environment of a population is changing drastically, for example, if temperatures are increasing or if rapidly mutating pathogens are causing an epidemic.

12.4 The Benefits of Sex Depend on the Evolutionary Context

Now let us consider a slightly different question-does reshuffling genes from different individuals *always* help a population become fitter? Consider a population where most (but not all) individuals have genomes that carry both beneficial mutations. These individuals can boast of strong muscles and sharp vision–the optimal trait combination in our imaginary example.

If such a genome recombines with a genome that lacks these good mutations, then the offspring might actually end up having only one of the good mutations! In this case, recombination actually has a negative effect on fitness as it breaks up optimal trait combinations. Reshuffling genetic material is thus a random process that sometimes acts to bring together beneficial mutations and sometimes to separate them. If most individuals in the population have only one or the other desired mutation, reshuffling will mostly help in the adaptation, while if most of the population already has the optimal trait combination, recombination will do more harm than good. As you see, sex is a double-edged sword.

In fact, we can imagine that an individual's ability to reshuffle its genome is itself controlled by a certain gene. If an individual with a super-combination of good mutations acquires a "no-recombination" mutation on this gene, then it will be able to pass on its perfectly adapted genome, including the no-recombination mutation, to all of its offspring. Its descendants will do the same and eventually outnumber the sexually reproducing families in the population because they avoid (harmful) recombination. In turn, when environmental conditions change and new trait combinations are needed, a mutation that restores the ability to recombine and have sex will be beneficial again and might spread in the population. Therefore, it is important to realise that the evolutionary benefits of recombination and sexual reproduction depend on how fit the population is to start with. Further, recombination can *itself* evolve over several generations like any other trait because it will be favoured by selection when new trait combinations are needed and disfavoured when it would mostly break up well-adapted trait combinations.

12.5 Dealing with Harmful Mutations

We already have one answer to our question about why recombination is advantageous. But there is another reason. Recall that mutations can change the structure and function of proteins. Thus, most mutations actually tend to be harmful–simply because there are more ways to break a protein (that has been almost perfected by millions of years of evolution) than to improve it. Now, think about what happens in an asexual population. Mutations (mostly harmful) arise at random sites of the genome of an individual. The individual passes on all these mutations to its offspring, who acquire a few more mutations (again mostly harmful) of their own. Thus, the offspring passes on an even higher number of harmful mutations to its offspring. In fact, the number of harmful mutations within any lineage keeps increasing at a slow but steady rate. In extreme cases, an asexual population may have so many harmful mutations that it could even go extinct! The only way for fitness to increase again is if one of the sites with a bad mutation now mutates back to the original good variant in at least one individual. Moreover, this mutation cannot occur in just any cell of the individual; it has to occur in the gametes in order to be passed on to the next generation. However, such "back mutations" are an unlikely source of improvement in a population unless it is enormous.

So let us see whether sex can offer a partial solution to this problem. The meeting of genomes from different individuals and their recombination seems to again be a superpower, as it allows for the reshuffling of parts of the genome. Since individuals who meet for mating most likely have bad mutations in different parts of their genomes, reshuffling can create combinations that inherit the good, mutation-free parts from each parent's genome. Offspring inheriting these combinations will be fitter than their parents. They will, of course, also be superior to other siblings who might have, unluckily, inherited the bad mutations of both parents. Therefore, these lucky offspring will leave more offspring of their own than less fit members of the population. Thus, sex, which involves the exchange and reshuffling of genetic material from different individuals, is also good for quickly removing bad mutations from populations, thereby increasing their overall fitness (see Fig. 12.3).

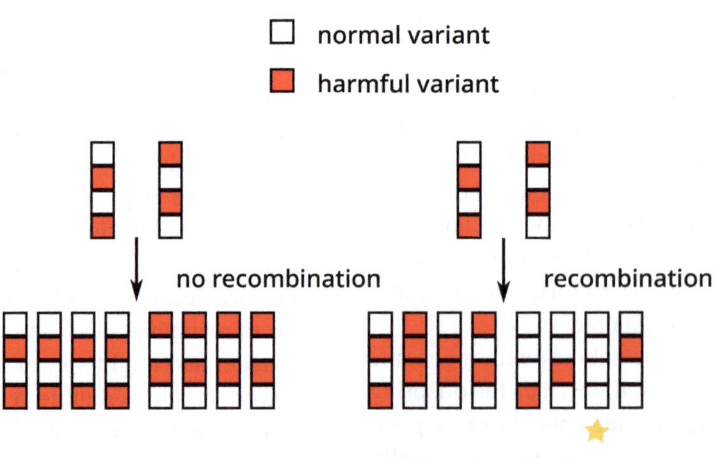

Fig. 12.3 **Sex and harmful mutations.** The first scenario represents an asexual population where recombination cannot reshuffle the two chromosomes. In this case, harmful mutations will be transmitted to the offspring. The second scenario shows that in a sexual population, recombination can reshuffle the two chromosomes and transmit to the offspring a chromosome with fewer harmful mutations (the fittest individual marked by a yellow star)

Population Size Matters!

An important assumption we have made so far is that mutations are unlikely to occur randomly at both of the required sites within a single genome. This is because the rate of mutation is very low. Thus, if a particular site in the genome mutates with probability 1/1000, then the chance that two specific sites would mutate is (1/1000) × (1/1000) or one in a million–a much lower number.

The chance of observing such a one-in-a-million event in a small population of 100 individuals is almost negligible. But what happens if the population is enormous? Mutations per individual per generation will still be rare, but since there are a lot of individuals, there is likely to be at least one individual among these who ends up with both required mutations, just by chance. This is true in general of any improbable event (such as winning a lottery or, in this example, a genetic lottery!). If just 100 people buy lottery tickets, it is very likely that no one will end up with the winning numbers. However, with a million people buying tickets, at least one person is likely to hit the jackpot! Thus, the larger the population, the higher the chance that it has at least one genome in which–going back to our example — both the strong muscle gene and good vision gene randomly mutated. This explains why recombination is very beneficial in small populations, where waiting for mutations is not feasible, but less so in large populations, where changes at the right sites have a reasonably high chance of arising. This might also explain why bacteria, whose population sizes reach billions, can afford to reproduce asexually and still be very viable.

However, even bacteria sometimes exchange genetic material via tiny rings of DNA called plasmids (see Chap. 6). This can be especially crucial when adapting to extreme environmental conditions, such as the high temperatures in hot water springs or high concentrations of antibiotics in the human body, which again hints at the importance of quick reshuffling during adaptation, even for those numerous bacteria.

Thus, the key benefit of sex is its ability to reshuffle and recombine different parts of the genome. Recombination allows offspring to have new combinations of genes not present in the parental generation. This can be especially important when the population is poorly adapted and the other source of change, mutation, is too unlikely to provide the genetic variation required for adaptation.

Sexual reproduction does open up a new set of challenges for a species, which require specialisation and adaptation — for example, how to choose a mate to produce the largest number of offspring and the fittest ones. Should one try to find as many mates as possible or just a single very good mate? Most animals and plants have evolved two very different strategies — one represented in males and the other in the females of a species. One could also ask why there are only two sexes and why most sexual species have independently evolved these two strategies. We will talk about this in Chap. 13.

DEFINITIONS

Meiosis: A type of cell division where a cell produces daughter cells that have half the number of copies of each chromosome compared to the parent cell

Recombination: The exchange of genetic material between pairs of chromosomes during meiosis

Sexual reproduction: The act of combining genetic material from more than one source

The twofold cost of males: The cost on populations of organisms where males take

Getting Rid of Harmful Mutations Without Sex

To estimate the likelihood of back mutations (that revert any site in the genome to its wild-type state), consider the human genome, which has 3 billion bases. Every generation, around 100 new random mutations occur in an individual. The chance that one specific site mutates in this individual is thus 100 in 3 billion, which is about 3×10^{-8}. If we consider a population of 1000 individuals, the chance that this site mutates in at least one of them is 1000 times higher: 3×10^{-5}. However, this is still an incredibly low probability. It takes a population size of at least ten million to make this probability of back mutation at any particular site a realistic number, like 0.3. Therefore, the size of the population will again determine whether the genetic changes required to become fitter can be achieved by random mutations alone. While large bacterial populations might be fine waiting for the right mutations, animals and plants that live in small populations need to use additional mechanisms to eliminate bad mutations from their genomes.

up resources, but do not take part in the resource-intensive task of producing the next generation

SUGGESTED READING

- Smith, JM. The evolution of sex. Vol. 4. Cambridge: Cambridge University Press, 1978.
- McDonald MJ, Rice DP, Desai MM. Sex speeds adaptation by altering the dynamics of molecular evolution. Nature. 2016 Mar 10;531(7593):233–6.
- Gibson AK, Delph LF, Lively CM. The twofold cost of sex: experimental evidence from a natural system. Evol Lett. 2017 May;1(1):6–15.
- Otto SP, Lenormand T. Resolving the paradox of sex and recombination. Nat Rev Genet. 2002 Apr;3(4):252–61.

Testing the Benefits of Sex in the Lab

Initially, hypotheses about the evolution of sex were mostly based on mathematical models and logical reasoning, quite similar to what we have discussed so far. However, scientists have since experimentally tested and even confirmed some of these theories. One such interesting experiment was carried out with baker's yeast *Saccharomyces cerevisiae*, which can reproduce both sexually and asexually.

Researchers took advantage of this feature to explore whether sexually and asexually reproducing yeast populations adapt differently to changing conditions. They put the yeast populations in a new environment and let them evolve for over 1000 generations. To determine how well the populations had adapted, they measured the growth rate of each population in the new environment at the beginning and at the end of the experiment. Both sexual and asexual populations improved their growth rate over 1000 generations, but the sexual ones experienced a notably larger improvement. What was going on in the genomes of the sexual yeast populations that made them better adapted by the end? The researchers looked into this using DNA sequencing.

One might expect that some good mutations arose in some yeast cells, became frequent in the population and made them better adapted. That's right–yellow lines in Figure B show that this is indeed the fate of good mutations in both the sexual and asexual populations: They increase in frequency over generations. What is different is that harmful mutations (blue lines) also increase in frequency in asexual populations, while in sexual populations, they eventually disappear. What could cause this difference? Well, harmful mutations should not spread unless they happen to be on the same genome as a very good mutation and "hitchhike" to high frequencies with them. Recall that asexuals do not have recombination, and so there is no way to decouple these two kinds of mutations-they will share the same evolutionary fate. When yeast has sex, though, it reshuffles parts of chromosomes, and harmful mutations eventually separate from good mutations. Thus, fitter combinations (with fewer harmful mutations) increase in frequency, while less fit ones eventually disappear.

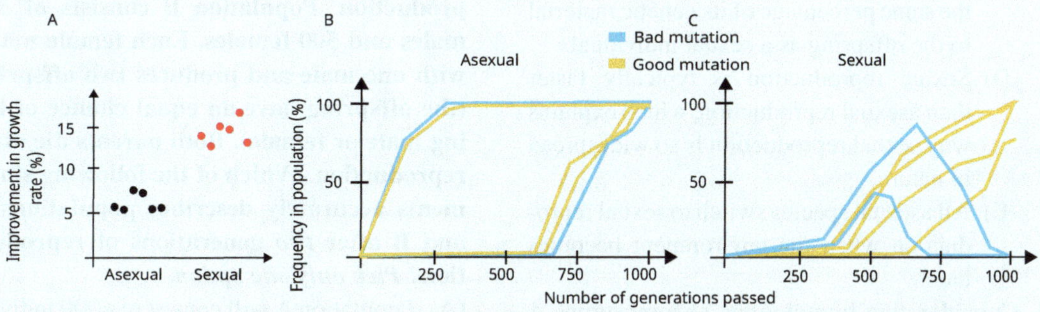

Fitness change and mutation fates in asexual and sexual populations. (**a**) Fitness change in asexual (black dots) and sexual (red dots) yeast populations after 1000 generations of adaptation. Frequencies of good (yellow) and bad (blue) mutations (percentage of individuals carrying them) are shown over time in (**b**) asexual and (**c**) sexual populations, respectively. (Figures adapted from McDonald, M., Rice, D. and Desai, M. Sex speeds adaptation by altering the dynamics of molecular evolution. Nature 531, 233-236 (2016). https://doi.org/10.1038/nature17143)

12.6 Questions

1. **Below you can see the genetic sequence of a certain stretch of a chromosome. The two rows represent paternally and maternally inherited chromosomes.**

   AACTGAAAGTCTCGA
   TACGGAAAGAATCGA

   During the formation of sperm cells, the two chromosomes undergo a process of recombination. Since the above sequences are quite short, imagine that there is *at most one* breakpoint within the sequence, at which the chromosomes exchange segments. Which of the sequences shown below might represent the same stretch of chromosome in the sperm cell?
   (A) TACGGAAAGAATCGA
   (B) AACTGAAAGAATCGA
   (C) AACTGAAAGTCTAAA
   (D) TACGGAAAGTCTCGA
   (E) TACGGAAAGAGTCTCGA
2. **Mark all correct statements.**
   (A) A sexually reproducing individual always produces offspring that are fitter than itself.
   (B) All living species have two mating types or sexes.
   (C) An asexual individual passes on almost the same percentage of its genetic material to the offspring as a sexual individual.
   (D) Sexual reproduction is typically faster than asexual reproduction, which explains why sexual reproduction is so widespread in nature.
   (E) All asexual species switch to sexual reproduction when the environment becomes harsh.
3. **Consider two homologous chromosomes *A* and *B* — one inherited from the mother and the other from the father. These two chromosomes undergo recombination during gamete formation to create a new chromosome *C*. Which of the following statements are true?**
   (A) The chromosome *C* will have more harmful mutations than either *A* or *B* individually.
   (B) The chromosome *C* may have certain mutations not present in either *A* or *B*.
   (C) The chromosome *C* will inherit exactly 50% of the beneficial mutations present in *A* and 50% of the beneficial mutations in *B*.
   (D) The physical length of *C* will typically be equal to the combined length of *A* and *B*.
   (E) The chromosome *C* can have either more or less harmful mutations than *A*.
4. **Which of the following factors make it more likely for a population to evolve sexual reproduction?**
   (A) The population consists of individuals who are far apart in space.
   (B) The population lives in an environment that varies a lot over time.
   (C) Individuals in the population carry a heavy "mutation load" (a large number of harmful mutations).
   (D) The population is extremely well adapted to its current environment.
   (E) All individuals in the population have exactly the same genetic sequence.
5. **Consider two populations A (sexual) and B (asexual). Population A consists of 1000 females. Each of these females produces two offspring asexually and then dies after reproduction. Population B consists of 500 males and 500 females. Each female mates with one male and produces two offspring (the offspring have an equal chance of being male or female). Both parents die after reproduction. Which of the following statements accurately describes populations A and B after *two* generations of reproduction? *Pick only one option.***
   (A) Population A will consist of 4000 individuals, all female. Population B will consist of 2000 individuals: 1000 males and 1000 females.
   (B) Populations A and B will have the same number of individuals after two generations of reproduction. This is because populations A and B started out with the same number of individuals, and in both, there are two offspring per female.

(C) Population A will consist of 2000 individuals, all female. Population B will consist of 1000 individuals: 500 males and 500 females.
(D) Population A will consist of 4000 individuals, all female. Population B will consist of 1000 individuals: 500 males and 500 females.
(E) Population A will consist of 4000 individuals, all female. Population B will go extinct because of the twofold cost of sex.

6. **Which of the following are examples of sexual reproduction?**
(A) A potato is kept out for a few weeks. New stems start to develop from the eyes of the potato. When this potato, with its newly formed stems, is planted in the ground, a potato plant develops.
(B) Pollen from the anther of a particular flower is deposited on the stigma of *the same* flower. Fertilisation occurs, and eventually, a seed-bearing fruit is formed.
(C) Certain species of white worms spontaneously fragment into a few pieces. Each of these pieces regenerates a new head and tail.
(D) Male and female fish shed egg and sperm cells in the water. The sperm fertilises the egg in the water, outside the female's body.
(E) Two single-celled individuals, with two sets of chromosomes each, fuse to form a cell with four sets of chromosomes. This cell then loses some of its chromosomes at random and returns to a state with two sets of chromosomes.

7. **Despite the advantages of sexual reproduction, there are some pitfalls. Figure 12.4 represents a diploid individual with two pairs of chromosomes, each carrying one locus of interest. In which of the situations described below would the next-generation offspring of this individual benefit more if it reproduced asexually than if it reproduced sexually?**
(A) The individual carries alleles A and B on two genes that together confer a strong selective advantage (e.g. strong ligaments combined with strong muscles in a lion). However, they are just marginally beneficial when inherited separately. Additionally, the two genes are located on two different chromosome pairs (as shown in the picture).
(B) The individual carries alleles (A and B) that, when inherited together, have a significantly negative effect on fitness; however, they do not have a negative effect when inherited separately. Additionally, the two genes are located on two different chromosomes (as shown in the picture).
(C) The environment is extremely variable, and environmental conditions change drastically from one generation to the next.
(D) The environment is extremely stable, and this individual has the highest fitness.
(E) Allele A is harmful, while allele B is beneficial to the individual.

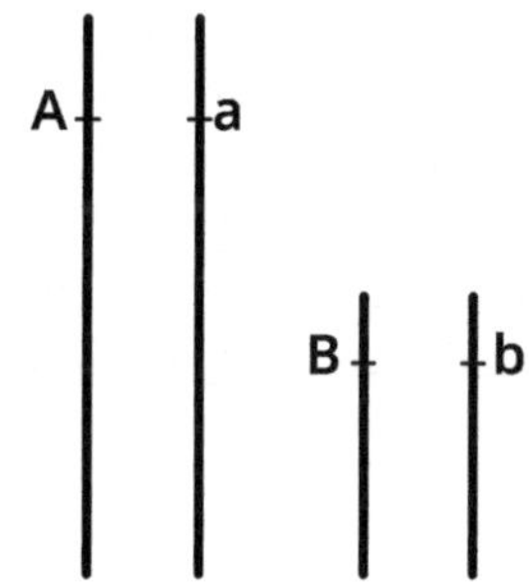

Fig. 12.4 Diploid individual with two pairs of chromosomes

8. **Humans have 22 pairs of homologous, autosomal chromosomes, and usually, each pair undergoes recombination (also called crossing-over) during meiosis, where the two homologous chromosomes exchange corresponding parts. What would happen if, in a small population of humans, suddenly all meiotic crossing-over stops, from this generation on, while everything else in the meiosis would otherwise proceed normally?**
(A) The children would be clones of their parents.

(B) The genetic variation in this population would dramatically rise.
(C) Reproduction would become impossible.
(D) This "no-crossing-over" population would be less likely to quickly adapt to a rapidly changing environment than a normal human population of comparable size.
(E) The number of distinct combinations of alleles present on the same chromosome would remain the same or decrease over time.

12.7 Project: When Is Sex Really Useful?

Materials

- Pen and paper
- calculator or table editor (Excel, Google Sheets)
- Three dice
- Online access to use the random pair generator at https://en.edub.sk/everyday-evolution

Part 1: Live and Prosper (in a Stable Environment)

Population A Consider a sexually reproducing population consisting of 100 individuals, 50 males and 50 females. The individuals mate randomly, and each female is able to produce, on average, two offspring in her lifetime. The generations replace one another (parent generation dies after laying eggs, before the offspring hatch), and the sex ratio of the offspring produced by sexual reproduction within each generation is about 1:1.

Follow this population for three generations and write down the numbers of males and females in each generation.

In the fourth generation, one of the females mutates and is able to reproduce asexually — she can now produce offspring on her own. This female still produces two offspring in her lifetime, but unlike the sexually reproducing females, the asexual female will produce two genetically identical daughters, who will also reproduce asexually (creating two daughters each) in their turn.

Follow the fate of this population for the next eight generations and write down the number of males, sexual females and asexual females in each generation in Table 12.1 (or create a similar table in a table editor, e.g. Excel or Google Sheets). If the population size is odd, for example, 31, then assume that there are 15 males and 16 females or vice versa — the numbers that are closest to a 1:1 sex ratio. Suppose that the population growth is not limited by environmental factors.

Population B For this population, everything is the same as for population A, but each female is able to produce, on average, three offspring in her lifetime. Still, the sex ratio across the whole population in each generation of sexually produced individuals is about 1:1.

Follow this population for three generations and write down the numbers of males and females in each generation.

Also in this population, an asexual female arises in the fourth generation. In this case, the asexual female is able to produce three asexual females during her lifetime, and each of these produces three asexual daughters and so on.

Follow the fate of this population for the next eight generations (or more if necessary to answer the questions) and write down the number of males, sexual females and asexual females in each generation in Table 12.2 (or create a similar table in a table editor, e.g. Excel or Google sheets). If the population size is odd, for example, 31, then have 15 males and 16 females or vice versa — the numbers that are closest to a 1:1 sex ratio. Assume that population growth is not limited by environmental factors.

Analysis

1. In which generation will the asexual individuals represent about half of the whole population A? Population B?
2. In which generation will the asexual individuals outnumber the sexual individuals in population A? In population B?
3. Is it possible that the sexually reproducing individuals will eventually disappear from the population, and the whole population will become asexual? If yes, specify the circumstances under which that would happen. If not, explain why.

Part 2: Adapt and Survive (in a Changing Environment)

Imagine a population of moths that spend their resting time sitting on tree trunks or building walls. The colour of their wings is coded for by two genes on one chromosome, each with two possible variants (alleles): dark or light. The more

Table 12.1 Population A

Generation	1	2	3	4	5	6	7	8	9	10	11	12
Males	50											
Sexual females	50											
Asexual females	0	0	0	1								
Population size	100											

Table 12.2 Population B

Generation	1	2	3	4	5	6	7	8	9	10	11	12
Males	50											
Sexual females	50											
Asexual females	0	0	0	1								
Population size	100											

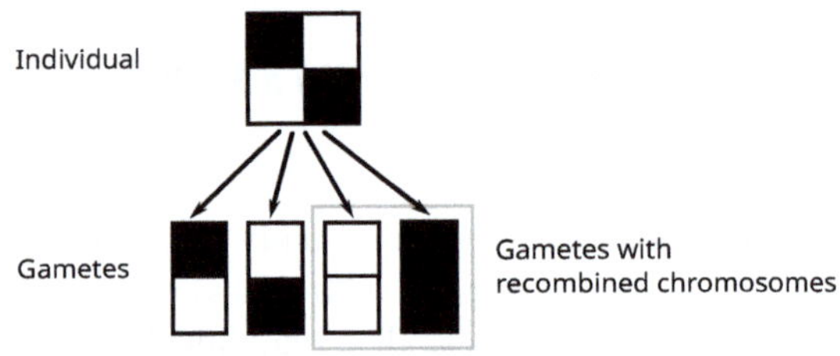

Fig. 12.5 **Production of gametes with recombined chromosomes**

dark alleles an individual has, the darker its wings are: An individual with four light alleles will be white, while an individual with four dark alleles will be the darkest, almost black.

For many generations, the white (all light alleles) or slightly grey (only one dark allele) individuals were most common in the population, since they would blend the best with the surfaces on which they rest. However, due to smog and pollution, the environment has started to change: Tree trunks and building walls are getting darker. The light-coloured individuals stand out against the darkening trunks and can be easily spotted by birds that eat them, while darker individuals blend with the surroundings and survive.

Scenario 1: Sexually Reproducing Population

Look at Fig. 12.5 and read the instructions describing how sexual reproduction works in this population.

1. The reproducing population consists of ten moths (generation 1).[2]
2. The genotype of each individual is represented by a square with four sites. The individuals are diploid; the two sites on the left represent one chromosome, and the two sites on the right represent the other chromosome. The top sites represent two alleles of the first gene, and the bottom sites represent the pair of alleles of the second gene. White sites represent white alleles, while dark sites represent alleles for dark colouration.
3. Each individual forms four gametes: two with chromosomes identical to its own and two with chromosomes that undergo recombination. To recombine chromosomes, take one allele from one chromosome and the other from the other chromosome, as shown below.
4. Individuals mate; each female can have multiple partners. By combining two gametes, a new individual is conceived. In total, 20 new fertilised eggs are produced and form generation 2. After laying their eggs, the parent generation dies.
5. Although all 20 individuals hatch, half of the moths are eaten by birds before they reach maturity. Only ten of the darkest ones, most suitable to the new environment, survive to adulthood and mate (Fig. 12.6).

[2]Of course, a real population would have hundreds of individuals. But we wanted to be nice to you, so ten should be enough for this project :)

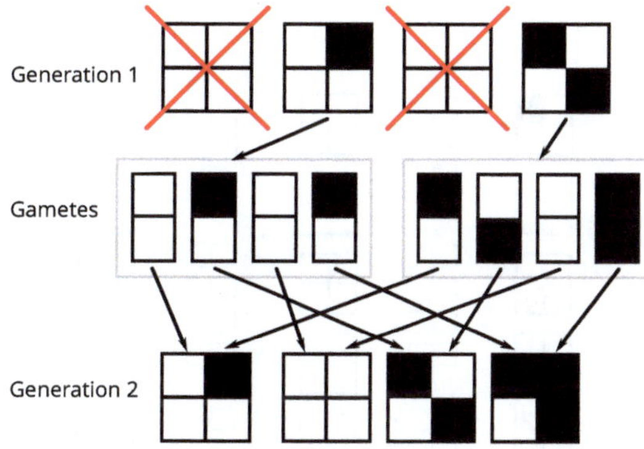

Fig. 12.6 **Example: Only half of the individuals survive and mate**

Your Task

1. Look at Figs. 12.7 and 12.8. The first row represents individuals of the first generation. Originally, there is very low variation within the populations, and almost all individuals are white or light grey. The ten lightest-coloured moths are eaten, and ten survive.
2. Each of the surviving individuals produces four gametes. Colour the chromosomes in gametes that each surviving individual produces (see example in Fig. 12.5).
3. Mating season! Randomly pair two gametes from the pool of 40 using the online pair generator. Go to our online simulator https://en.edub.sk/everyday-evolution and look for *random-pair-generator* to create 40 offspring at once. By combining two gametes, a new individual is conceived. Colour individuals of the following generation according to the chromosomes they inherited (see example in Fig. 12.6).
4. For each individual, roll three dice. If you get three sixes, a new mutation arises. Colour one more site in black and highlight the mutant in the picture.
5. Again, ten of the lightest individuals are eaten, and only ten individuals survive until reproductive age. If several of the individuals are exactly the same, pick any of these, such that in total, ten survive. The surviving individuals form a new generation.
6. Repeat steps 2-5 to simulate the following nine (or more, if you like) generations!

Scenario 2: Asexual Population

1. The population consists of ten asexual individuals: During reproduction, each individual produces two identical copies of itself.
2. Because of the predation in a darkening environment, the darker the individual, the higher the chance for its survival. Only the ten darkest ones survive.
3. From time to time, a new mutation arises, changing a light allele to a black one.

Your Task

1. Look at Fig. 12.9. The first row represents individuals of the first generation — the population is exactly the same as the sexual population in the first generation. Almost all individuals are white or light grey. The ten lightest-coloured moths are eaten, and ten survive.
2. Each individual produces two offspring: Colour two individuals in the next generation identically to their parent.
3. For each individual, roll three dice. If you get three sixes, a new mutation arises. Colour one more site black and highlight the mutant in the picture.
4. Cross out the ten lightest-coloured individuals from the population. If several individuals are identical, some may survive, while others may be eaten. Surviving individuals form a new generation.
5. Let the remaining individuals reproduce and continue modelling the next generations by repeating steps 1 to 4.

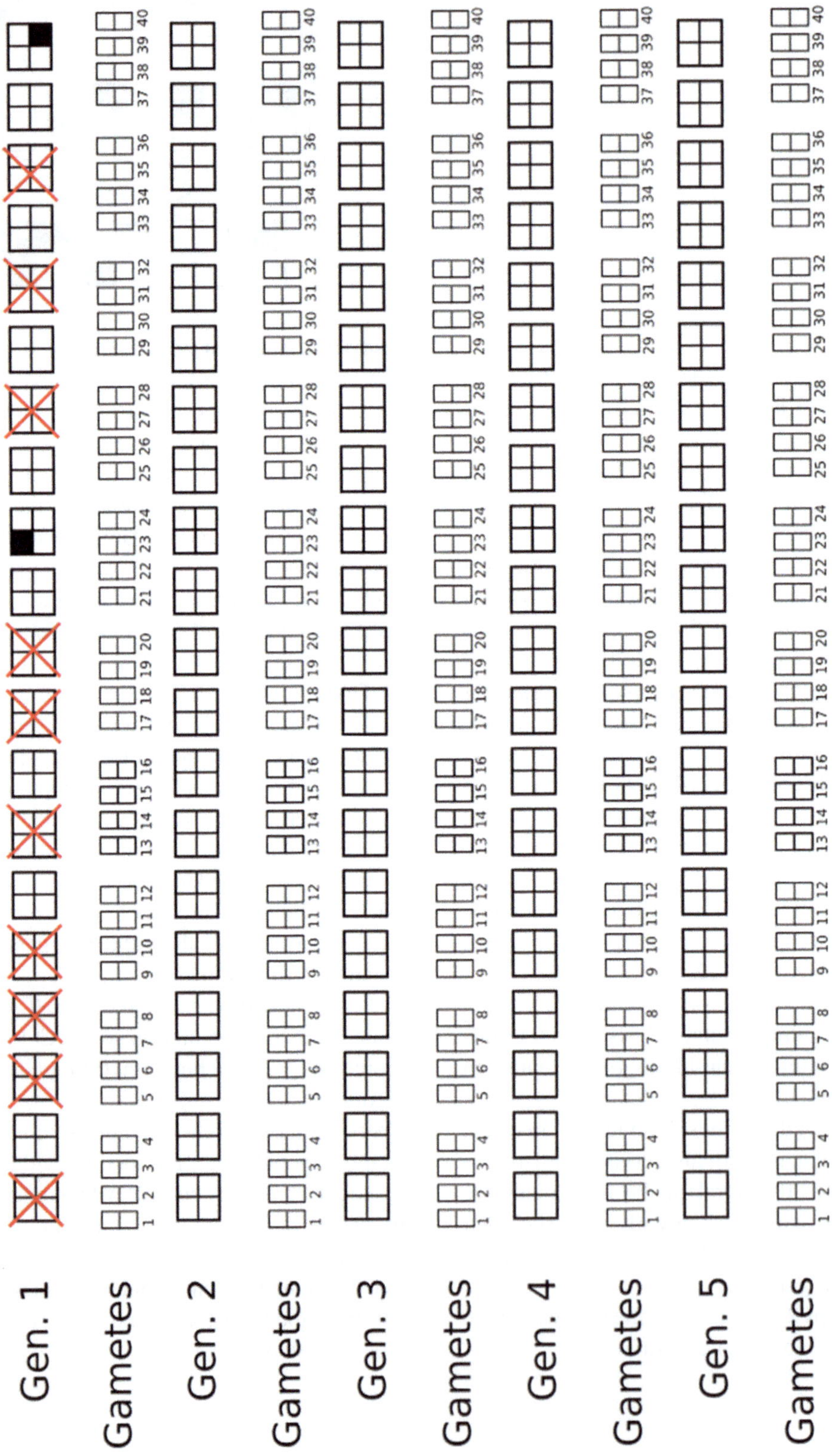

Fig. 12.7 Sexually reproducing population (generations 1–5)

Fig. 12.8 Sexually reproducing population (generations 6–10)

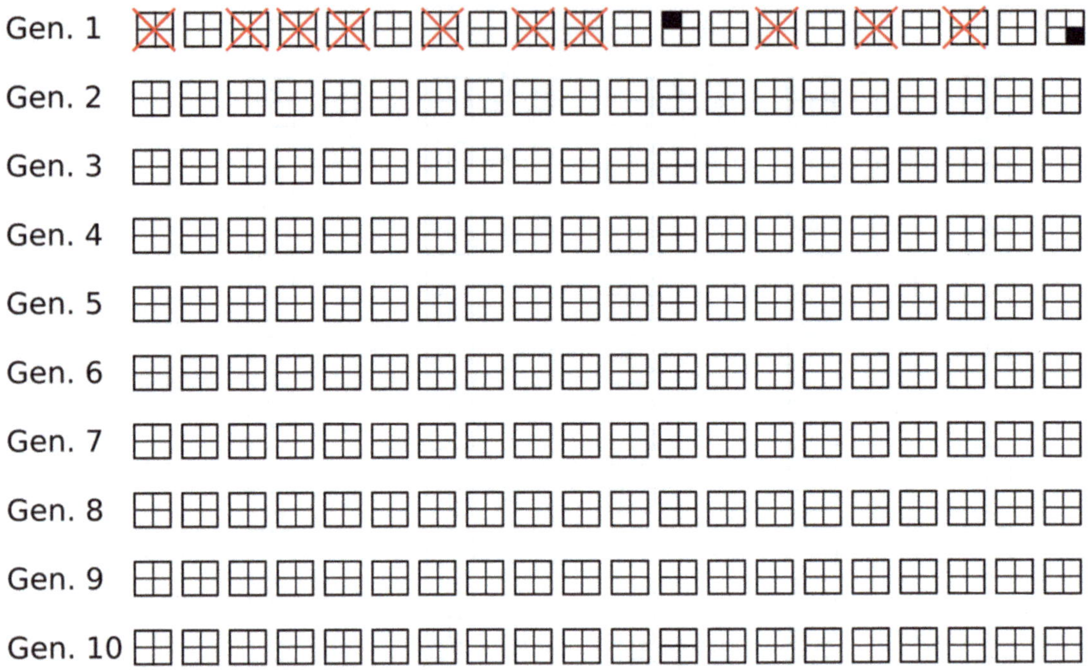

Fig. 12.9 Asexually reproducing population

Analysis

1. For both populations, show the changes in colour across generations by plotting the number of dark alleles in each generation.
2. Which population (sexual or asexual) adapted faster to the environment?
3. How would you expect the adaptation process to be different if there were a population of asexuals 100 times larger than the one modelled?
4. Would you expect your answer to question 2 to be different if both populations had one perfectly adapted individual (with all four dark alleles) in generation 1?

This project was developed by Kristína Hudáková and Barbora Trubenová.

The Causes and Consequences of Sex Differences 13

Gemma Puixeu Sala

▶ When we look around, we see that many species, like humans, have two (sometimes very distinct) sexes — but have you ever wondered why? In this chapter, we will walk together through the steps that led to the evolution of females and males and the implications of having these two particular mating types.

One of the main puzzles in evolutionary biology is: Why is sexual reproduction so widespread, even though it is much more biologically costly than clonal reproduction? Indeed, in contrast to asexual reproduction, where each adult individual can on their own produce offspring genetically identical to themselves, in sexual populations, only half of the individuals (the females) bear offspring, which have only half of their DNA. Besides, having to mate is costly in terms of time and effort, and the reshuffling of genetic material during sex can break up beneficial allelic combinations. Then, why does sex prevail? Precisely because of its ability to **recombine** different genomes, which can bring beneficial mutations together and unlink them from deleterious ones, allowing for fitter combinations to arise, ultimately speeding up adaptation. Because both asexual and sexual reproductions have advantages and disadvantages, the most adaptive strategy is context-dependent. Sex is often more prevalent in smaller populations with more changeable environments, where the pressure to adapt is high (see Chap. 12).

Baker's yeast *Saccharomyces cerevisiae*, for example, is a diploid organism that usually reproduces asexually by mitotic cell division but switches to sex under stressful conditions. To have sex, each yeast cell generates four haploid gametes, two of mating type **a** and two of mating type $\boldsymbol{\alpha}$. To give rise to a new individual, two gametes, one of each type, need to fuse. Sounds familiar, right? This is similar to what happens when a mammalian sperm fuses with an egg and gives rise to a zygote. However, there is one fundamental difference between the sexual reproductive systems of this yeast and mammals: Mammals have females and males, while yeast do not.

13.1 What Are Females and Males?

Why is it that yeast does not have females and males, whereas mammals do? What features define females and males? Let's take a moment to think about it.

G. P. Sala (✉)
Institute of Science and Technology Austria, Klosterneuburg, Austria

B. Trubenová et al. (eds.), *Everyday Evolution: Unveiling Life's Hidden Patterns*, https://doi.org/10.1007/978-3-032-07050-0_13

There is a surprisingly simple and universal definition: **Females** are the mating type that produce larger gametes, and **males** are those that produce smaller ones.[1] We can see this clearly in humans: Human egg and sperm cells are, respectively, the largest and smallest cells in the body, with egg cells being around 50 times larger than sperm cells. The average man produces roughly 525 billion sperm cells over a lifetime, of which around 250 million are released in a single ejaculation. Women, however, are born with around 1 million eggs, of which 300,000 remain after puberty. Of these, only 500 will be released during ovulation over the course of her reproductive lifetime.

This form of sexual reproduction, in which the formation of the zygote involves the fusion of two gametes of *different* sizes, is known as **anisogamy**. All anisogamous species have two (and only two, as discussed below) *sexes*: females, which produce larger gametes (eggs) in the female gonads (ovaries), and males, which produce smaller gametes (sperm) in the male gonads (testes in animals and anthers in plants). In contrast, **isogamy** involves the union of two gametes of similar sizes, which is the case for baker's yeast: The two types of gametes generated after meiosis all come in similar sizes, so there are neither males nor females in yeast.

In species with anisogamy, the two mating types can be found in separate individuals or within a single individual. In the case of mammals, we can talk about female and male *individuals*, since each individual produces just one of the two types of gametes — a condition known as **gonochory** in animals and **dioecy** in plants (Fig. 13.1). But, in other species, such as slugs and most flowering plants, each individual produces both types of gametes, a condition known as **cosexuality**. In plants, there are two types of cosexuality: **monoecy**, in which a single plant produces separate female and male flowers (like hazels or maize), and **hermaphroditism**, in which a single plant produces flowers with both female and male functions (like lilies). In animals, all cosexuals are hermaphrodites since they only have a single gonad of each type (unlike plants, which often have many flowers).

Fig. 13.1 Dioecious plants. In dioecious plants, male and female reproductive organs are found on separate individual plants. For example, willow trees are dioecious. (Figure based on Thakur, A., Sharma, P. et al. "Hybrid Seed Production", Elements of Plant Breeding, IIP Series, Volume 3, May, 2024, Page no.63-73, https://www.doi.org/10.58532/nbennurch221)

Hermaphroditic individuals, therefore, have both female and male *components* (Fig. 13.2). For example, think of a lily (Fig. 13.3). Have you ever touched those brown worm-like structures at the tips of the long filaments and gotten stains on your hands? Those are the anthers (male gonads), each of which contains hundreds of millions of pollen grains that carry the male gametes. The female reproductive structure is a thicker stalk-the carpel — at the bottom of which is the ovary, containing the eggs. By cutting the ovary, dozens of structures containing eggs can be seen with the naked eye.

Thus, we say that lilies have both a male component and a female component because the two types of gametes that need to fuse to generate a zygote are of (very!) different sizes. A similar picture appears, for example, in the hazel, with the difference that female and male components are found in separate flowers (Fig. 13.2). Simply

[1]When we talk about *females* and *males*, we explicitly refer to the biological dimension of sex, which differs from the cultural aspects underlying *women* and *men* as identities, as well as other forms of gender expression in humans, which we discuss later on in the article.

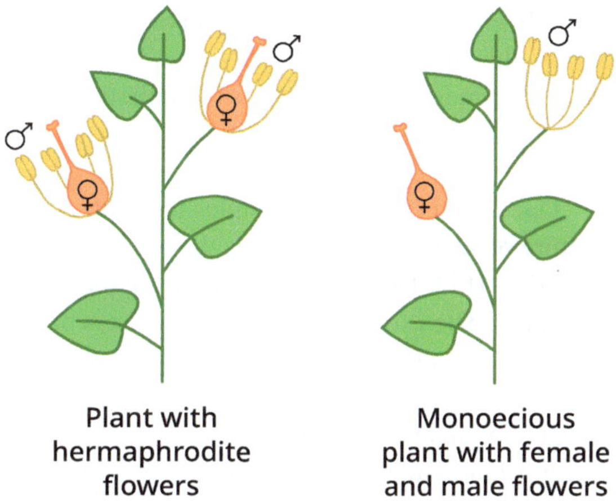

Fig. 13.2 The two types of cosexuality. Hermaphroditism with female and male reproductive parts in a single flower and monoecy, with separate female and male flowers in a single plant. (Figure based on Thakur, A., Sharma, P. et al. "Hybrid Seed Production", Elements of Plant Breeding, IIP Series, Volume 3, May, 2024, Page no.63-73, e-ISBN: 978-93-6252-234-4, https://www.doi.org/10.58532/nbennurch221)

Fig. 13.3 The lily flower. Both female (carpel) and male (anthers) components are clearly visible

looking at gamete sizes, we can identify male and female individuals (in gonochoric species) or components (in hermaphrodite species) in *all* anisogamous sexual species. Using other traits to identify the sexes can be misleading because they do not always align with a specific gamete size. For example, we know that it is *male* seahorses that undergo "pregnancy", a female condition in many other species, because they are the mating type that produces smaller gametes.

13.2 The Evolution of Anisogamy

Now that we know what makes a female a female and a male a male, the next question we might ask is: Why are there two sexes in the first place? We learned that asking *why there are two sexes* is the same as asking *why gametes come in two different sizes*. Why, then, did anisogamy evolve?

Let's take a population of colonial green algae (e.g. volvox) that reproduces sexually by spawning two different types of gametes in the water. The two types, "+" and "-", are produced by two different types of colonies. To give rise to a zygote, two gametes of opposite types need to meet, which happens more often if the algae produce more gametes: there is *selection for increased gamete number*. At the same time, a zygote is more likely to successfully develop into an adult if it is bigger because it can accumulate more resources; thus, there is also *selection for increased gamete size*. However, suppose the algae only have a limited budget for gamete production. In that case, selection cannot lead to an increase in both the size and number of gametes in every individual at the same time. Thus, there are two selective pressures *in conflict with each other*. A possible solution to this size-quantity trade-off is that one of the colony types produces fewer but bigger gametes, while the other evolves to produce more but smaller gametes. And *voilà*: anisogamy, and with it, the two sexes evolve!

The anisogamous green algae in our example are **dioecious** since the two gamete types are produced by different individuals. But if we imagine another example in which isogamous "+" and "-" gametes were initially produced by a single organism, the same selection process would have led to hermaphroditism. Actually, dioecy can evolve from hermaphroditism and vice versa, and scientists do not know which one evolved first.

Interestingly, almost all multicellular sexual organisms have female and male components or individuals. Scientists believe that this feature evolved many times independently. Nevertheless, there also exist organisms with two isogamous mating types (without males and females, as is the case with yeast) and even with multiple mating types, as Box: (Multiple) Mating types explains.

13.3 Sexual Dimorphism

At this point, another question might arise: What does anisogamy have to do with the (sometimes very large) differences between the sexes that we commonly observe across species?

Sexual dimorphism is the technical term for differences between females and males that go beyond reproductive organs. These differences can be related to a variety of characteristics (such as hormonal levels, body size, colour, behaviour, lifespan or disease risk) and may be of different magnitudes (from very mild to extremely accentuated). For instance, to get an idea of the extent of sexual dimorphism in mammalian physiology, a study looked at sex differences in over 230 traits using almost 15,000 mice. They found significant sexual dimorphism in 10% of qualitative (categorical) and 57% of quantitative (continuous) traits.

One of our favourite examples to illustrate the nature of sexual dimorphism is the superb bird of paradise (Fig. 13.4). It is extremely sexually dimorphic: The external plumage of males is black, whereas females are brown and white. During the mating season, after carefully preparing a "dance floor", a male first attracts a potential female with a loud call. When he has her attention, he unfolds his turquoise breast shield, transforming his frontal view into a spectacular ellipse-shaped creature that rhythmically dances back and forth. The female might agree to mate with him (they reject an average of 15–20 potential suitors before accepting!), after which he will repeat the ceremony to try to mate with other females. The female will lay the fertilised eggs and take care of the offspring after they hatch. The fact that males

(Multiple) Mating Types

A mating type is a group of individuals that generate a specific type of gamete and thus cannot mate among themselves. Instead, they need to mate with individuals of another mating type that generate compatible gametes. As such, **a** and ***α*** are mating types in baker's yeast, colonies producing "+" or "-" gametes are mating types in volvox algae, and individuals producing eggs or sperm (females and males) are mating types in all anisogamic sexual species. In all of these examples, there are only two mating types. But there also are organisms with *multiple mating types*, in which there are more than two types of organisms generating mutually compatible gametes. For example, ciliates, a group of protozoa, present various cases of multiple mating types: Within a single genus, *Euplotes*, the number of mating types ranges from five to 12. A more exciting example is the fungus *Schizophyllum communae*. It has four mating genes: $A\alpha$, $A\beta$, $B\alpha$ and $B\beta$, each with 9, 32, 9 and 9 different alleles, respectively. A **mating genotype**, or an allelic combination of mating genes, is made up of a combination of four alleles, one from each type of mating gene. Thus, the total possible number of mating genotypes can be calculated by multiplying together the number of choices for each allele: $9\times\ 32\times\ 9\times\ 9 = 23{,}328$. Since any two different mating genotypes are compatible, this species has 23,328 mating types! Importantly, even though there are organisms with more than two mating types, there is *no organism with more than two sexes*. This is because multiple sexes would imply having different mating types generating gametes of different sizes, a condition that is not evolutionarily stable. Indeed, all known examples of organisms with multiple mating types are isogamous.

Fig. 13.4 **Example of sexually dimorphic species.** Male (behind) superb bird of paradise trying to persuade a female to mate with him. (Scholes and Laman (2018), Distinctive courtship phenotype of the Vogelkop Superb Bird-of-Paradise Lophorina niedda Mayr, 1930 confirms new species status. PeerJ 6:e4621; https://doi.org/10.7717/peerj.4621. CC-BY-4.0 https://creativecommons.org/licenses/by/4.0/)

of this species invest more energy in mating and females in parental care is not by chance. Because of anisogamy, the production of a single gamete is more expensive for females than for males. Since sperm is so "cheap", the number of offspring a male has (his fitness) increases with the number of times he mates. By contrast, females are mostly constrained by the resources they need to make the egg[2] and take care of it as it develops. For example, pregnancy, typically a female task, is extremely costly. First, the mother must provide the developing fetus with a large amount of resources, in many cases for an extended period of time (640–660 days in elephants!). Second, because females cannot have more offspring during this period, pregnancy limits the number of descendants a female can have, thus limiting her fitness.

Anisogamy, therefore, generates an **asymmetry in the reproductive costs between females and males**, which implies that the two sexes will maximise their fitness using different strategies: Females are often more choosy about whom they mate with and are usually more involved in postmating events (e.g. pregnancy, birth, parental care), while males typically spend more energy on attracting or competing for mates or sperm production. These sex differences in reproductive interests lead to **sex-specific selection**, which leads to the evolution of sex differences. The strength of sex-specific selection and the traits on which it acts can vary a lot across species. This leads to the extensive diversity in sexual dimorphism in the natural world. But, crucially, there is one fundamental sex difference that is common to all species and is the starting point for the evolution of the rest: anisogamy.

Importantly, while the series of causal evolutionary events described above (known as the sexual cascade) could broadly explain why females and males look different, they provide a very simplistic view of the evolution of sex differences. Many studies are increasingly showing that reality is much more complex, with females and males also displaying characteristics typically associated with the other sex. Indeed, there are examples across many species that show that females benefit from multiple mating and that males can be choosy and good fathers, too. This variation in sex-associated traits is very important to consider when trying to understand the implications of sex, which is important from both a biological and sociocultural perspective. See the last section for more on this discussion.

[2]This can be very large. For example, the kiwi egg takes up about 20% of the mother's body!

Sex Determination Systems

Various **sex determination systems** can be found across different organisms. One of these is **genetic sex determination**, which typically relies on **sex chromosomes**, a specific set of chromosomes that contains the genes that trigger sex-specific developmental programs. There are three types of chromosomal sex determination systems: (i) **male heteromorphic**, in which males have two different sex chromosomes — XY — while females have the same — XX. The Y chromosome is much smaller than the X, containing only a few genes. Examples include all mammals, Coleoptera (beetles) and many Diptera (for example, fruit flies). (ii) **Female heteromorphic**, in which females have two different sex chromosomes — ZW — while males have the same — ZZ. Examples include birds, many reptiles and Lepidoptera (including butterflies). (iii) **Homomorphic**, in which the two sex chromosomes only differ in the sex determination genes but look exactly the same otherwise. This type is prevalent in amphibians and common in fish. Another type of genetic sex determination system is **haplodiploidy**, in which females are diploid (originating from fertilised eggs) and males are haploid (originating from unfertilised eggs). Examples include ants and bees (Hymenoptera).

On the other hand, some species have **environmental sex determination**. For example, crocodile eggs incubated at warmer temperatures will produce females; otherwise, males will hatch. In wrasse (a kind of fish) colonies, when the only male dies, the largest female becomes a male. Finally, there are **hermaphrodites**, which have both male and female functions, so sex determination occurs in each gonad type.

13.4 Sex Determination and Differentiation

Now we know *why* females might differ from males: It is a consequence of sex-specific selection forces. But have you ever wondered *how* they can be so different when both share almost exactly the same genome? The answer lies in the *almost*. In fact, in mammals, the only element we need to predict the sex of an individual is the presence of a Y chromosome,[3] more specifically, a gene it contains: *SRY* or sex-determining region Y. In XY individuals, this gene is expressed during early development and acts by "switching on" a developmental program that will turn the yet undifferentiated gonads into testes. If *SRY* is absent (in XX, but also XY individuals with mutated *SRY*), an alternative developmental program that generates egg-producing ovaries is executed instead. The development of sex-specific gonads is called **primary sexual differentiation** and, in mammals, **is genetically determined** already in the zygote (Fig. 13.5). However, there are also other sex determination systems, which we will learn about in Box: Sex Determination Systems.

The expression of other sex-associated traits (in humans: body hair, breasts, odour, voice,...) is called **secondary sexual differentiation**. It starts in puberty and depends on hormones produced primarily by the gonads. Ovaries and testes secrete oestrogens and androgens (of which testos-

[3]This simplistic take on sexual development does not consider intersexuality — a condition in which an individual is born with reproductive or sexual anatomy that does not fit the typical binary definitions of female and male. Intersexuality is common in humans (with an estimated incidence of about 1.7% of the population, similar to the proportion of red-haired individuals) and can involve differences at the genetic, gonadal, genital or hormonal level. For example, an XY individual with a mutation in the *SRY* gene might develop ovaries and express female-like hormones, or a genetically female individual with an excess of androgen production (due to e.g. congenital adrenal hyperplasia) might present masculinised genitalia.

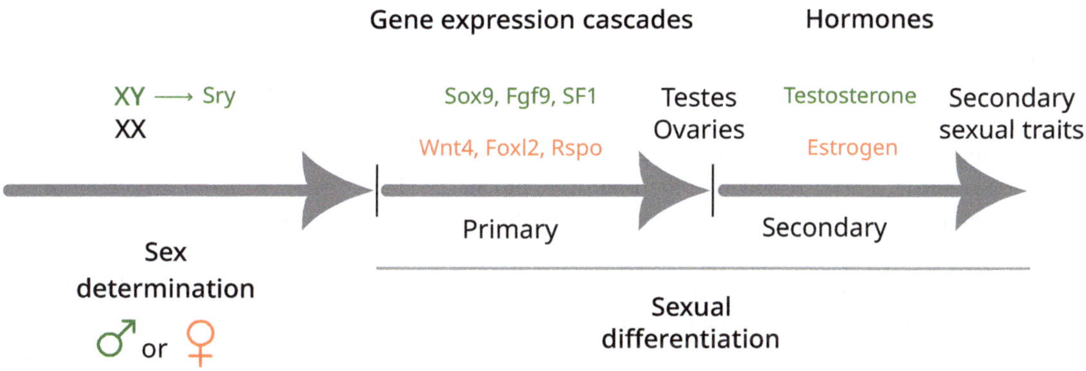

Fig. 13.5 Human sex determination and differentiation. Sex in mammals is determined by the presence of the gene *SRY* in the Y chromosome. This gene initiates a gene expression cascade leading to the development of testes or male primary sexual differentiation; in the absence of *SRY*, an alternative set of genes is activated, leading to the development of ovaries (female primary sexual differentiation). Testes and ovaries primarily produce different types of sexual hormones, which orchestrate the differentiation of male and female secondary sexual traits

terone is a type), which lead to the development of female- and male-associated secondary sexual characters, respectively, by affecting the expression of genes across the genome. This means that many sex differences that we observe arise because the *same* proteins (the majority of which are encoded by genes lying outside sex chromosomes) are produced in different quantities in males and females in response to varying hormone levels. Both types of hormones are present in both sexes at different but often overlapping levels. This means that both sexes have both types of hormones, but typically at different average levels.[4] This quite effectively generates sexual dimorphism, i.e. *quantitative sex differences in secondary sexual characters* from the *qualitative primary sex condition*. For example, in humans, higher testosterone levels typically mean more facial hair since the hormone causes hair follicles to grow on the face. There are obvious differences in facial hair between males and females, but also among males (from very patchy to very dense beards) and females, some of whom have facial hair. Sexual dimorphism in facial hair, as well as in other secondary sexual traits driven by hormonal levels, is, therefore, *quantitative* (can take a continuous range of values) rather than *qualitative* (a yes/no trait).

[4]E.g. testosterone is typically higher in males and oestrogen in females, but there is a lot of interindividual variation, with some individuals having hormonal levels that are more typical for the other sex.

13.5 Sexual Dimorphism as a Resolution to Sexual Conflict

Sex-specific selection can cause the levels of sexual dimorphism to evolve over time. In particular, if the expression of a specific trait becomes controlled by a sex-specific developmental program, this trait will evolve to be sexually dimorphic. This is often what happens in the process of sexual conflict resolution.

Let's depict the process of sexual conflict resolution with an example. More colourful fish are more visible to predators and, therefore, have lower survival rates. However, in many species, such as guppies, colourful males are more attractive to females and more successful in finding partners. This is not reciprocal: Male guppies do not seem to show a preference for colourful females. If a mutation (let's call it *golden*) that generates yellow pigmentation in the tail appears in a population of guppies, it will be *selected for* in males since it gives them a mating advantage, but *selected against* in females, since it only makes them more visible to predators and thus harms them. A situation in which a trait that is shared by both sexes is selected differently in each sex is called **sexual conflict**. How can sexual conflict be resolved?

First, if the *golden* mutation "jumps" (due to an accidental error during meiosis) to the non-recombining region of the Y chromosome, which is only present in males, it will be inherited directly with male-determining factors and only be expressed in males. Second, a mutation that makes the expression of *golden* responsive to high levels of androgens might arise. Such a mutation would be positively selected as it would cause *golden* to be more expressed in males (just like other male secondary sexual traits) and thus increase their mating success without compromising female fitness. This can happen even if the gene is found outside the sex chromosomes and is therefore shared by both sexes.

Either way, yellow tail colouration would become a male-associated secondary sexual trait, resolving the instance of sexual conflict and increasing the overall levels of sexual dimorphism in this guppy population. This example illustrates that sexual dimorphism typically evolves as a mechanism to resolve sexual conflict and that it does so by linking the expression of the trait under conflict to mechanisms leading to sex determination or female/male sexual differentiation.

13.6 Parental Conflict and Imprinting

As hinted earlier, another aspect in which sexes typically differ, and therefore also have conflicting interests, is **parental care**. Given what we learned about anisogamy and reproductive costs, we would expect females to take over most of parental care. But this is not always true, as we discussed earlier. Indeed, mammals present an extreme case of female-biased parental care since females host and feed the developing offspring during pregnancy, nurse them after birth and, in some cases, take care of them for several years (e.g. up to 6-7 years in orangutans). Males, instead, assist the females only in 5-10% of mammalian species. Nevertheless, paternal care is common in other species, particularly in fish and many bird species (for example, it is male emperor penguins that incubate the egg, while the female forages for food, and male spotted sandpipers and kiwi birds that take over the nursing of the chicks almost entirely on their own after the females leave the nest). Thus, there is a lot of variation in the time and intensity of parental care across species.

Parental conflict emerges when the interests of the mother and the father over parenting do not match. For example, mammalian fathers want their offspring to elicit all the care they can get from their mothers since more maternal care increases the offspring's chances of survival. However, it is in the mother's interest to provide equal care to *all* her offspring (including future ones) and not only to the ones coming from a specific father. Since she only has limited resources, she wants her offspring to be more independent and to take less of her resources to leave enough for their (half-)siblings. Increased demand for care (e.g. increased begging of offspring for food) and social bonding in the offspring is caused by high levels of oxytocin, also known as the "love hormone". A gene that increases oxytocin production in the offspring might thus be beneficial for fathers but not for mothers.

One way to fight the battle of parental conflict is via **imprinting** or **parent-specific gene expression** (more details in Box: Genomic Imprinting and Epigenetics), in which the expression of some genes in the offspring depends on which parent they inherited the respective copy from. In our example, a gene that increases oxytocin production might be maternally imprinted (silenced in the maternally transmitted copy and expressed in the paternal), and a gene decreasing it, paternally imprinted instead. Interestingly, there are known human diseases associated with mistakes in the imprinting of oxytocin-related genes: The loss of certain pro-oxytocin paternally expressed factors leads to children with Prader-Willi syndrome, which is marked by asocial or autistic behaviour. On the other hand, the loss of certain anti-oxytocin maternally expressed factors leads to Williams syndrome, which is characterised by a hyper-sociable phenotype.

Genomic Imprinting and Epigenetics

Individuals inherit two copies of each gene-one from their mother and the other from their father. Normally, both copies are active, but in some cases, only one is turned on, and this depends on which parent it came from. This phenomenon is called **genomic imprinting** or **parent-specific gene expression**. We say that a gene is **maternally imprinted** if the allele from the mother *is not* expressed, whereas the allele from the father *is*. In particular, the mother turns the gene off in the oocytes during their development. For paternally imprinted genes, the situation is the same, but with roles reversed. This "turning off" occurs by attaching a small molecule (a methyl group) to the segments of DNA corresponding to that particular gene, which are then interpreted by gene expression machinery as "do not express" signs. The phenomenon of imprinting evolved about 150 million years ago in a common ancestor to all mammals. In humans, only a small percentage of genes undergo imprinting. We do not know why some are imprinted and others are not — although parental conflict might explain the presence of imprinting in some genes (as explained in the main text). Receiving both copies of an imprinted gene from the same parent due to an error in meiosis (a phenomenon known as **uniparental disomy**) can lead to developmental disorders.

Genomic imprinting is a type of **epigenetic** modification, i.e. a modification that changes the regulatory state or expression of the DNA but does not change the DNA sequence itself. Epigenetic modifications can be transmitted to daughter cells after division (just like genetic modifications) but are less stable and more easily reset to the ancestral state. For example, have you ever wondered why skin cells are so different from eye or blood cells despite sharing the same set of genes? They have different epigenetic states that activate different groups of genes. These epigenetic states are configured during development, when cells differentiate into their respective types, and are then passed on from mother to daughter cell.

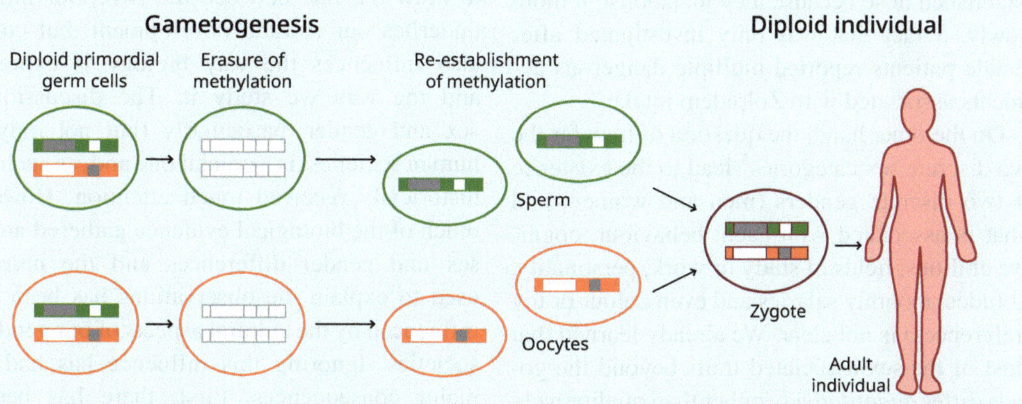

Genomic imprinting in humans. Imprinting marks (methyl groups, indicated in grey) are erased and re-established during female (orange) and male (green) gametogenesis. Those methylation patterns are then maintained during an individual's lifetime. (Figure based on Ishida M, Moore GE. The role of imprinted genes in humans. Mol Aspects Med. 2013 Jul-Aug;34(4):826-40. https://doi.org/10.1016/j.mam.2012.06.009. Epub 2012 Jul 4)

13.7 A Note on (the Study of) Sex and Gender

We feel that a discussion on sexual dimorphism should also include some reflection on the way human societies are shaped by the ways in which we study and view sex differences. Although humans are one of the least sexually dimorphic primates (quite surprising, right?!), it is evident that human females and males *do* differ. This is important to acknowledge, particularly for medical reasons, since there are sex-related differences in the prevalence, age of onset, mortality rate and specific clinical manifestation of many diseases. For example, autoimmune diseases like multiple sclerosis or inflammatory bowel disease are more common in women, while the incidence of neurocognitive disorders such as autism is higher in men. Since both sexes have distinct metabolisms, they also differ in treatment requirements. Indeed, until recently, only male cells and individuals were used in many clinical and preclinical trials, leading to the release of multiple drugs with a higher risk of side effects in women. Zolpidem, a drug used to treat sleep problems, is one such example. Women should take half the initially established dose because they metabolise it more slowly, a fact that was only investigated after female patients reported multiple dangerous accidents associated with Zolpidem intake.

On the other hand, the question of how far the two discrete sex categories[5] lead to the existence of two discrete genders (men and women, and what is associated with each: behaviour, cognitive abilities, fields of study or work, personality, attitudes, monthly salaries and even colour or toy preferences) is not clear. We already learned that most of the sex-associated traits beyond the gonads differ *quantitatively* rather than qualitatively between females and males. For example, even though there are typically female and male brain features, if we take a brain at random, we would not be able to reliably say whether it belongs to a female or a male. One study looked at different types of brain features in thousands of individuals and classified them as "typically female" or "typically male" depending on whether they were mostly found in one sex or the other. Then, they asked whether those two types of traits can be used to effectively classify brains based on sex. They found that the vast majority of human brains possess a combination of typically female and male traits, indicating that there is no such thing as "gendered brains", and that everyone has an intersex brain instead. Additionally, some neurologists argue that the differences in brain structure between females and males do not account for the differences in attitudes and behaviours between the two genders. Instead, they suggest that these sex differences in brain features may largely reflect the gendered stereotypes conditioning our development from a very young age.

In general, sex refers to biological and physiological differences and categorises individuals into females or males, while gender is a social construct that shapes roles, behaviours and identities associated with femininity, masculinity or other gender identities. While it is assumed that the first is mostly shaped by biology and the second by culture and society, it is difficult to draw the line between the two: Our biology underlies our cultural development, but culture also influences the way biology is expressed and the way we study it. The discussion of sex and gender, particularly (but not only) in human societies, is a crucial one and, as such, has historically received much attention. However, much of the biological evidence gathered around sex and gender differences and the narrative used to explain the observations has been very influenced by the cultural aspects of our gendered societies. Ignoring this influence has had two major consequences. First, there has been a bias towards describing systems and species with patterns of sex-specific phenotypes and behaviours that reflect our gendered expectations (e.g. a particular focus on depicting aggressivity and competition as male-specific traits and choosiness, passivity and parental care as female-specific traits, while largely overlooking species where these patterns do not apply). Second, the generation of biased biological evidence has been

[5] It is in itself questionable that there are only two discrete categories for sex. The existence of intersex individuals challenges this view, as discussed above.

used, in turn, to justify the gendered societies we live in.

Fortunately, this is starting to change. Many scientists are currently taking a critical view on the study of the biological evidence of sex and gender while acknowledging these societal biases, producing a more colourful picture that accounts for the beautiful diversity associated with sex and gender in biology.

DEFINITIONS

Mating type: A genetic system in some organisms defining compatible partners for sexual reproduction.
Sex determination: The biological process that determines whether an individual should be a female or a male. In most species, sex is genetically or environmentally determined.
Sexual conflict: Differences in evolutionary interests between males and females, which may cause antagonistic coevolution.
Sexual dimorphism: Differences in size or appearance between the two sexes of an organism, beyond the differences in their sexual organs.
Anisogamy: A form of sexual reproduction that involves the fusion of two gametes, which differ in size.
Isogamy: A form of sexual reproduction that involves the fusion of two gametes that are of the same size.

SUGGESTED READING

- Joel D et al. Sex beyond the genitalia: The human brain mosaic. Proc Natl Acad Sci U S A. 2015 Dec 15;112(50):15468-73.
- Bachtrog D et al; Tree of Sex Consortium. Sex determination: why so many ways of doing it? PLoS Biol. 2014 Jul 1;12(7):e1001899.

13.8 Questions

1. ***S. communae* has 923,874 different:**
   (A) Mating types
   (B) Sexes
   (C) Sex chromosomes
   (D) Gametes of different sizes
   (E) Combinations of different alleles of sex genes
2. **Edible snails (*Helix pomatia*) are a typical example of hermaphroditism. Which of the following statements are correct?**
   (A) There are no male and female snail individuals.
   (B) One snail individual has both female and male reproductive organs.
   (C) Each individual snail produces gametes of two different sizes.
   (D) There are snail females and males; however, they all look the same.
   (E) Snails cannot mate with other individuals and only fertilise themselves instead.
3. **Differences in secondary sexual characteristics between males and females arise because**
   (A) These characteristics are only influenced by genes located on sex chromosomes.
   (B) The expression of some genes located on the autosomes is affected in a sex-specific manner.
   (C) Males and females carry different alleles at genes that influence these characteristics.
   (D) Females and males have completely different sets of hormones, which switch on/off the expression of genes influencing these characteristics.
   (E) Females and males have the same hormones but at different levels, which can change the expression of genes influencing these characteristics.
4. **In human populations with Northern European ancestry, red-green colour blindness is found in about one in 12 males and one in 200 females. Which of the following hypotheses are consistent with this observation?**
   (A) Colour blindness is caused by the excessive expression of a gene that would normally be maternally imprinted.
   (B) Colour blindness is caused by high levels of androgens (e.g. testosterone).
   (C) Colour blindness is due to a dominant mutation in the non-recombining region of the Y chromosome.
   (D) Colour blindness is due to a dominant mutation in the recombining region of the Y chromosome.
   (E) Colour blindness is due to a recessive mutation in the X chromosome.
5. **For the time being, it is not possible for same-sex couples to have children of their own, even if scientists manage to fuse same-sex gametes in the lab. What reasons might there be?**
   (A) Two sperm cells fused are too small to give rise to a viable zygote.
   (B) The set of autosomal genes contained in the cell resulting from fusing two same-sex gametes is different from that of a heterogametic fertilisation.
   (C) Two same-sex gametes have equal imprinting states, which leads to an excess/loss of expression in some genes and can cause embryo inviability.
   (D) As both of the gametes are of the same sex, the resulting gamete is haploid.
   (E) The cell resulting from the fusion of two same-sex gametes is tetraploid (with four sets of genes).
6. **Sequential hermaphrodites are organisms that change their sex during their lifetime. One such example is the bluehead wrasses, which are fish living in the western Atlantic Ocean. Large aggressive males of this species defend territories with harems of females. These males have blue heads and stripes, while the females are yellow. When the male dies, a large female of the group turns into a male. Interestingly, there are also small yellow young males. They look like females and can hence sneak into big males' territories to mate. Researchers found that gene expression in the brains of these yellow males resembles that of females. Which of the following statements is/are true?**
   (A) Bluehead wrasses can self-fertilise.

(B) Big blue-headed males produce small gametes. Females and small yellow males produce large gametes.
(C) Small yellow males produce gametes that are intermediate in size to the ones produced by large males and females.
(D) When a fish switches from female to male, the size of the gametes it produces changes.
(E) Big blue-headed males and small yellow males carry different sets of genes.

7. **Which of these arguments could explain why we don't observe more than two sexes?**
(A) Because gametes that are neither as big as possible nor as numerous as possible do not have any strong advantage for the fitness of the individual.
(B) More than two sets of genetic information would be needed from parents, eliminating the advantages provided by simple sexual reproduction.
(C) If more than two parents contributed genetic information, organisms would be polyploid and not able to survive.
(D) If there were three or more sexes, the two sexes with the most suitable gamete size would outcompete the others, eventually reducing the number of sexes to two.
(E) We do observe more than two sexes-for instance, in Volvox.

8. **What can cause sexual dimorphism to evolve?**
(A) Differences in the reproductive interests between males and females.
(B) Males and females achieve a higher fitness with different strategies.
(C) A drive to avoid potential predators by forcing them to adapt to preying on two different targets.
(D) Differences in the anatomy of gametes.
(E) The need to reduce the potential for competition among males and among females.

13.9 Project: About Eggs and Sperm

In this project, we will examine the evolution of eggs and sperm, specifically the evolution of anisogamy. As we explained in the article, the production of gametes is quite costly — organisms need time and resources to make them. In this project, you will simulate six fictional species of marine algae that form gametes of different sizes. You will investigate how the gamete number and size affect the chances of fertilisation and the survival success of the new zygote. Notice that the total investment in the gametes is the same in all cases. The cost of producing gametes is represented by the surface area of the square further below, which is divided into four large (8×8 cm), 16 smaller (4×4 cm) or 64 really small (2×2 cm) squares, representing the gametes.

Material

- About ten sheets of paper of two different colours. We will refer to them as green and yellow, but feel free to use any combination. You can print the mesh in Fig. 13.6 to assist you with cutting.
- Dice.

Hypothesis

All of these algae species have different gamete sizes, written in Table 13.1. Look at them and write a testable hypothesis about the probability of fertilisation and the survival of the zygote. Which reproductive strategy is the best in your opinion?

Simulation

Species A

1. Cut out four green and four yellow gametes-squares of 8×8 cm. Bend or fold them a bit with your fingers to prevent them from sticking together.
2. Put all of them into a small container — a bowl or something.
3. Stir, mix and shake thoroughly. Make sure gametes do not stick to each other, ruffle and fluff them as much as possible!
4. Now just throw everything out of the bowl on the floor from at least 1 m height. Don't be gentle — simulate nature! Just pour it out. This may take a bit of practice: Make sure they don't end up all lumped together. Try it a few times before starting your experiment to calibrate the best throw, and then, use the same way throughout the experiment.
5. Look at the spilt gametes. Do any of the green and yellow gametes overlap or at least touch? If yes, fertilisation occurs! Write the number of newly formed zygotes in Table 13.1.
6. However, the environment is harsh. Cold currents and a lack of nutrients can affect the zygote's survival. Only if the zygote contains enough nutrients does it have a chance to live. For each of the zygotes, roll the dice one by one. If the size of the zygote — the sum of the sizes (areas) of both parental gametes — is larger than ten times the number you rolled, the zygote survives. For instance, if you roll 4, the zygote survives only if the sum of gamete sizes is more than 40 cm^2.
7. Count the number of surviving zygotes and write them into Table 13.1.
8. Repeat steps 2-7 three times and write the numbers into Table 13.1. Calculate the average number of formed and surviving zygotes for this species.

Species B-F

Repeat this simulation for all the other five species. Make sure you cut the correct number of the right-sized gametes! Write down the numbers of formed and surviving zygotes for each trial in Table 13.1 and calculate the averages.

Analysis

1. In Table 13.1, indicate which gamete colour represents sperm and which one eggs in which species (or write it down).
2. In which species did you get the most fertilisations? In which one is the least?

Fig. 13.6 Gametes of various sizes

Table 13.1 Fertilisation and survival

Species	A	B	C	D	E	F
Green gamete size	8 × 8	4 × 4	2 × 2	8 × 8	4 × 4	2 × 2
number	4	16	64	4	16	64
Yellow gamete size	8 × 8	4 × 4	2 × 2	2 × 2	8 × 8	4 × 4
number	4	16	64	64	4	16
Potential zygote size	128	32	8	68	80	20
Trial 1: Zygotes formed						
Zygotes survived						
Trial 2: Zygotes formed						
Zygotes survived						
Trial 3: Zygotes formed						
Zygotes survived						
Mean: Zygotes formed						
Zygotes survived						

3. In which species was the percentage of the surviving zygotes the highest?
4. Make a graph of the dependence of the number of surviving zygotes on the size of the zygote.
5. Overall, which species would you consider to be the most successful one? Which one is the least successful? What do you think will happen with these species?
6. Discuss any discrepancies between your initial hypothesis and the observed results.
7. Find at least three interesting examples of extreme anisogamy among plants or animals. Introduce each of the organisms briefly: If possible, give the sizes and/or ratios of their gametes, as well as the estimated size of each type of gamete. Include your sources of information.

This project was developed by Kristína Hudáková and Barbora Trubenová.

14 The Secret World of Plant Sex

Katarína Boďová and Melinda Pickup

▶ How do plants choose a mate? Unlike animals, plants can't get up and move around to find a potential mating partner. In most plant species, each plant has male and female reproductive organs. Some plants ensure they have a mate by pollinating themselves—they self-pollinate. Yet many plants can only mate with a different plant of the same species—they cross-pollinate. These plants attract animals, such as bees, which then carry their pollen, or rely on wind to disperse their pollen to other plants. When pollen arrives, they need to solve another problem: Is this pollen mine or from another plant? In this chapter, we will learn why many plant species cross-pollinate and become familiar with some ingenious incompatibility systems they use to prevent self-pollination.

If we find ourselves in an alpine meadow in summer, we might look around and admire the incredible floral diversity that surrounds us. Even within a single environment, flowers vary dramatically in colour, shape and form (Fig. 14.1). But what drives the evolution of this amazing diversity? The answer lies in plant mating. Let's have a closer look at how plants mate, why many of them avoid self-pollination (selfing), and let's explore the elegant systems that have evolved to prevent selfing from occurring.

14.1 Plant Mating: Selfing vs. Outcrossing

Plants can have sex in three main ways (Fig. 14.2): First, they can self-pollinate, meaning that their own pollen is transferred to their own stigma (also called **selfing**). Second, they can cross-pollinate, meaning that pollen from other plants is transferred to their stigma (also called **outcrossing**). Third, they can use a mix of selfing and outcrossing (**mixed mating**).

For outcrossing species, specific floral morphologies—shape, size and colour—can facilitate pollen transfer by increasing the chances that vectors—animals or wind—move pollen between plants (Fig. 14.3).

In exchange, the plants offer nectar and/or pollen to the animal as a reward. The evolution of these win-win plant-pollinator interactions (see Fig. 14.3 for some examples) is an ingenious solution to the problem of being immobile yet

The author Melinda Pickup was deceased at the time of publication.

K. Boďová (✉)
Department of Mathematical Analysis and Numerics, Faculty of Mathematics, Physics, and Informatics, Comenius University in Bratislava, Mlynska Dolina, Bratislava, Slovakia
e-mail: katarina.bodova@fmph.uniba.sk

B. Trubenová et al. (eds.), *Everyday Evolution: Unveiling Life's Hidden Patterns*,
https://doi.org/10.1007/978-3-032-07050-0_14

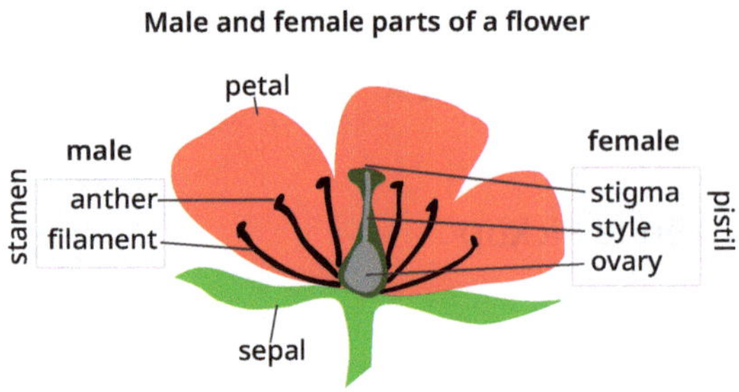

Fig. 14.1 Parts of a flower in hermaphroditic plants. A closer look at the flowers of most plants reveals that they contain both male and female reproductive organs, making them hermaphroditic. The pistil includes the ovary that produces female gametes, and the stamen includes the anther, which produces pollen grains that contain the male gametes. After the pollen has landed on the stigma, it starts to germinate and create a pollen tube that grows through the style to reach the ovary, where fertilisation occurs

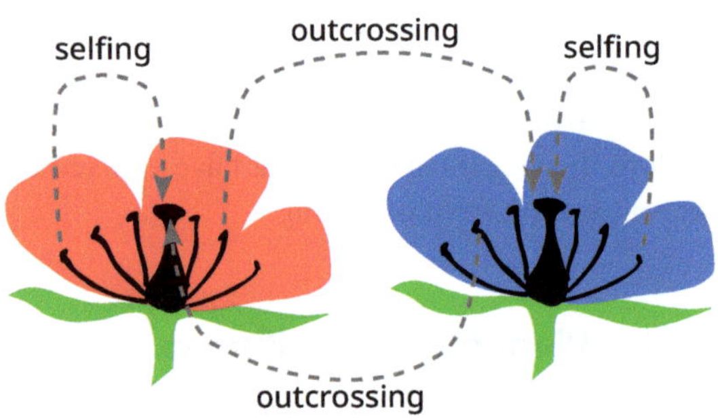

Fig. 14.2 Selfing (A) and outcrossing (B) in flowering plants. The pollen is transferred from the anther to the stigma of the same or another plant

needing to find a mate. Plants have evolved different morphologies that are specifically adapted to different vectors. For example, some plants are specialised for wind-mediated pollination, others are especially attractive to bees, and yet other plants rely on butterflies or birds to visit them and collect and disperse their pollen (see Fig. 14.3). In this way, they reduce competition with each other for pollinators and increase the chance that their pollen is carried to another plant of their species. This specialisation is one of the reasons why we observe such a huge diversity in flowers.

However, if a plant is going to self-fertilise, it does not require adaptations to convince a pollen vector to move pollen from one individual to another. So, what do these flowers look like? As you might expect, the flowers of selfing species tend to be small and white (inconspicuous), offer little to no pollinator rewards like nectar and have a reduced distance between their anther and stigma—traits collectively known as the **selfing syndrome**. These traits evolved because making large, colourful flowers with sugar-rich nectar to reward pollinators is costly for the plant, and it brings no benefit to plants that self-pollinate. Furthermore, bringing the anthers and stigma close together increases the likelihood that their own pollen will reach their own stigma.

Selfing brings an obvious advantage: complete independence and the certainty of having a mate—**reproductive assurance**. Among plant taxa, selfing has indeed evolved many times from outcrossing, which leads us to ask: Under what conditions might selfing and the selfing syndrome evolve?

Changes in the environment, such as pollinator decline, can provide the selection pressure for selfing to evolve. But can this be a rapid process? Experimental studies have shown that selfing can evolve within a few generations in response to pollinator decline or loss (see Box: Rapid evolution of selfing after pollinator loss). But then why do not all plants self-pollinate? It has been suggested that, despite the short-term advantage of selfing through reproductive assurance, selfing may be disadvantageous in the long term for two reasons. First, self-fertilisation quickly leads to homozygosity at many loci. With increased homozygosity, recessive deleterious mutations can be expressed, leading to lower plant fitness known as inbreeding depression (see Box: The Cost of Selfing).

Second, selfing generally decreases genetic variation within populations. This means that selection has less variation to act on, which reduces the ability of selfing populations to adapt to environmental changes and increases their risk of extinction. This highlights one of the greatest benefits of sex between genetically different individuals (see Chap. 12): the shuffling of genes and creation of new genetic combinations as the raw material for evolution.

Fig. 14.3 Floral diversity and pollination systems in plants. Plants pollinated by animal vectors have flowers with attractive colours and shapes, while wind-pollinated plants have simple or plain flowers. (Photo courtesy of David L. Field, reproduced with permission)

For species with separate sexes, such as most animals and some plants, having males and females ensures that mating occurs between genetically different individuals. However, as discussed above, most plants are hermaphrodites, meaning they have both male and female reproductive organs and the potential for self-fertilisation. Therefore, many plants have evolved elaborate mechanisms to ensure that their gametes can only fuse with gametes of unrelated individuals (i.e. neither themselves nor another plant that is a close relative). These mechanisms are known as genetically based self-incompatibility systems, meaning that pollen from an individual is incompatible with its own stigma, and thus, the male gamete cannot fertilise the female gamete produced by the same plant (see Fig. 14.4). So, how do plants distinguish which pollen is their own and which is from a genetically different plant?

They do this by maintaining multiple **mating types**, i.e. distinct combinations of alleles in a specific region of the genome. As we will see below, many species have intricate molecular mechanisms to ensure **self-incompatibility**, i.e. successful fertilisation only if pollen is transferred between flowers with different mating types.

Rapid Evolution of Selfing After Pollinator Loss

The evolution of selfing from outcrossing is common across the plant kingdom. How this process occurs and the conditions that may trigger this evolutionary transition have intrigued evolutionary biologists since Charles Darwin. *Mimulus guttatus*, or the monkeyflower (Figure A below), is a bee-pollinated, mostly outcrossing wildflower that is common across North America.

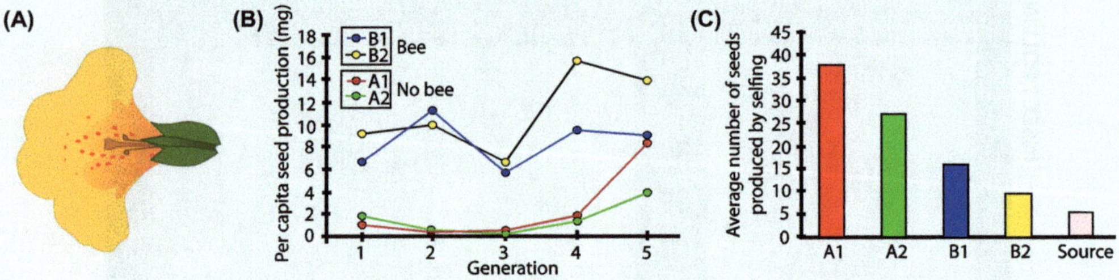

Rapid evolution in *M. guttatus*. (A) *M. guttatus* flower, (B) per capita seed production for each population and (C) average number of seeds produced by selfing. (Adapted from Figures 1-3 in Roels & Kelly 2011. Sarah A. Bodbyl Roels, John K. Kelly, Rapid evolution caused by pollinator loss in *Mimulus guttatus*, Evolution, Volume 65, Issue 9, 1 September 2011, Pages 2541-2552, https://doi.org/10.1111/j.1558-5646.2011.01326.x)

To test if selfing could rapidly evolve in this species, researchers evolved populations for five generations under two distinct treatments: (i) abundant bee pollinators and (ii) no pollinators. The populations (A1 and A2 in Figure B) without pollinators showed a large reduction in seed production in the early generations. Here, they produced, on average, less than 25% of the seed of the populations with pollinators (B1 and B2). However, from generation 3 onwards, the number of seeds produced by A1 and A2 started rising again. Figure C shows how the average number of seeds produced by selfing increased in populations A1 and A2 over five generations of selection. These rapid phenotypic changes included not only increased seed production in the absence of pollinators but also a shorter distance between the anthers and stigma, making self-pollen transfer easier. Researchers were able to establish that these changes were, in part, due to rapid allele frequency changes in regions of the genome that affect anther-stigma separation and pollen viability. This experiment provides a fascinating example of how traits associated with the "selfing syndrome" can evolve rapidly following the loss of mutualistic plant-pollinator interactions.

14.2 Self-Incompatibility and Mating Types

Recognition systems that prevent self-fertilisation have evolved to ensure that a plant mates only with a genetically different plant and not with itself (see Fig. 14.4). The recognition systems underlying self-incompatibility are widespread, with these systems found in at least 100 plant families and 40% of plant species. For plants, there are two main self-incompatibility systems: one where a plant can recognise and reject its own pollen (self-recognition) and one in which it recognises and accepts all other mating types but not its own (non-self-recognition). A common feature of both these self-incompatibility systems is that they are determined by a single part of the genome called the *self-incompatibility locus*

The Cost of Selfing

The harmful effects of inbreeding were first documented in detail and quantified by Charles Darwin, who carried out careful crossing experiments on 57 plant species. He compared the fitness of seeds produced by self-fertilisation vs. cross-fertilisation and concluded that "cross-fertilisation is generally beneficial, and self-fertilisation injurious" (The Effects of Cross and Self-fertilisation in the Vegetable Kingdom—Darwin 1876). These results from plants prompted him to be concerned about marriage between relatives, which was common in various human populations at the time! We now understand more about the genetics underlying this reduction in fitness following self-fertilisation or mating among related individuals (see figure below).

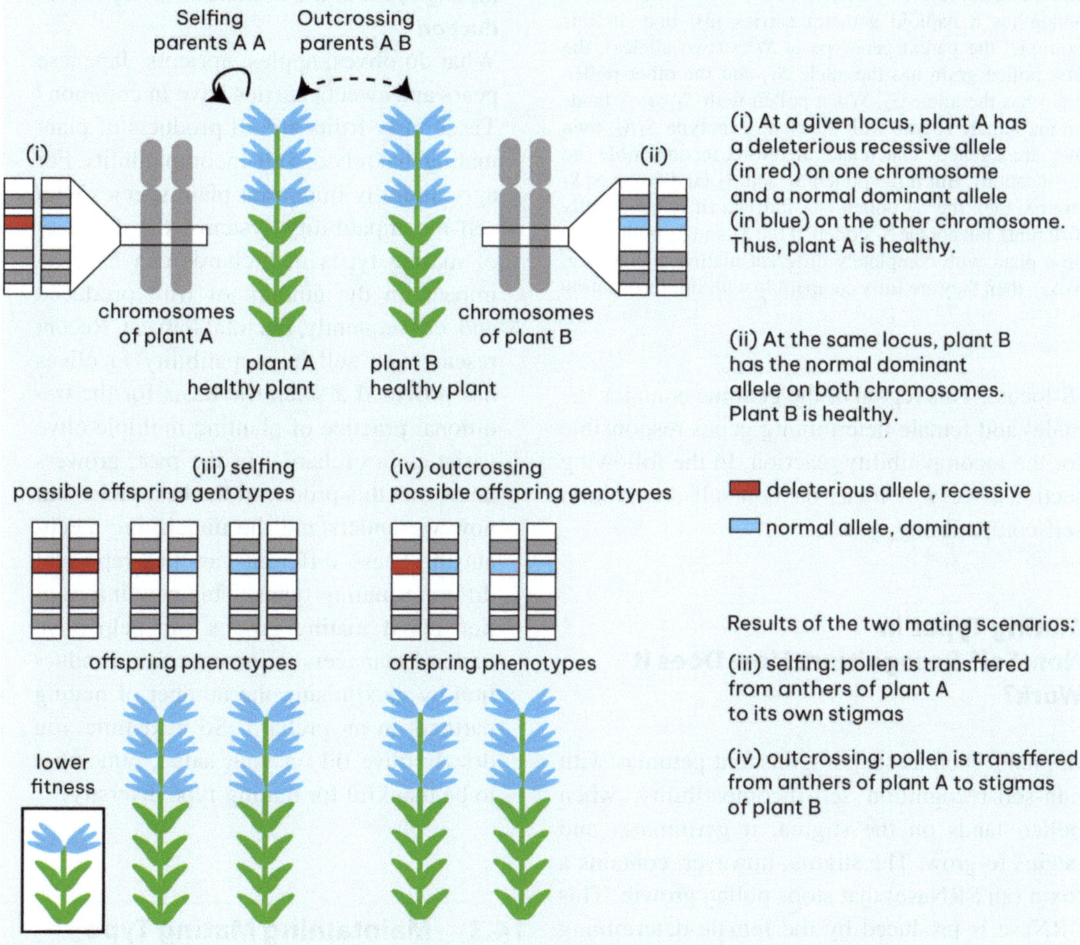

Fitness of selfed vs. outcrossed offspring. Self-fertilisation can lead to decreased fitness due to the expression of recessive deleterious mutations. While in the outcrossing case (parents A, B), all offspring are heterozygous and healthy (iv), and self-pollination (parents A, A) may lead to homozygous offspring that carry two copies of the deleterious allele and thus have reduced fitness (iii)

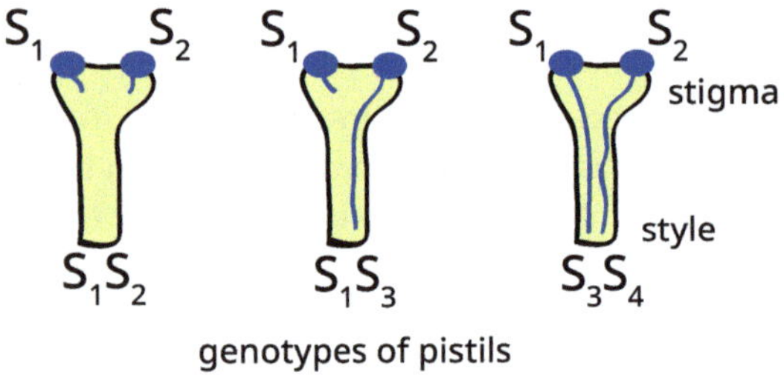

Fig. 14.4 Pollen growth in self-incompatible plants. The S-locus is very diverse and can have a large number of possible alleles S1, S2, S3, S4,... although any one diploid individual only carries two of these alleles, and pollen (as a haploid gamete) carries just one. In this example, the parent genotype is S_1S_2 (two alleles), the first pollen grain has the allele S_1, and the other pollen grain has the allele S_2. When pollen with S_1 or S_2 lands on the female stigma with the same genotype S_1S_2, then they are a genetic match and, therefore, incompatible (no fertilisation). But if the pollen S_1 and S_2 land on an S_1S_3 stigma, then the S_2 pollen can fertilise (it is genetically different), but not the S_1 pollen. If the S_1 and S_2 pollen land on a plant with completely different mating alleles (e.g. S_3S_4), then they are fully compatible with full fertilisation

(S-locus). This region of the genome contains the male- and female-determining genes responsible for the incompatibility reaction. In the following sections, we will focus on non-self-recognition self-compatibility systems.

Mating Types in Non-Self-Recognition: How Does It Work?

In plants such as snapdragons and petunias with non-self-recognition self-incompatibility, when pollen lands on the stigma, it germinates and begins to grow. The stigma, however, contains a toxin (an SRNase) that stops pollen growth. This SRNase is produced by the female-determining genes of the self-incompatibility locus (S-locus). The pollen, in turn, produces "antidotes" to neutralise the toxins, allowing pollen tubes to resume growing. These antidote proteins are produced by a set of male-determining (F-box) genes of the S-locus that are expressed only in the pollen. Crucially, in self-incompatibility systems, pollen contains antidotes to all toxins except those produced by the self-stigma. This interaction between products of the female- and male-determining genes means that pollen can fertilise when it lands on a stigma that does not belong to the same plant, but not when it lands on the plant's own stigma (see Box: Non-Self-Recognition Self-Incompatibility: A Closer Look). These plants utilise a toxin-antidote system to ensure that they only mate with genetically distinct plants.

Mating Types in the Orchard: The Key to Production

What do olives, apples, apricots, Japanese pears and sweet cherries have in common? These tasty fruits are all products of plant mating that rely on self-incompatibility. For agriculturally important plant species with self-incompatibility systems, the diversity of mating types in orchards can have an impact on the amount of fruit produced and, consequently, the total harvest. Recent research on self-incompatibility in olives has provided a scientific basis for the traditional practice of planting multiple olive varieties in orchards. In the past, growers knew that this produced the best yield. But now we understand the underlying mechanism: These different varieties represent different mating types. This new information about mating groups can help olive orchard managers to increase fruit production by maximising the number of mating partners in an orchard. So next time you drizzle olive oil on your salad, remember to be thankful for mating-type diversity!

14.3 Maintaining Mating Type Diversity

As we have seen in the previous section, in non-self-recognition systems, the male and female genes work together to determine recognition, so a particular combination of male and female genes forms a mating type. Plants that rely on these recognition systems usually have an astonishing diversity of mating types. This diver-

Non-Self-Recognition Self-Incompatibility: A Closer Look

Each pollen contains a collection of male-determining F-box genes in the self-incompatibility locus (shown as boxes of different colours). These genes determine whether the pollen can fertilise a plant or not. The principle is simple: Pollen must contain antidotes to all types of toxins (represented by circles of different colours) produced by the stigma of the receiving plant, in order to fertilise this plant. However, the pollen does not carry an antidote to its own toxin and thus cannot fertilise the plant it came from. The figure shows three different plants that can be fertilised by either one or none of the self-incompatible pollen types of the plant on the left.

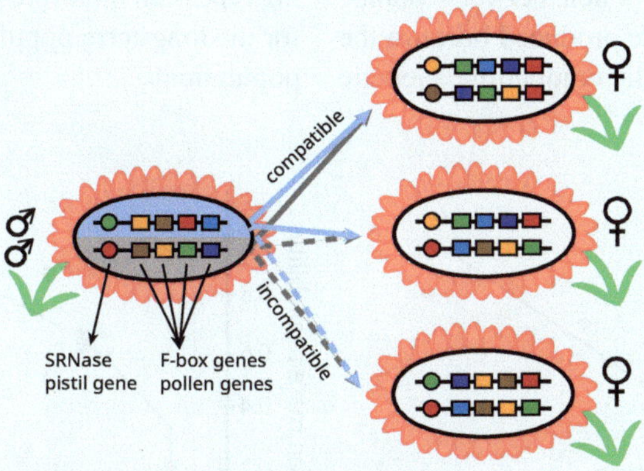

S-locus genes determine compatibility in the non-self-recognition system. Note that the male and female symbols do not indicate individuals who are male and female (since both individuals are hermaphrodites), but individuals who receive pollen (i.e. an individual in its female role) and an individual who donates pollen (i.e. an individual in its male role)

sity is generated by a unique form of selection known as **balancing selection**, which favours the maintenance of multiple alleles or genetic combinations at a particular locus in a population. In self-incompatibility systems of plants, balancing selection occurs because rare mating types, which are present at low frequencies in the population, have an advantage over common types that are present at high frequencies. This is referred to as **negative frequency-dependent selection**. Remember that a rare mating type has lots of potential mates, since most individuals in the population are genetically dissimilar to it at the S-locus. Thus, rare alleles increase in frequency in the next generation. By contrast, common types are selected against because they are unable to mate with many others in the population, since they can't mate with other plants that have the same common allele. Therefore, these alleles decrease in frequency in the next generation. Over many generations, this process tends to even out the frequencies of different alleles by causing common ones to decrease and rare ones to increase in frequency. In general, rare alleles have a high probability of being lost from a population simply due to the random variations in the number of offspring between individuals. However, in self-incompatibility systems, the mating advantage that comes with being rare partially protects low-frequency variants from being lost by chance.

Balancing selection is believed to play an important role in many other systems and may be instrumental in maintaining diversity at disease resistance genes, for example, at the major histocompatibility (MHC) locus in animals, which contains many immune response genes. When a

The Loss of Mating Types: A Pressing Issue in Plant Conservation

For small or declining populations of self-incompatible plants, the loss of genetic diversity in mating types can pose a significant problem for conservation. This is because, even with an adequate number of plants in the population, if there are too few mating types left, the remaining plants will have very few individuals with which they can successfully mate. This results in a situation where pollinators may transfer pollen between plants, but little to no seeds are produced because the plants are not genetically compatible. Genetic rescue, which involves introducing new mating types, can provide two benefits for these small populations with low genetic diversity. First, it can increase mate availability by increasing the number of mating types in the population. Second, offspring of parents with different mating types often have higher-than-average heterozygosity and are thus usually fitter than their parents. Genetic rescue of mating types can therefore provide a double bonus for the long-term population viability of small populations.

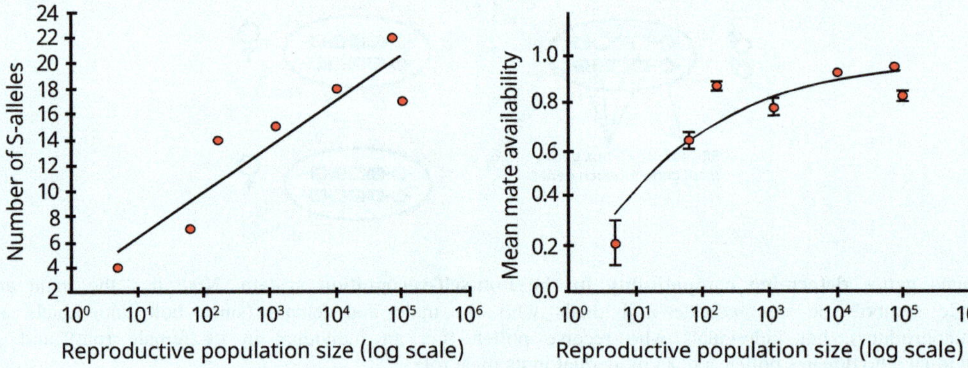

S-alleles and mate availability. Number of S-alleles and mean mate availability for seven populations of *Rutidosis leptorrhynchoides* as a function of population size. (Adapted from Figure 1 in Young & Pickup 2010. Young, A.G. and Pickup, M. (2010), Low S-allele numbers limit mate availability, reduce seed set and skew fitness in small populations of a self-incompatible plant. Journal of Applied Ecology, 47: 541–548. https://doi.org/10.1111/j.1365-2664.2010.01798.x)

new allele arises at the resistance gene, it can confer a strong selective advantage to its carrier simply because existing pathogens are unlikely to have evolved ways of evading it. As explained above, this process causes rare alleles to increase in frequency, leading to high genetic diversity in immune response genes.

How Do Different Mating Types Evolve?

One of the key questions in the evolution of mating type diversity for self-incompatibility is: How does a new mating type evolve, given that this requires mutations in both the female and male components? Non-self-recognition systems provide an interesting example to understand the conundrum. Imagine that there is a change in the female side (stigma) first. It then produces a new toxin for which no other pollen has an antidote, so mating can't occur. This means that a change in the male side (pollen) must occur first, causing the antidote to appear, and then, it waits for a corresponding change in the female side (stigma). But how could this change become prevalent in the population if the corresponding female toxin that it can neutralise is not yet there? Is there a particular order of mutations that is more likely to

The Cube

Vincenzo Natali's classic sci-fi horror film *Cube* (1997) features a group of strangers who mysteriously find themselves in a cube-shaped room. The room has a door on each of the six sides and ingenious death traps hidden in some of the rooms behind them. The survival of the main characters depends on how well they tackle the challenges while traversing through the labyrinth of cube-shaped rooms and searching for a way out.

Similar to this movie classic, the non-self-recognition mating system provides a labyrinth of pathways that may lead to a variety of outcomes, ranging from the arrival of a new self-incompatible mating type to the death of multiple existing mating types. The main evolutionary mechanisms at work are mutation, selection and random genetic drift. The evolution of new self-incompatible mating types requires multiple mutations in both male- and female-determining genes of the S-locus in a particular order.

The mutations in RNase and F-box genes can be visualised as transitions between the edges of a cube, where a successful arrival of a new mating type (with a new SRNase) corresponds to a completed path from the lower left front corner to the upper right back corner of the cube. While some of the intermediate states (red colour nodes) are self-incompatible, others are self-compatible (blue colour nodes). Arriving mutations may thus increase or decrease the fitness of an individual. In this aspect, the non-self-recognition system provides a similar set of traps to the rooms as in the sci-fi movie, since taking the wrong mutation path can lead to the elimination of the mating type.

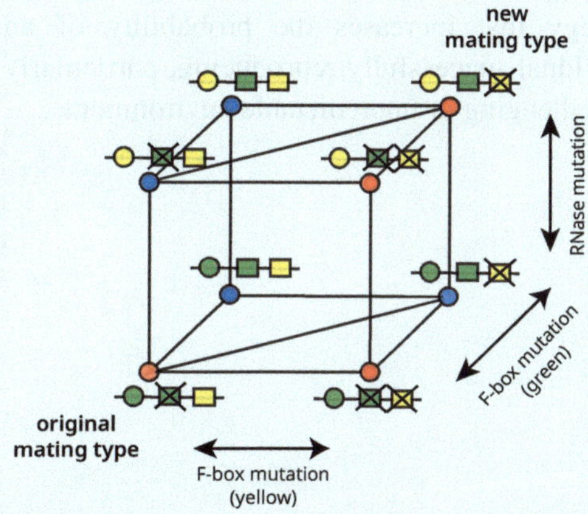

Evolution of a new mating type in the non-self-recognition system. Interestingly, some paths are more likely than others, and some are not feasible at all. However, quantifying this difference is a difficult problem that requires knowledge of both biology and advanced mathematics! (Adapted from Figure 2 in Bod'ová, K. et al. (2018) Evolutionary Pathways for the Generation of New Self-Incompatibility Haplotypes in a Nonself-Recognition System. Genetics, Volume 209, Issue 3, 1 July 2018, Pages 861–883, https://doi.org/10.1534/genetics.118.300748)

create a new mating type? Solving this complex evolutionary problem requires knowledge of both biology and advanced mathematics, underscoring the importance of an interdisciplinary approach to addressing fundamental questions in biology. With this type of collaboration, maybe we could also solve another outstanding question in these systems: How did the initial two mating types evolve in the first place?

DEFINITIONS

Cross-fertilisation/outcrossing: The fusion of gametes from two different parents, leading to offspring with genetic material from both.

Self-fertilisation/selfing: The fusion of male and female gametes produced by the same individual organism.

Self-incompatibility: Mechanism in flowering plants that prevents self-fertilisation and mating among genetically related plants.

Self-recognition: The ability of an organism to distinguish its own cells and molecules (self) from those of another organism (non-self).

Non-self-recognition: The ability of an organism to identify and respond to foreign cells, molecules or pathogens.

Balancing selection: A type of natural selection that maintains genetic variation within a population, often through mechanisms like heterozygote advantage or frequency-dependent selection.

Reproductive assurance: Any mechanism or strategy that increases the probability of an individual successfully reproducing, particularly in challenging or unpredictable environments.

SUGGESTED READING

- Roels SA, Kelly JK. Rapid evolution caused by pollinator loss in Mimulus guttatus. Evolution. 2011 Sep;65(9):2541–52.
- Young AG, Pickup M. Low S-allele numbers limit mate availability, reduce seed set and skew fitness in small populations of a self-incompatible plant. Journal of Applied Ecology. 2010;47:541–548.
- Takayama S, Isogai A. Self-incompatibility in plants. Annu Rev Plant Biol. 2005;56:467–89.
- Busch JW et al. Pollinator loss causes rapid adaptive evolution of selfing and dramatically reduces genome-wide genetic variability. Evolution. 2022 Sep;76(9):2130–2144.

14.4 Questions

1. **If an individual carries only male or only female reproductive organs, but not both, then the plant species is called dioecious. Examples include several *Cannabis* species, especially those that are used to obtain drugs (the drugs are obtained from the female flowers). Which of the following statements are true?**
   (A) Cannabis plants can reproduce in two different ways: by self-pollination or by outcrossing.
   (B) Since Cannabis is dioecious, it cannot be wind-pollinated.
   (C) To cultivate Cannabis over several generations, one needs male and female plants.
   (D) Dioecious plants can display sexual dimorphism, i.e. male and female plants have different physical characteristics on average.
   (E) Female Cannabis plants can only receive pollen from the same plant.
2. **Look at the two plants in Fig. 14.5. Circles designate SRNases (expressed in pistil), while boxes designate F-box genes (expressed in pollen). Which of the following scenarios is possible?**
   (A) Pollen from plant 1 lands on plant 2 and fertilises it.
   (B) Pollen from plant 2 lands on plant 1 and fertilises it.
   (C) Pollen from plant 1 lands on plant 2 but cannot fertilise it.
   (D) Pollen from plant 2 lands on plant 1 but cannot fertilise it.
   (E) Pollen from some other plant fertilises both plants above.

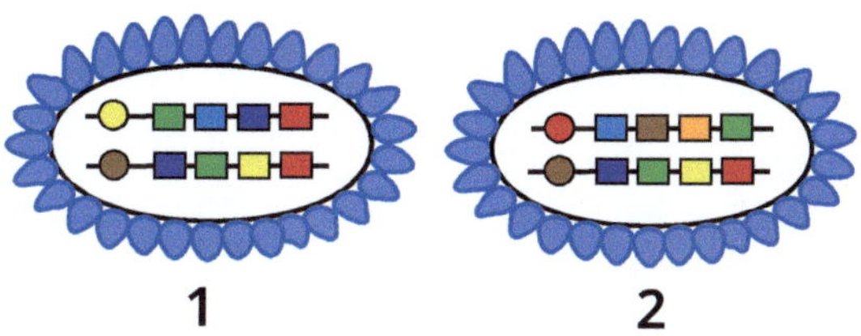

Fig. 14.5 Schematic of two plants with different genotypes

3. **A plant that can only reproduce by self-pollination is heterozygous at a given locus, i.e. one chromosome carries allele A and the other allele a. What is the percentage of type aa, Aa and AA descendants of this plant after three rounds of reproduction?**
   (A) AA: 0.4375; Aa: 0.125; aa: 0.4375
   (B) AA: 0.25; Aa: 05; aa: 0.25
   (C) AA: 1; Aa: 0; aa: 1
   (D) AA: 0.333333; Aa: 0.333333; aa: 0.333333
   (E) AA: 0.375; Aa: 0.25; aa: 0.375
4. **Which of the following statements are true? A bee-pollinated plant species with a genetic self-incompatibility system**
   (A) Can survive in an environment in which bees are less abundant by evolving a shorter distance between the anther and the stigma.
   (B) Is likely to go extinct in a patchy habitat where individual plants are very far from each other.
   (C) Can only be successfully fertilised by pollen from genetically dissimilar plants from another species.
   (D) Is likely to have larger, more ornate flowers than a sister species which reproduces via self-fertilisation.
   (E) Will typically be homozygous at more loci than a sister species which reproduces via self-fertilisation.
5. **Three seeds of *Antirrhinum*, which is a genetically self-incompatible plant, end up on a distant island, where there are no other *Antirrhinum* plants. Based on the genotypes of the seeds at the S-locus, in which case are the seeds *most* likely to establish a new *Antirrhinum* population on the island?**
   (A) Seed 1: S_1S_2; seed 2: S_1S_2; seed 3: S_1S_2
   (B) Seed 1: S_1S_2; seed 2: S_1S_2; seed 3: S_1S_3
   (C) Seed 1: S_1S_2; seed 2: S_1S_2; seed 3: S_3S_4
   (D) Seed 1: S_1S_2; seed 2: S_3S_4; seed 3: S_5S_6
   (E) Seed 1: S_1S_2; seed 2: S_1S_3; seed 3: S_4S_5
6. **Is the establishment of a new *Antirrhinum* population impossible in any of the above cases?**
   (A) A
   (B) B

(C) C
(D) D
(E) E

7. **Consider a new mutation that has recently arisen in a snail population. In which of the following scenarios is the mutation likely to be under negative frequency-dependent selection, i.e. in which scenarios does the mutation have an evolutionary advantage which decreases as the mutant becomes more common in the population?**
   (A) Mutant snails have a shell pattern which is very distinct from that of wild-type snails. Wild type snails are the preferred food of a crab species common in that area.
   (B) Mutant snails have a shell pattern which is very distinct from that of wild-type snails, which are avoided by crabs (since the wild type is poisonous).
   (C) The mutant is very small and can slip into the many rock crevices that are inaccessible for the larger wild type and for crabs (who feed on the wild type).
   (D) The mutant is better able to resist infection from an extremely fast-evolving bacterial strain that causes heavy mortality in the wild type.
   (E) Male mutant snails are more attractive to females, who preferentially mate with them instead of wild-type males.

14.5 Project: How Many Mating Types Can Evolve?

In this activity, you will model the population of fictional plants with multiple mating types and explore how natural selection in combination with random drift affects their survival and abundance. You will also introduce novel mating types, which will emulate the mutation process.

It may be fair to say that the analogue of the popular singer Lady Gaga, within the music industry, is a plant named *Oenothera* within plant genetics. Its popularity exploded when *Oenothera lamarckiana* was used by Hugo de Vries to rediscover Mendelian laws and continued when a tiny population of *Oenothera organensis* (with less than 1000 individuals in New Mexico) and its genetically coded self-incompatibility became a source of inspiration for one of the greatest geneticists, Sewall Wright. Eighty years ago, Wright formulated a simple explanation of the genetics of self-incompatibility alleles that we will look at in this project in more detail.

In Wright's model, the individuals in a population of flowering plants have multiple mating types, where a new type may be introduced by mutation in the S-locus (self-sterility locus). The mating rules ensure that plants mate with a genetically distinct individual. The way it works is that the male pollen, which carries a single copy of the genome (haploid), is able to fertilize a plant when both S-alleles (the pistil on which pollen lands is diploid) of the pollen-receiving plant differ from the pollen allele. An immediate consequence of this rule is that all plants in a population which obey these rules are self-incompatible.

In this experiment, you will compare the fate of two populations which start with different numbers of S-alleles (Fig. 14.6).

Materials

- Print-out of initial populations (Fig. 14.7)
- Several copies of print-outs of template individuals (Fig. 14.8)
- Colouring templates for individuals of both populations (Fig. 14.9)
- Scissors and a bowl
- Yellow, green, red, blue and black pencils
- Printed table and figure templates

Hypothesis

Read the instructions for running the simulation below. Formulate a clear and testable hypothesis considering the following:

1. How will the number of different S-alleles in a population affect its survival?
2. How will the number and proportion of alleles change over the generations?
3. What will be the fate of a mutation (a new S-allele) suddenly appearing in a population?

Instructions for Running the Simulation

Preparation

1. In both populations, your initial sample at generation 1 will consist of ten plants. (For better results, you may work with more individuals (e.g. 15 or 20), but that is more time-consuming. If you have a friend to help you, it is worth trying.) Each plant will have a mating type that is a combination of two alleles.
2. Cut out rectangles representing individuals of the first generation (print-outs further below). **Population A** contains four different alleles (blue, green, yellow and red) in equal proportions. **Population B** contains three different alleles (blue, green and red) in unequal proportions (see Fig. 14.6).
3. Keep the two populations separated. You will run the simulation first for population A and then for population B.
4. Cut out white rectangles representing individuals of future generations (print-outs below) and set them aside. You will need them to generate individuals for subsequent generations.

Modelling a Single Generation

1. Fold the coloured rectangles with the colored sides facing out. Put all the rectangles into a bowl and shuffle thoroughly. These are the plants of generation 1.

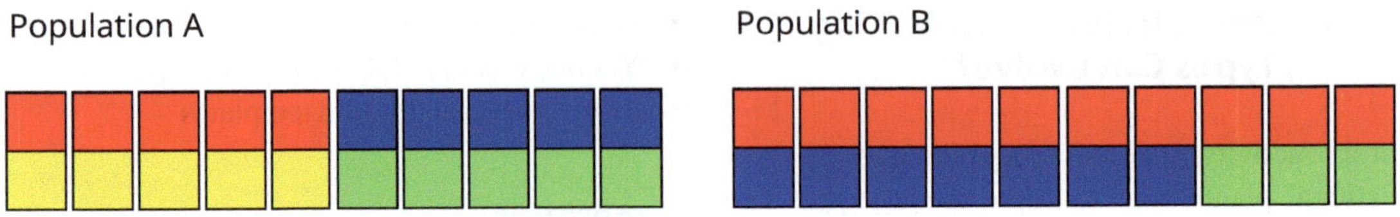

Fig. 14.6 Genotypes of individuals in initial populations A and B

2. The plants are maturing and are just about to reproduce. Each plant forms a diploid pistil represented by a two-colour rectangle and haploid pollen, represented by each one-colour square (one side of the folded rectangle). Rectangles in the bowl represent all pistils and all pollen in the population.
3. Draw a random plant from the bowl and place it on a table, still folded. The upper colour that you can see represents haploid pollen—a single allele.
4. Draw again a random plant from the bowl and place it on a table, unfolding it. It represents the diploid pistil on which the pollen grain lands.
5. If the colour of the pollen allele is identical to at least one of the pistil colours, put the plant back into the bowl. **Note this event as an unsuccessful fertilisation attempt.**
6. If the pollen colour is different from both pistil alleles, fold the **pistil** rectangle again and put it on a table with closed eyes. The colour on top represents a haploid egg that will be fertilised. Note the genotype of this new individual by colouring the first rectangle in the second generation in the colouring template.
7. Return the two mated plants back into the bowl and shuffle them.
8. Repeat the process to generate a new generation of plants. Each time a new individual is created, note its genotype in the colouring template. Note down also the number of unsuccessful fertilisation attempts.
9. Modelling of the next generation is finished, when you successfully produce **ten new individuals OR** when you have had **15 unsuccessful fertilisation attempts**. After 15 unsuccessful fertilisation attempts, the mating season for this generation is over, and the next generation will consist only of the individuals created until then.

Modelling Subsequent Generations

1. Put aside the rectangles representing the old generation.
2. Create the new generation by colouring new white rectangles according to the genotypes listed in the colouring template. You can recycle some rectangles from the previous generation if you want to save yourself some colouring.
3. Fold the newly generated rectangles as before, put them into the bowl, and shuffle them.
4. Repeat the steps describing the formation of a single generation, modelling five generations in total.
5. Before you start mating the individuals of the fifth generation, pick one individual at random, place it on a table, and mutate the allele that ended up on top-change it to black (repaint it or make a new individual).
6. Place this mutated individual back into the population. Note this change also in the Fig. 14.9 (in the fifth generation).
7. Continue running the simulation as before, until you have in total of ten generations.
8. If you encounter a situation when all mating attempts are unsuccessful, report it and explain why it happened. Return back two generations and continue from there. Mark this event in the table and mention it in the Analysis.
9. If the new allele (black) has vanished from the system, introduce the same mutation again in a subsequent generation and mark this in the table.

Recording and Visualising Results

1. After running the simulation for both populations A and B, use the data recorded in the colouring template to complete Tables 14.1 and 14.2.

Fig. 14.7 Cut-outs: Mating types for population A in the two top rows and B in the two bottom rows

Fig. 14.8 Cut-outs: Mating type templates for the next generations

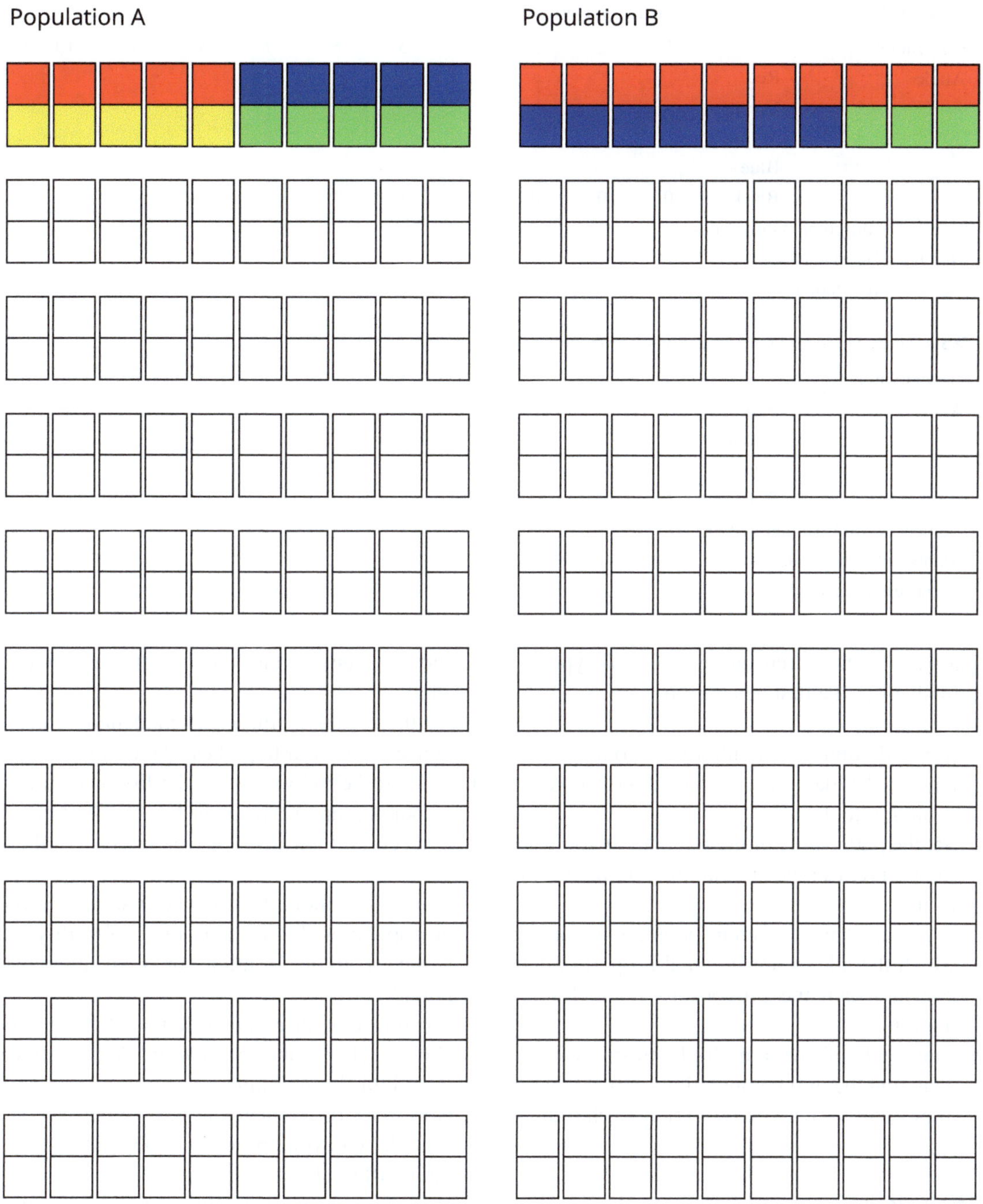

Fig. 14.9 **Colouring template for populations A and B**

2. For each population, create a graph showing the number of alleles and the number of individuals through the course of the ten modelled generations. You may use the provided graph template in Fig. 14.10 or create your own graphs.

Running a Computer Simulation for Many Individuals

We have prepared a computer simulation for you, which runs exactly the same procedure as you did. You can run it for many individuals at once, and

Table 14.1 Modelling ten generations in population A

Generation		1	2	3	4	5	6	7	8	9	10
Allele	Red										
	Green										
	Yellow										
	Blue										
	Black	0	0	0	0	1					
Number of different mating types											
Number of individuals		10									
Unsuccessful matings											

Table 14.2 Modelling ten generations in population B

Generation		1	2	3	4	5	6	7	8	9	10
Allele	Red										
	Green										
	Blue										
	Black	0	0	0	0	1					
Number of different mating types											
Number of individuals		10									
Unsuccessful matings											

you can set the initial number of genotypes of individuals in the first generation.

1. You will find the simulation on our website among the data: http://en.edub.sk/everyday-evolution/tools.
2. Set the parameters of the population you want to observe, run the simulation, and record the results.
3. Try at least two different populations, one in which the alleles are in equal proportion and one in which the proportions of alleles are uneven.
4. Run each of them multiple times, without changing the initial populations. Note down in a few words what you observe in each trial.

Analysis

1. Summarise in a few sentences what you observed in both paper-modelled populations A and B **before** the introduction of the mutated allele.
2. Summarise in a few sentences what you observed in both paper-modelled populations A and B **after** the introduction of the mutated allele.
3. Do the results of the paper-modelled simulation support your hypothesis? Explain.
4. Look at the individuals of the first generation in population A. What is the probability of a successful fertilisation for a given pollen grain?
5. Look at the individuals of the first generation in population B. What is the probability of a successful fertilisation for a given pollen grain?
6. Imagine that there are just two different S-alleles left in the population. What effect would this have? Explain.
7. How many different genotypes (with respect to S-allele) can you observe in a population with three different S-alleles?
8. How many fully compatible partners does each genotype in such a population have?
9. Explain the effect of the number of different genotypes on the fertilisation success rate.
10. Did you observe any homozygotes in any of the modelled populations? Explain.
11. Summarise in a few sentences what you observed in the computer-modelled popu-

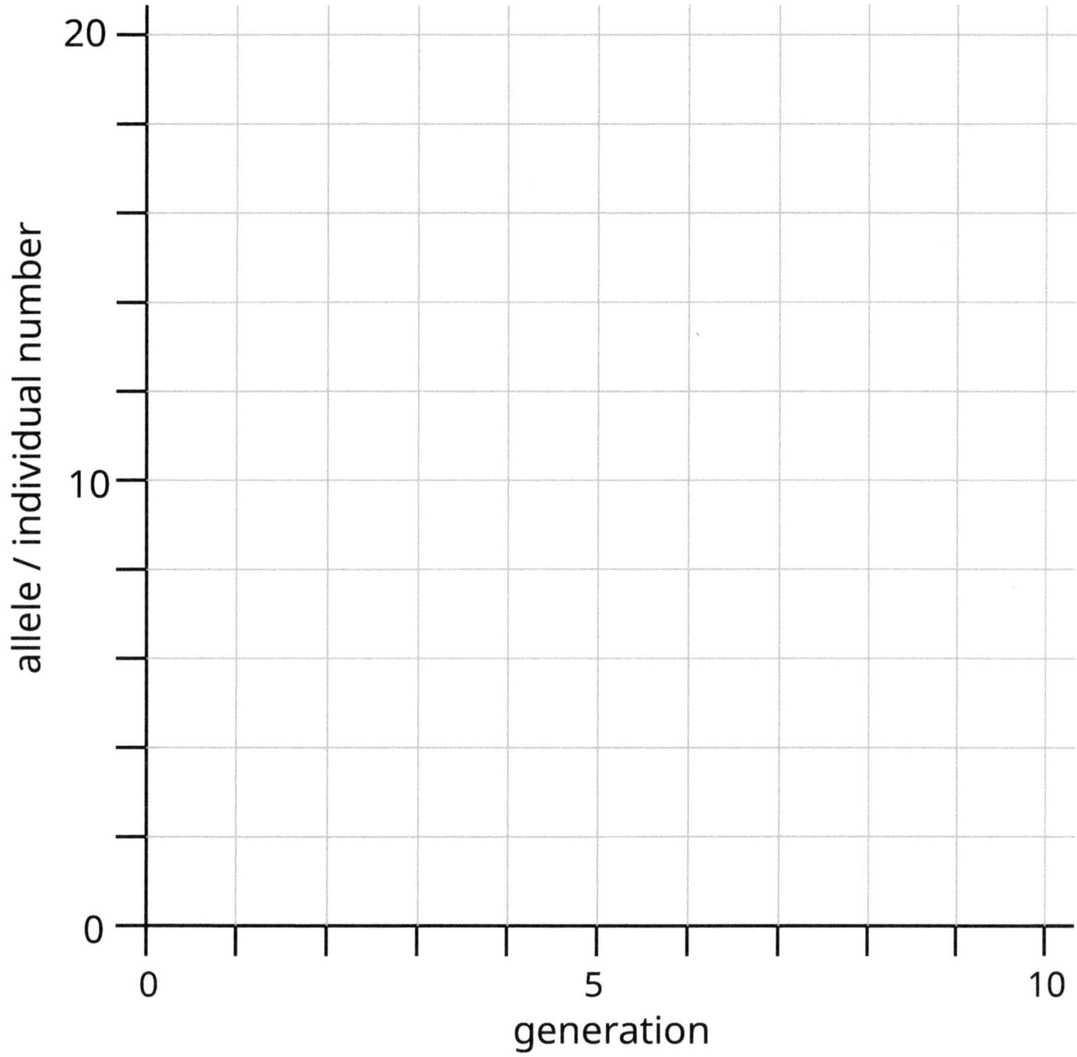

Fig. 14.10 **Allele and individual number in time**

lations when running multiple simulations for the same initial population. Were there any differences between the two starting populations?

This project was developed by Kristína Hudáková, Katarína Boďová and Barbora Trubenová.

Origin of Eukaryotes

15

Kristína Hudáková

► All living organisms are composed of one or more cells, which can be classified as either prokaryotic or eukaryotic. For more than half of the history of life on Earth (about two billion years), the Earth was inhabited only by simple, single-celled organisms similar to today's bacteria, known as prokaryotes. Although there are still millions of prokaryotes in any given environment, including on and within the human body, the organisms we are most familiar with — animals, plants, and fungi — are composed of eukaryotic cells, the other cell type.

The origin of the eukaryotic cell is considered one of the most critical events in the history of life on Earth (see Box: Major evolutionary transitions). It gave rise to organisms made of cells substantially different from those before and eventually led to the evolution of sexual reproduction, multicellularity, specialised cells, and complex organisms such as ourselves. But how did these complex cells evolve?

K. Hudáková (✉)
eduB, Bratislava, Slovakia
e-mail: kristina.hudakova@edub.sk

It is now generally accepted that eukaryotic cells evolved from a symbiotic[1] relationship between ancient Bacteria and Archaea, which are the two groups of prokaryotes and two of the three currently recognised domains of life. The third domain—Eukarya—evolved from the merging of these two domains, and encompasses all organisms composed of eukaryotic cells. The oldest microfossil evidence dates the origin of eukaryotes to about 1.5 billion years ago (Figure 15.1).

Both bacteria and archaea are unicellular prokaryotes that are abundant in every habitat. They have a similar cell structure but differ in the chemical composition of their cell walls and membranes. Bacterial cell walls are made of peptidoglycan; archaeal cell walls may be made of a similar compound, pseudopeptidoglycan, but also of other polysaccharides and glycoproteins. Archaea often inhabit extreme environments, such as hot springs or extremely acidic, alkaline, or saline waters. Additionally, archaea are important symbionts that thrive in anaerobic conditions, including the digestive tracts of cows, termites, and even humans. Bacteria are known to cause numerous diseases; in contrast,

[1] To be consistent with the terminology in published scientific articles on this topic, we use the term symbiosis for a mutually beneficial interaction in this chapter.

B. Trubenová et al. (eds.), *Everyday Evolution: Unveiling Life's Hidden Patterns*,
https://doi.org/10.1007/978-3-032-07050-0_15

The Major Evolutionary Transitions

The increase in complexity of Earth's life forms can be explained as the result of several changes in the way information is stored and transmitted. These changes are known as major evolutionary transitions, each representing a new level of complexity. The most important transitions are:

- **From individual genetic elements to organised genomes**: Origin of complex genomes and the organisation of genetic material into chromosomes.
- **From RNA-based life to life dependent on DNA, RNA, and proteins**: Origin of the genetic code, ribosomes, and translation.
- **From prokaryotes to eukaryotes**: Origin of membrane-bound organelles within cells. The genome is located in the cell's nucleus. Transcription and translation occur in different cellular compartments.
- **From asexual clones to sexual populations**: Origin of sexual reproduction, life cycles with alternating haploid and diploid phases, formation of gametes and zygotes.
- **From unicellular to multicellular organisms**: Origin of tissues and cell differentiation.
- **From solitary organisms to social groups**: Origin of cooperative behaviours, animal societies (such as ant and termite colonies), and the evolution of complex signalling and social learning.
- **From animal to human societies**: Origin of language and socio-cultural evolution.

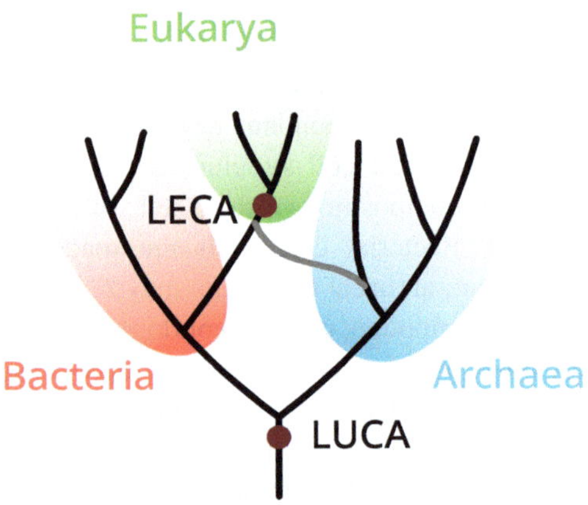

Fig. 15.1 Evolutionary tree. This simple evolutionary tree shows the relationships of the three domains, the last universal common ancestor (LUCA) and the last eukaryotic common ancestor (LECA). Drawing based on Moody, E.R.R., Álvarez-Carretero, S., Mahendrarajah, T.A. et al. The nature of the last universal common ancestor and its impact on the early Earth system. Nat Ecol Evol 8, 1654–1666 (2024). https://doi.org/10.1038/s41559-024-02461-1

archaea have not yet been clearly identified as pathogens.

15.1 The endosymbiotic theory

The first comprehensive theory explaining the origin of eukaryotic cells was proposed in 1976 by Lynn Margulis. She described the evolution of photosynthesis in prokaryotes living in the anaerobic conditions of the early atmosphere, as well as the subsequent evolution of cellular respiration in bacteria put under selection pressure by the increasing amount of oxygen in the atmosphere.

Margulis suggested that the eukaryotic cell evolved in a series of steps. The first step occurred when an anaerobic prokaryote engulfed another prokaryotic cell capable of using atmospheric oxygen. Instead of being digested, the engulfed cell survived, forming a symbiotic relationship with the host that eventually led to a single organism. The engulfed cell was the ancestor of the mitochondrion.

A key feature of all eukaryotic cells is the presence of a nucleus that divides by mitosis. Margulis suggested that the nucleus and mitosis also originated from an endosymbiotic event, which may or may not have occurred simultaneously

Prokaryotic and Eukaryotic Cells

All cells share some basic features: they are bounded by a cell membrane, filled with semi-fluid cytoplasm, have genetic material in the form of DNA, and have ribosomes that synthesise proteins according to the information provided by the DNA.

Several features unique to eukaryotes allow us to distinguish them from prokaryotes: The main genetic material of eukaryotes is surrounded by a nuclear envelope — this feature alone is so important that it is considered both *necessary* and *sufficient* to define an organism as a eukaryote. In addition, eukaryotes have an extensive endomembrane system stretching from the endoplasmic reticulum (which is in direct contact with the nuclear envelope) to the Golgi body, to vesicles capable of fusing with the cell membrane. Then, all eukaryotic cells have mitochondria, and many have chloroplasts. Finally, the cytoskeleton of eukaryotic cells enables movement of the cell itself and of cellular components (e.g., organelles, vesicles, and chromosomes) within the cell.

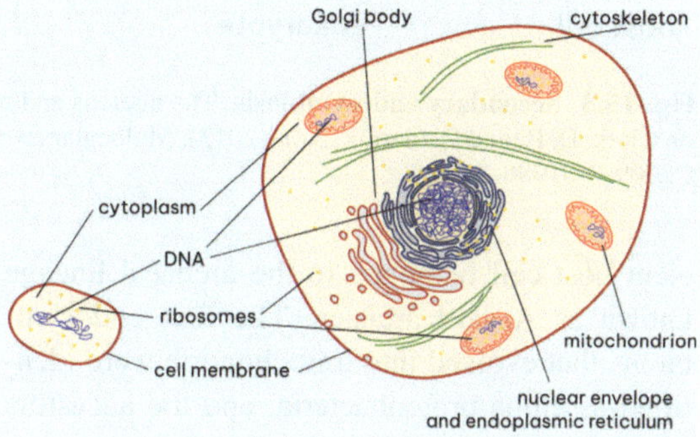

Prokaryotic and eukaryotic cell

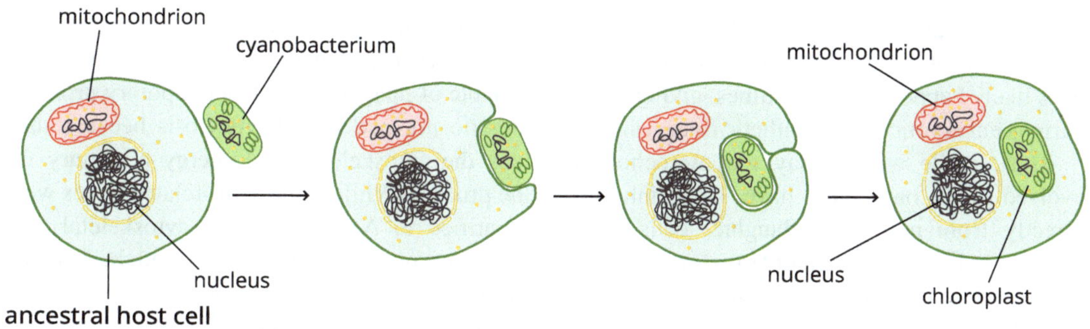

Fig. 15.2 Primary endosymbiosis. Primary endosymbiosis resulted in the acquisition of chloroplasts. Mitochondrion and nucleus are already present in the host cell. Figure based on Clark, D. P., and Pazdernik, N. J. (2013). Molecular Evolution. Molecular Biology, 812–853. https://doi.org/10.1016/C2009-0-01986-2

with the acquisition of the mitochondrial ancestor. The cytoskeleton necessary for the proper distribution of the genetic material could have come from a motile (or "swimming") partner. Later, some of these cells acquired photosynthetic endosymbionts, which became the ancestors of chloroplasts (Figure 15.2).

All the above-mentioned events are cases of **primary endosymbiosis** because the internalised cells were prokaryotes. These ideas are consistent with the observation that all eukaryotic cells contain mitochondria (though these are substantially reduced in some lineages) and reproduce by mitosis, but not all of them contain chloroplasts.

Although it took considerable time for the scientific community to accept these ideas, the symbiotic origin of eukaryotic cells is now a widely accepted theory, particularly with regard to the origin of mitochondria and chloroplasts. Molecular genetic evidence suggests that the an-

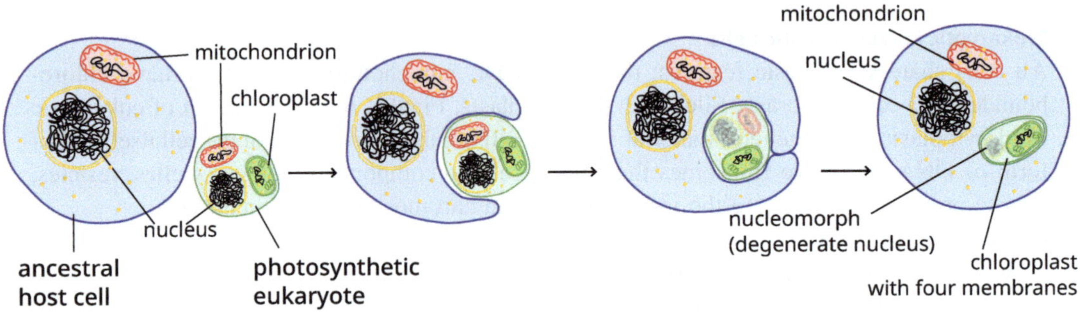

Fig. 15.3 **Secondary endosymbiosis.** The nucleus and mitochondrion of the endosymbiont degenerate. Figure based on Clark, D. P., and Pazdernik, N. J. (2013). Molecular Evolution. Molecular Biology, 812–853. https://doi.org/10.1016/C2009-0-01986-2

cient host cell belonged to the archaeal lineage known as Asgard archaea. The first endosymbionts that evolved into mitochondria were identified as alpha-proteobacteria, and the ancestors of chloroplasts were cyanobacteria.

Some chloroplasts are surrounded by additional membranes, which is considered to be evidence of **secondary endosymbiosis** — the process by which a photosynthetic eukaryote (which already contains a chloroplast) becomes an endosymbiont of another eukaryote (Figure 15.3). This likely happened several times, giving rise to various eukaryotic photosynthetic lineages.

Thus, while we can say that mitochondria evolved only once and have been inherited directly from mother to daughter cells in all eukaryotic lineages (**vertical transfer**), the story for chloroplasts is different. While it is accepted that chloroplasts also evolved just once, they could be transferred either by **vertical transfer**, passed on to daughter cells, or acquired through **horizontal transfer** by another symbiotic event.

The key evidence supporting the endosymbiotic theory was provided by the discovery of DNA molecules within mitochondria and chloroplasts. Both organelles transcribe their own genes and translate them on their own ribosomes to produce their own proteins. However, the genomes of these organelles have been greatly reduced. The ribosomes found in mitochondria and chloroplasts are more similar to bacterial ribosomes than to those of eukaryotic cells. Additionally, the fact that both mitochondria and chloroplasts can move around in the cell, grow, and reproduce independently suggests that their ancestors may have existed as independent organisms.

Inside out or outside in?

Although the chimeric nature of the eukaryotic ancestor is no longer in question, various aspects of the merger remain unclear. An important question is whether the ancient archaean host was capable of phagocytosis. Contemporary prokaryotes do not perform phagocytosis because they lack the cytoskeleton and energy necessary for the process. While some actin-related genes were identified in Asgard archaea, cytoskeletal proteins have not yet been observed, and the function of these genes remains unproven.

In 2014, Baum and Baum proposed an alternative mechanism to phagocytosis for both the acquisition of mitochondria and the origin of the eukaryotic nucleus and the endomembrane system. In their paper, the authors introduced their model as "inside-out", in contrast to the traditional models of endosymbiosis, which they referred to as "outside-in". Traditionally, it was assumed that the endosymbiont entered the ancient host cell via phagocytosis, and that the eukaryotic nucleus subsequently formed within this phagocytosing cell, also (or at least partially) as a result of phagocytosis. Baum and Baum's suggestions are based on the ability of extant archaea to produce extracellular protrusions (extensions of their cells outside of the cell walls) and the general

Chloroplasts and Mitochondria

Both chloroplasts and mitochondria produce ATP — the form of energy that cells use for all their activities. In chloroplasts, during photosynthesis, light energy powers the electron transport chain located within the thylakoid membrane, producing ATP. This ATP is then used to create energy-rich organic molecules in the Calvin cycle, which takes place in the stroma. Oxygen is released, and carbon dioxide is consumed in the process. In mitochondria, during cellular respiration, energy-rich organic molecules are broken down in the Krebs cycle in the mitochondrial matrix. This powers the electron transport chain located within the internal membrane, which produces ATP, which directly powers cellular work. Oxygen is consumed, and carbon dioxide is released in this process.

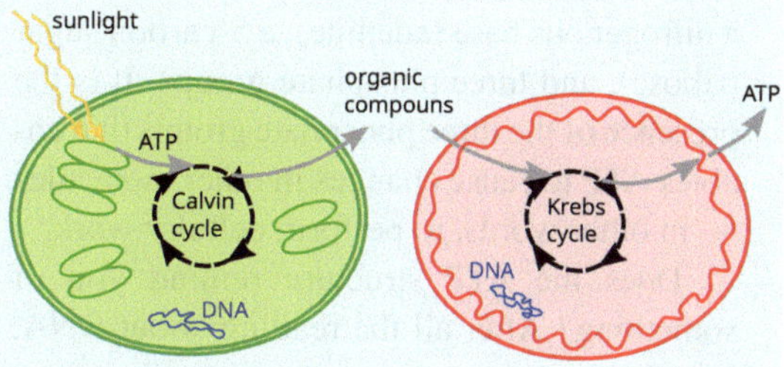

Chloroplast and mitochondrion.

lack of evidence for phagocytosis in prokaryotic cells.

In particular, the authors suggested that the complex organisation of the eukaryotic cell emerged when a prokaryotic cell extended extracellular protrusions. These protrusions immobilised the symbionts outside the original cell boundary, and eventually fused back around the captured symbionts and outside of the original cell membrane. This process led to the formation of a membrane-bound network, giving rise to the endoplasmic reticulum and other structures of the eukaryotic endomembrane system.

The eukaryotic nucleus was formed within the original prokaryotic cell, enclosed by the original cell membrane, which eventually became the nuclear envelope. The "inside-out" mechanism allows for *gradual* internalisation of the symbiont, with functional intermediate steps that may have provided advantages for both organisms.

In 2020, a research team led by H. Imachi studying anaerobic prokaryotes on the ocean floor isolated an archaeon *Promethearchaeum syntrophicum*, which was found to be an obligate symbiont. Its cells were very similar to the intermediate steps of endosymbiont acquisition proposed by the "inside-out" model (Figure 15.4). The cells lacked internal membrane-bound compartments but had a distinct cell body from which long extracellular protrusions and smaller blebs extended. The scientists studying the organism suggested that these extensions allowed for the exchange of material with extracellular symbionts.

Based on their findings, Imachi and colleagues suggest that the eukaryotic ancestor was an anaerobic Asgard archaeon that relied on an aerobic partner (alpha proteobacterium) to survive in aerobic conditions. The process of **eukaryogenesis** might have been slow, with a closer association between the partners evolving over many generations, rather than a fast event such as the engulfment of one cell by another through phagocytosis.

Based on observations of the extant members of the Asgard archaea, the eukaryotic ancestor was probably small and lacked the energy and cell machinery (cytoskeleton) necessary to engulf another cell in such a momentous event as phagocytosis.

Though the inside-out mechanism seems very plausible, subsequent endosymbiotic events likely involved phagocytosis. The ancestors of chloroplasts were probably phagocytised, particularly during secondary endosymbiotic events. The energy necessary for cytoskeletal movement was likely provided by the already-present mitochondria.

ATP

Adenosine triphosphate (ATP) is often referred to as the "energy currency" of the cell. Cells use this molecule to perform work — for example, to move using their flagella, replicate their DNA, or to transport molecules across the cell membrane. The ATP molecule consists of a nitrogenous base (adenine), a 5-carbon sugar (ribose), and three phosphate groups. It is the presence of the three phosphate groups that enables ATP to cause changes in other molecules — in other words, to perform cellular work.

Does the ATP structure remind you of something? After all the reading about DNA, it probably does.

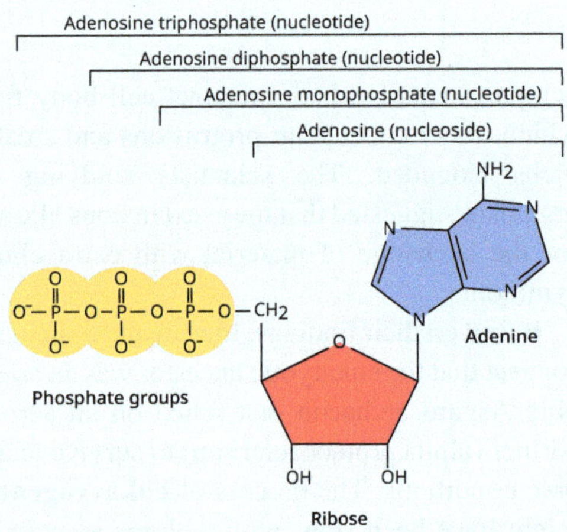

ATP molecule.

A **nucleoside** consists of a nitrogenous base (adenine (A), cytosine (C), guanine (G), thymine (T), and uracil (U), and a five-carbon sugar (ribose or deoxyribose). These five bases are called the primary nitrogenous bases (nucleobases). A **nucleotide** consists of a nucleoside and one to three phosphate groups. Depending on the number of phosphate groups attached, the nucleotide may be called a nucleoside monophosphate, a nucleoside diphosphate, or a nucleoside triphosphate. The nucleotides present in an RNA or DNA molecule are nucleoside monophosphates. The ATP molecule is a nucleoside triphosphate.

When a DNA or RNA strand is being formed, nucleoside triphosphates align with the template strand according to the base-pairing rule. As the phosphodiester bond forms between the neighbouring nucleotides, the two phosphate groups are removed. The nucleotides to be incorporated into the new strand bring with them the energy needed to form the bond in the form of triphosphates! This is one of the reasons a new nucleotide is always added by its 5' end to an exposed 3' end of a previous nucleotide — the triphosphate is attached to the 5' carbon.

Now, it becomes obvious that there is not only ATP in the cell but also GTP, CTP, TTP, and UTP. While less known, the guanosine triphosphate (GTP) is also involved in cellular processes as a source of energy, like ATP. The other nucleotides are used less as energy sources but are involved in various pathways, for example, as cofactors in enzymatic reactions.

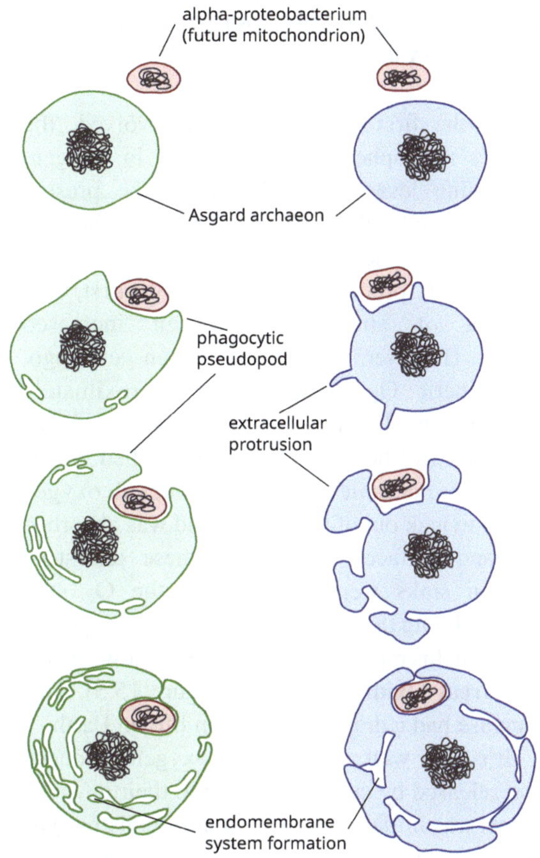

Fig. 15.4 Models of endosymbiosis. Comparison of the "outside-in" (left) with the "inside-out" (right) models of endosymbiosis. The endomembrane system, including the nuclear envelope, will be formed by layers of membrane around the nuclear DNA. The drawing is based on the illustration in Baum DA, Baum B. An inside-out origin for the eukaryotic cell. BMC Biol. 2014 Oct 28;12:76. https://doi.org/10.1186/s12915-014-0076-2

Evolution of endosymbiosis — observed

One classic example of the observed evolution of endosymbiosis was the result of a stroke of luck. In 1966, some unknown bacteria[2] infected a culture of the unicellular eukaryote *Amoeba proteus* cultivated by Kwang W. Jeon. This was unusual: amoebae do not typically feed on bacteria, but can digest them if bacteria happen to invade. These bacteria clearly survived, multiplied within the amoebae and killed *almost* the entire culture — at first. However, a few of the infected amoebae survived. Scientists discovered that these survivors still harboured bacterial cells in surprisingly high numbers (approximately 40,000 bacteria per amoeba), but the bacteria were no longer killing their hosts. The surviving amoebae were able to reproduce, establishing new colonies consisting entirely of amoebae-containing bacteria. Scientists maintained and studied these colonies and discovered some interesting facts: If the bacteria within amoebae were killed, either by antibiotics or by increased culture temperatures, the host cells soon died as well — the amoebae had become dependent on their endosymbionts for survival! Moreover, when these bacteria were isolated from the amoebae and introduced to bacteria-free amoebae, they quickly established symbiotic relationships with their new host cells. These amoebae then developed dependence on the bacteria in just ~ 200 generations (18 months). The symbiosis changed the bacteria as well — they were no longer as virulent.

[2]They were called X-bacteria then; now they are known as *Candidatus* Legionella jeonii.

15.2 Oxygen, the nucleus, and reasons for sex

When the first living organisms evolved, the Earth's atmosphere was very poor in oxygen, containing less than 0.001 % of the present atmospheric level (PAL). This changed when the first photosynthetic organisms — cyanobacteria — started producing oxygen. At first, the amount of atmospheric oxygen increased slowly. However, around 2.45 billion years ago, atmospheric O_2 suddenly rose approximately 10,000 times. Initially, the oxygen produced was absorbed by the oceans and the seabed rocks. Once these oceanic sinks were saturated, oxygen began to leak out of the oceans and was absorbed by land surfaces. Only after these terrestrial oxygen sinks were filled could the O_2 start accumulating in the atmosphere. The short period in which this rapid change occurred is now termed the **Great Oxidation Event** (Figure 15.5). This increase had a dramatic effect on life on Earth.

It is now well accepted that oxygen produced and released by cyanobacteria into their environments was initially toxic to other organisms. The toxicity stemmed from the high concentrations in which oxygen was produced, combined with other environmental factors. These factors include ultraviolet radiation and desiccation, which converted the produced oxygen into **reactive oxygen species (ROS)**. ROS were harmful to vital biological molecules, such as nucleic acids and enzymes. Mechanisms to detoxify such forms of oxygen were scarce in an anaerobic world, so many groups of organisms died. Some anaerobic prokaryotes survived, restricted to oxygen-poor environments, such as the ocean floor.

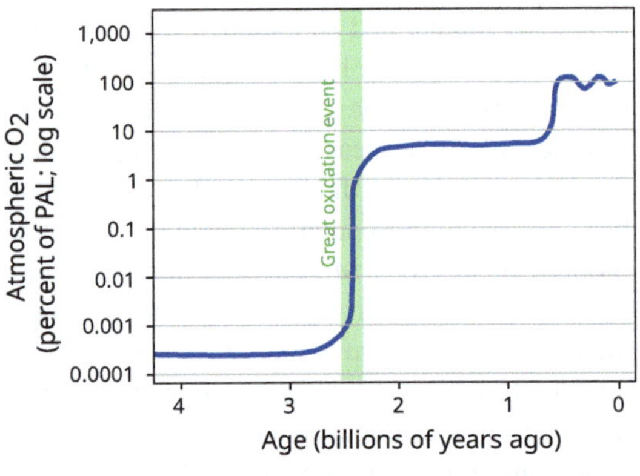

Fig. 15.5 Increase of O_2 in the Earth's atmosphere. The green column indicates the dramatic increase in atmospheric O_2 which occurred about 2.45 billion years ago. Adapted from Kump, L. The rise of atmospheric oxygen. Nature 451, 277–278 (2008). https://doi.org/10.1038/nature06587

However, some organisms survived in aerobic conditions by adapting. They not only evolved the ability to protect themselves from oxygen's toxicity but also began to utilise this abundant oxygen for energy extraction from organic molecules in a process called aerobic respiration. The evolution of aerobic respiration was a pivotal metabolic breakthrough. This process is estimated to generate significantly more ATP — approximately 16 times more — than anaerobic fermentation.

Therefore, acquiring the capacity to metabolise oxygen would provide a significant advantage for the host cell. However, the process of cellular respiration itself generates reactive forms of oxygen within the cell, which can increase the frequency of mutations in both the endosymbiont and the host cell.

The combination of these factors (ROS formation both in the environment and within the endosymbionts, UV radiation and unstable environmental conditions) exerted a strong selection pressure. This pressure drove the movement of genetic material deeper into the cell and its protection with additional membrane layers. The evolution of key eukaryotic features — including the endomembrane system, nucleus, and sexual reproduction — can be explained as a result of this selection pressure.

Scientists believe that the genome of eukaryotes expanded significantly early in their evolution, increasing the total DNA content and gene number. This expansion can be partially attributed to the endosymbiont. Subsequently, the expanded genome underwent duplication, which provided significant advantages: it could mask the effects of harmful mutations, enable the repair of damaged DNA, and allow mutations to accumulate in duplicated genes, potentially leading to novel functions. The advantages of possessing an "extra" copy of the genome explain why many organisms have life cycles with a dominant diploid stage. On the other hand, increased genome size poses several challenges: cells must regulate gene expression to ensure suitable levels of gene products within the cell and, more importantly, ensure that the genetic material is properly divided between daughter cells. These challenges were likely overcome thanks to the energy provided by the endosymbiont.

The basic meiotic mechanisms likely evolved from an earlier, simpler DNA repair mechanism present in the archaeal host, primarily in response to a high mutation rate. Subsequently, sexual reproduction with diploid-haploid cycles proved to be highly efficient in eliminating deleterious mutations.

Sexual reproduction is an elaborate process that depends on complex cell machinery and an adequate energy supply. Again, the maintenance of the complex machinery and the availability of energy were likely made possible thanks to the aerobic symbiont. It appears that sex evolved simultaneously with the evolution of eukaryotic traits, largely in response to the same array of selection pressures: the rise in environmental oxygen levels, reactive oxygen species produced by the aerobic endosymbiont, and other environmental stresses such as UV radiation and desiccation.

The Genes Behind Meiosis

Evidence that meiosis evolved from prokaryotic DNA repair mechanisms comes from studying the bacterial *RecA* protein and its homologs. It is an enzyme responsible for DNA repair and homologous recombination during bacterial transformation. There are two homologs of this protein in eukaryotes — *Rad51* and *DMC1*.

The *Rad51* protein is involved in repairing DNA double-strand breaks, both in somatic cells and during meiosis. The *DMC1*, however, is active only during meiosis, in which it plays a key role in the recombination of homologous chromosomes. Because recombination between homologous chromosomes is necessary for their proper segregation into daughter cells, the evolution of *DMC1* may have contributed directly to the evolution of sexual reproduction in eukaryotes.

RecA was most likely transferred from the mitochondrial genome to the nuclear genome of the ancestral eukaryote, where it duplicated. The functions of the two duplicate genes differentiated during subsequent evolution. The presence of *DMC1* in many protists, including the earliest diverging lineages, supports the idea that meiosis evolved very early in eukaryotic evolution and that the **last eukaryotic common ancestor** (LECA) was sexual.

15.3 Genetics of the merger

Compared with free-living bacteria, the genomes of mitochondria and chloroplasts are highly reduced. Mitochondria have lost up to 99% and chloroplasts up to 95% of their genes compared with their independent ancestors. This corresponds to a decrease from several thousand genes to approximately 100–200, or even fewer. For example, most chloroplasts have lost genes for the production of peptidoglycan, the main component of the bacterial cell wall, which is essential for survival. When living inside another cell, the protective peptidoglycan layer is no longer necessary for survival, so the genes for it are no longer needed.

However, despite the low number of genes, the abundance and variety of proteins present in mitochondria and chloroplasts are similar to those in free-living bacteria. How is that possible?

Many proteins specific to mitochondria and chloroplasts are encoded by genes in the nuclear genome. These genes are transcribed in the nucleus, translated within the host cytoplasm, and the resulting proteins are transferred into these organelles with the help of a specific transit peptide. This method of delivering newly synthesised proteins to specific locations within the eukaryotic cell is unique to chloroplasts and mitochondria. The underlying mechanism is quite complex, requiring proteins encoded by both the nuclear and the organelle genome. Interestingly, the proteins must cross two membranes surrounding the chloroplasts or mitochondria. To achieve this, they remain in an unfolded state (as a long strand of amino acids) and are only processed and folded once inside the organelle.

How did proteins essential for mitochondria and chloroplasts come to be encoded by nuclear genes? Two explanations have been suggested: First, it is possible that both cells involved in the ancient symbiosis already possessed a gene for a certain protein. As the relationship between the two symbionts strengthened, the protein acquired the ability to be imported into the endosymbiont. The endosymbiont could then produce the protein itself *and* also receive it from the host nucleus. Under such conditions, the endosymbiont's gene could have been lost without negatively affecting its processes.

Another possibility is that some genes originally present in the endosymbiont ancestor were transferred into the host genome. This could have happened, for example, when an old endosymbiont degraded within the cell, and random parts of its genome escaped degradation and became integrated into the nuclear genome. As in the previous example, this would result in both the nuclear and organelle genomes coding for the same protein. Consequently, the endosymbiont gene could be lost without any functional impairment. Selective forces, primarily driven by the toxicity of reactive oxygen species, then favoured the relocation of endosymbiont genes from the region with a high concentration of reactive oxygen to one protected by the nuclear membrane. Thus, the same pressure acted to both reduce the endosymbiont genome and enhance the host genome.

So, why did mitochondria and chloroplasts retain any genes at all? The retained genes primarily code for the proteins of their respective electron transport chains (located in the inner mitochondrial membrane or the thylakoid membrane in chloroplasts) and ribosomal proteins. The smooth functioning of the electron transport chains and production of ATP requires a precise balance in the amount and distribution of specific membrane proteins. Therefore, it is advantageous for mitochondria and chloroplasts to be able to synthesise these proteins independently of the nucleus. The localisation of these genes and their products within the same cellular compartment allows for rapid and efficient regulation.

15.4 Evolutionary importance of (endo)symbiosis

Traditionally, the diversification of life is attributed to natural selection acting on the diversity produced by random mutations and recombination. Over longer timescales, this may lead to the reproductive isolation of related groups under differing selection pressures. This view is often represented by a branching, tree-like diagram, in which each branching point signifies an event where an ancestral population split, eventually leading to new species. However, Lynn Margulis, who described the key steps in early eukaryotic evolution, also argued that symbiosis itself is a pivotal evolutionary force. While natural selection requires variety — typically sourced from random mutations and gene shuffling during sexual reproduction — symbiosis brings together distinct genomes, making genetic changes through gene transfer more likely. This process can lead to novel genetic combinations for natural selection to act upon, suggesting that over very long time scales, symbiosis and related gene transfer may be as important a source of variety as random mutations, hybridisation, or ploidy changes. Consequently, the traditional "tree of life", which depicts organisms as gradually branching off from a common ancestor, becomes more complex; branches may merge, transforming the picture into a web of life rather than a tidy tree, at least close to its root.

Green Animals

When learning about Cnidarians, you likely encountered the green hydra (*Hydra viridis*). You probably also know why these animals are green: they live in symbiotic relationships with unicellular *Chlorella* algae. But have you seen photosynthesising *slugs*? Unlike the green hydra, this is not a symbiotic relationship between two entire organisms. The sea slug *Elysia chlorotica* eats the algae and, remarkably, keeps only their chloroplasts alive and functioning — within its own animal cells! The cells of the green alga *Vaucheria litorea* are the slug's main food source. When the slug feeds on them, most algal cell contents are digested, but the chloroplasts are retained. The chloroplasts are taken up by cells lining the slug's extensively branched digestive system. They are then kept alive and functioning for the lifetime of the slug (about a year). These chloroplasts neither multiply within the host cells nor are passed on to the offspring when the slugs reproduce. Therefore, each individual animal depends on the *V. litorea* cells it consumes for its chloroplast supply. The *E. chlorotica* can rely on the products of photosynthesis provided by the chloroplasts when it is unable to find and eat fresh algae.

In all photosynthesising organisms, chloroplasts depend on the nuclear genome to code for the proteins necessary for their function. The long-term maintenance of functioning chloroplasts within the slug's cells was, therefore, a puzzle. This was resolved when scientists discovered several algal nuclear genes within the slug's own genome. These genes were even found in newly hatched slugs, which had never fed on *V. litorea*. Evidently, some genes from the algal nuclear genome were incorporated into the slugs' chromosomes and are passed from parent slugs to their offspring. Thus, baby slugs are prepared to take care of their chloroplasts even before they take their first algal meal.

Scientists were even able to observe the slugs produce chlorophyll: they fed the slugs food containing an amino acid labelled with radioactive carbon and observed that the carbon was incorporated into chlorophyll molecules. However, it has also been observed that chlorophyll production occurred only when the slugs were kept in the light. This is one of the rare observed cases of successful gene transfer across eukaryotic lineages.

Sea slug *Elysia chlorotica.* Patrick Krug in Cai, H., Li, Q., Fang, X. et al. A draft genome assembly of the solar-powered sea slug Elysia chlorotica. Sci Data 6, 190022 (2019). https://doi.org/10.1038/sdata.2019.22.CC-BY-4.0

DEFINITIONS

Endosymbiosis: A symbiotic relationship in which one organism lives inside the body or cells of another organism, with both typically benefiting from the association.

Evolutionary Transition: A fundamental change in the organization or complexity of life, typically characterized by the emergence of a new level of individuality or a novel mechanism for storing and transmitting genetic information.

Reactive Oxygen Species (ROS): Highly reactive molecules containing oxygen, often formed as a natural byproduct of oxygen metabolism. Due to their high reactivity, ROS can damage cellular components like DNA, proteins, and lipids.

Great Oxidation Event (GOE): A period approximately 2.45 billion years ago when atmospheric oxygen (O_2) levels rose dramatically and permanently, primarily due to the widespread photosynthesis of cyanobacteria.

Last Eukaryotic Common Ancestor (LECA): The hypothetical most recent common ancestor of all living eukaryotes.

SUGGESTED READING

- Baum DA, Baum B. An inside-out origin for the eukaryotic cell. BMC Biol. 2014 Oct 28;12:76.
- Imachi H et al. Isolation of an archaeon at the prokaryote-eukaryote interface. Nature. 2020 Jan;577(7791):519-525.
- Moody ERR et al. The nature of the last universal common ancestor and its impact on the early Earth system. Nat Ecol Evol. 2024 Sep;8(9):1654-1666.
- Richards TA et al. Reconstructing the last common ancestor of all eukaryotes. PLoS Biol. 2024 Nov 25;22(11):e3002917.
- Cai H et al. A draft genome assembly of the solar-powered sea slug Elysia chlorotica. Sci Data. 2019 Feb 19;6:190022.

15.5 Questions

1. **Rise in atmospheric oxygen concentration was the driving force for the evolution of eukaryotes. Why?**
   (A) Before the Great Oxidation Event, there were already some ancestral eukaryotes resistant to toxic oxygen chemical states. Since no prokaryotes had this advantage, most of them died and eukaryotes became prevalent, thus had more time to evolve.
   (B) When atmospheric oxygen concentration rose, bacteria that were already endosymbiotic could survive, because their host cells would protect them from the reactive oxygen states, whereas solitary bacteria were exposed to the atmospheric oxygen, which was toxic.
   (C) Aerobic respiration, which evolved in oxygen-rich environments, enabled organisms to produce ATP more efficiently. These organisms could then afford to maintain a larger genome and distribute it properly to daughter cells.
   (D) The cyanobacteria engulfed by other cells diversified into mitochondria, chloroplasts and other eukaryotic organelles.
   (E) Ancient and relatively simple mechanisms of DNA protection and repair evolved and most likely led to the origin of sexual reproduction.
2. **Which of the following statements is/are true?**
   (A) Chloroplasts were acquired before mitochondria by endosymbiosis.
   (B) Photosynthesis evolved earlier than aerobic respiration.
   (C) Many of the proteins needed for the function of mitochondria and chloroplasts are coded in the nuclear DNA, but not in the mitochondria or chloroplasts.
   (D) Mitochondria and chloroplasts kept most of their genetic material in their own DNA.
   (E) The last eukaryotic common ancestor contained both mitochondria and chloroplasts. Some eukaryotic lineages then lost chloroplasts and became heterotrophs.
3. **Suppose you identify the following structures in a cell: cell membrane, DNA, mitochondria and ribosomes. What type of cell could this be?**
   (A) An anaerobic eukaryote.
   (B) An aerobic eukaryotic cell.
   (C) An anaerobic archaean or bacterial cell.
   (D) A eukaryote or archaeon, but not a bacterium.
   (E) A bacterium or an archaeon, but not a eukaryote.
4. **Many protist parasites (e.g. of the genus *Trypanosoma*, *T. brucei*, which cause sleeping sickness) have highly reduced mitochondria and lack chloroplasts. These facts can be explained by which of the following?**
   (A) Within the host body, there is darkness and a lack of oxygen, which created a selective pressure to move all the mitochondrial and plastid genes to the nucleus and to get rid of the organelles.
   (B) The ancestors of these protists branched off from the last eukaryotic common ancestor before the acquisition of mitochondria and chloroplasts.
   (C) The parasites get energy-rich organic compounds from their hosts, which they then break down by anaerobic processes within their cells. This means the ancestors of these parasites could have afforded to lose both chloroplasts and mitochondria.
   (D) In dark and oxygen-poor conditions within the host body, the chloroplasts and mitochondria did not provide any advantage for the parasite, so the loss of chloroplasts and mitochondria did not affect its survival.
   (E) These protists are on their way to extinction because it is not possible for a eukaryote to survive without mitochondria.
5. **The free-living protist *Paramecium bursaria* lives in a symbiotic relationship with green algae of the genus *Chlorella*. Both the *P. bursaria* and the *Chlorella* algae can survive on their own, but when the *P. bursaria* comes into contact with the *Chlorella* algae, it ingests them, but keeps them alive within their**

cell. When the *P. bursaria* lives in a well-lit location, the *Chlorella* cells perform photosynthesis, supplying the host with glucose and oxygen. In return, the *Paramecium* provides nitrogen compounds and protection.

***P. bursaria* can lose its endosymbiotic *Chlorella* algae in two ways: when it is in darkness for an extended period of time and the *Chlorella* cells die, or when there is no prey (e.g. bacteria) and the starving *P. bursaria* digests its endosymbionts. In which of the following scenarios would the endosymbiotic interaction resume?**

(A) Abundant light, abundant bacterial prey, abundant *Chlorella* algae, scarce *Paramecia* in the environment.
(B) Abundant light, scarce bacterial prey, abundant *Chlorella* algae, abundant *Paramecia* in the environment.
(C) No light, abundant bacterial prey, abundant *Chlorella* algae, scarce *Paramecia* in the environment.
(D) No light, scarce bacteria prey, abundant *Chlorella* algae, abundant *Paramecia* in the environment.
(E) Abundant light, abundant bacteria prey, scarce *Chlorella* algae, abundant *Paramecia* in the environment.

6. **Which of the following are the well-supported aspects of the eukaryotic origin?**
(A) A photosynthetic prokaryote was engulfed by a heterotrophic prokaryote, enabling the host cell to survive in an oxygen-poor atmosphere.
(B) An archaeon acquired an internal aerobic partner, which later evolved into a mitochondrion.
(C) Members of two different prokaryotic lineages formed a symbiotic relationship; the aerobic one of the two evolved into a mitochondrion.
(D) Members of two different prokaryotic lineages formed a symbiotic relationship; the aerobic one of the two evolved into a chloroplast.
(E) An anaerobic photosynthetic archaeon became engulfed within a bacterial cell capable of cellular respiration.

7. **Which of the following statements about prokaryotes and eukaryotes are correct?**
(A) Organisms with prokaryotic cells are classified either as Bacteria or Archaea, while all organisms with eukaryotic cells are classified within the domain Eukarya.
(B) Ribosomes, DNA and cell membrane are found in all cells, whether prokaryotic or eukaryotic.
(C) Prokaryotes use RNA as their main genetic material, in contrast to eukaryotes, which rely on DNA.
(D) In eukaryotes, DNA is found in the nucleus, ribosomes and mitochondria, while in prokaryotes it is found only in the nucleus and chloroplasts.
(E) The process of photosynthesis evolved in prokaryotes. Eukaryotes became capable of photosynthesis by acquiring a photosynthetic endosymbiont.

8. **The increase of O_2 in the Earth's atmosphere dramatically changed the environment for living organisms. Which of the statements below correctly describe(s) some of the related events and processes?**
(A) The increase in the amount of oxygen caused the extinction of many unicellular organisms.
(B) The increase of oxygen enabled the evolution of photosynthetic pigments in eukaryotes.
(C) Organisms which evolved cellular respiration obtained an advantage since they could use the oxygen to obtain energy from organic compounds.
(D) The dramatic increase of O_2 in the atmosphere was caused by early photosynthetic eukaryotes, which evolved from a symbiotic interaction between green algae and archaea.
(E) The evolution of photosynthetic pigments in the ancestors of red and green algae helped to protect these organisms from the toxic effects of oxygen.

15.6 Project: When I grow up, I want to be a scientist!

In this project, you will develop your scientific career. You will work on two independent projects and experience the harsh life of a scientist.

Part I: Ph.D. in plantology

You are just starting your Ph.D. in a lab of a famous Plant Professor, and you need to plan your grand experiment. You have spent the last year reading everything there is to read about chloroplasts, mitochondria, and plants, and you know all there is to know about them:

The inheritance patterns of **genes located within chloroplast and mitochondrial DNA** differ from the patterns of inheritance of genes located within the nuclear DNA. This is connected with how the mitochondria and chloroplasts are inherited:

- A typical eukaryotic somatic cell contains only one nucleus, but many mitochondria and chloroplasts, which means it contains many copies of mitochondrial and chloroplast genomes as well.
- When such a cell divides by mitosis or meiosis, **the chloroplasts and mitochondria are distributed randomly** to daughter cells.
- When gametes fuse together to produce a zygote during sexual reproduction, the zygote usually gets the chloroplasts and mitochondria **just from the female gamete**, not from both. This is because the egg cells are large with all the necessary material for the zygote. In contrast, the sperm cell is usually reduced to the minimum.

Figure 15.6 shows possible scenarios of the inheritance of mitochondria with a newly acquired mutation. The same pattern applies to the inheritance of chloroplasts.

In your experiment, you will be working with the plant *Mirabilis jalapa*, which can have three types of shoot colouration: green, patchy (green and white), and white (see Figure 15.7).

The white cell colour is caused by a mutation in **the chloroplast DNA**, which makes the chloroplasts unable to get green (they are colourless). **A cell is white if it contains only mutated chloroplasts.** If a cell contains only normal green chloroplasts or a mixture of green and colourless chloroplasts, it will appear standard green. The patches of a patchy plant may vary in size: parts of leaves, whole leaves, or even whole branches.

Your experiment

You planted six seeds of *M. jalapa* plant and get one white, two green, and three patchy plants.

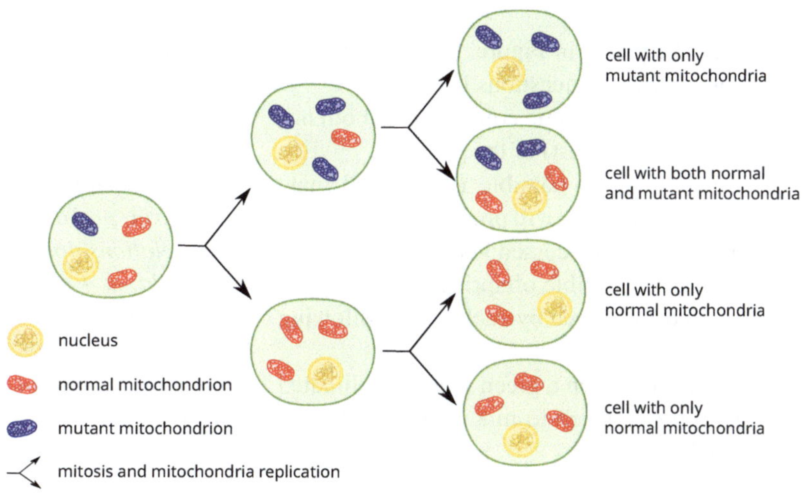

Fig. 15.6 Cell division. When a cell, which contains both normal and mutant mitochondria, divides, it produces a collection of cells (a tissue or an organ) varying in their mitochondria composition: some cells will have only normal mitochondria, some only mutant, and still other cells will have various mixtures of both

Fig. 15.7 **Three possible colourations of *M. jalapa* plants**

Fig. 15.8 **The six growing *M. jalapa* plants**

1. For each plant, draw the chloroplast composition of the zygote the plant has grown from.

The white plant dies soon. The green and patchy plants grow quickly and healthily, soon producing flower buds. The colour of the *M. jalapa* flowers is coded by **one nuclear gene**, for which two alleles exist with an **incomplete-dominance inheritance** pattern. Homozygotes have either red or white flowers, while heterozygotes have pink flowers.

Both your green plants produce pink flowers, two of the patchy plants produce red flowers, and one patchy plant produces white flowers.

2. Colour the plants in Figure 15.8. It will help you to answer the questions below.
3. Write down the genetic composition (both nuclear and chloroplast) of the plants.
4. Describe or draw possible female and male gametes these plants will produce (with respect to the two traits discussed here).
5. Describe the possible offspring that will grow from seeds produced by each of the following crosses.
   a. a pink flower on a green plant is pollinated with pollen taken from a white flower on a patchy plant.
   b. a pink flower on a green plant is pollinated with pollen taken from a red flower on a patchy plant.
   c. a white flower on a patchy plant is pollinated with pollen taken from a pink flower on a green plant.
   d. a red flower on a patchy plant is pollinated with pollen taken from a pink flower on a green plant.

Part II: Evil bunnies in your professorship

After moving 17 times across different countries, you finally became a respected professor at a respectable university. Now, you are expected to understand things that no one else does and come up with something amazing. You spent the last 10 years of your life thinking about chloroplasts, mitochondria, prokaryotes, eukaryotes and complicated relationships between them, day and night. Last night, at 4 am, you had an epiphany, and it all came to you, with astonishing clarity, in all its simplicity. It all made sense. You grabbed a pen, drew it on a piece of paper, and then you could finally sleep.

The next morning, you woke up and saw that all of your beautifully drawn pictures and schemes were torn, cut and chewed on by your vicious bunny that is refusing to look into your eyes. Look at the remnants of your notes and try to reconstruct the grand theory of the evolution of eukaryotic organisms!

On the following pages, you can observe various samples of unicellular prokaryotes and eukaryotes that you collected and observed under the microscope over the past 10 years.

1. Cut out the organisms and carefully study their anatomy.
2. Organise them to explain their evolution as they diversified or merged. Keep in mind that the tree of life branches, that branches may merge, and that some organisms are older (in evolutionary terms) than others.
3. Draw arrows to indicate the ancestors and descendants.
4. Annotate the pictures, label the structures.
5. **Create a clear, organised scheme.**
6. Describe your scheme, explain the steps, and provide observations that support your ideas in a few sentences.
7. If you see more possibilities, discuss them.

We would love to see your creation! If you would like to share it, please visit https://en.edub.sk/everyday-evolution and submit it through our website. Thank you!

This project was developed by Barbora Trubenová and Kristína Hudáková.

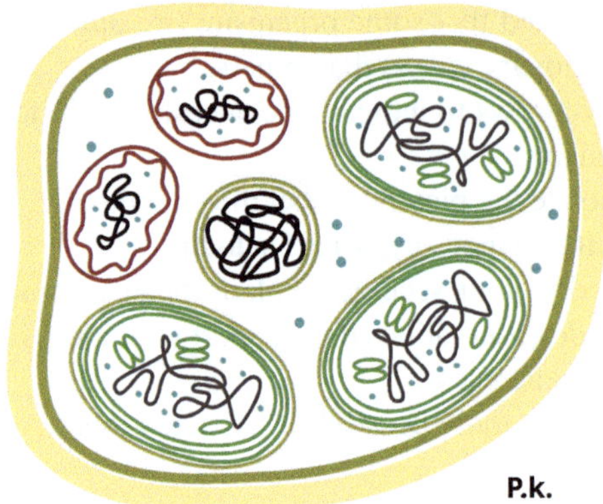

P.k.

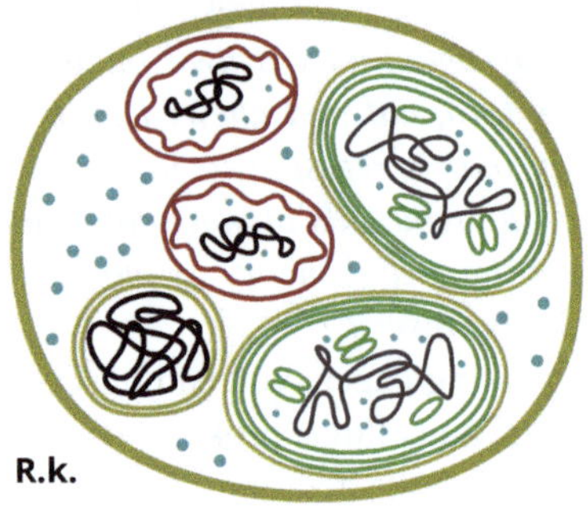

R.k.

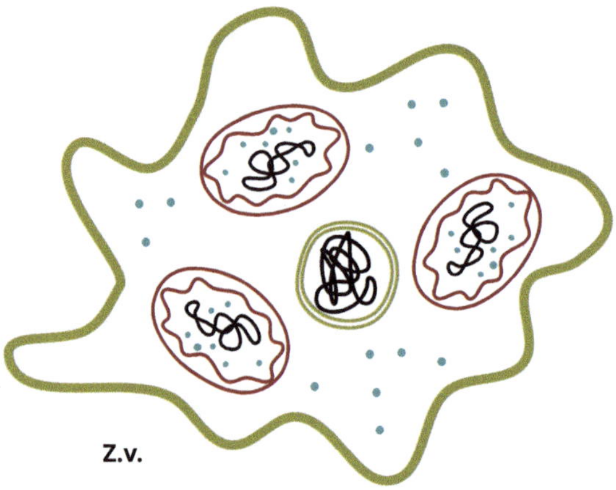

Z.v.

E.y.

U.a.

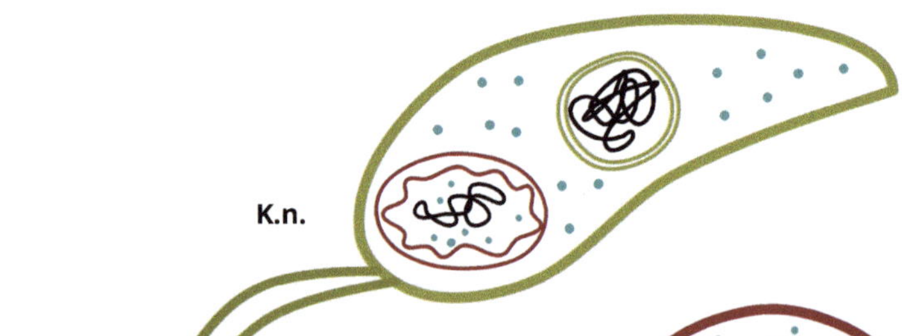

K.n.

U.i.

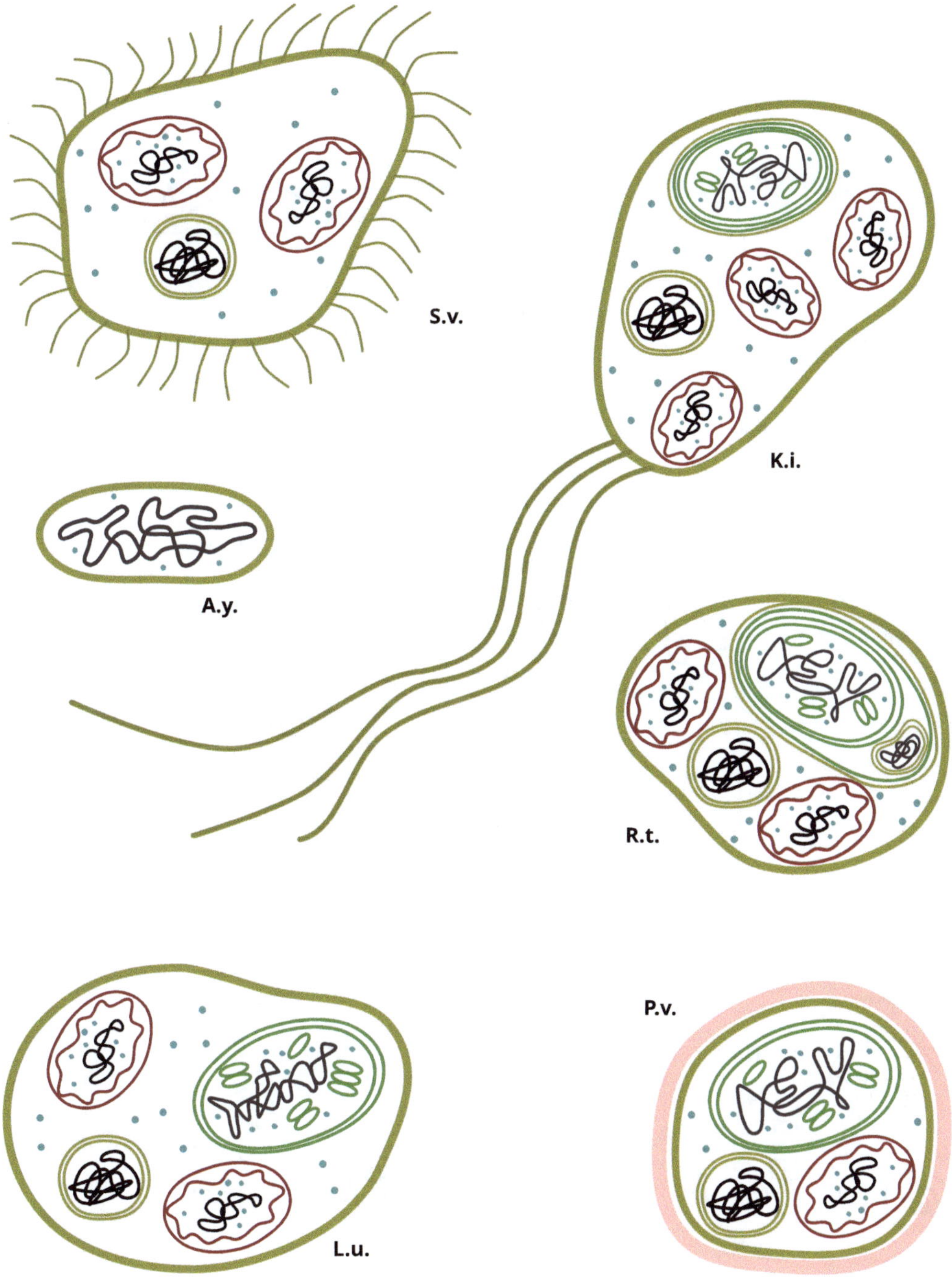
S.v.
K.i.
A.y.
R.t.
L.u.
P.v.

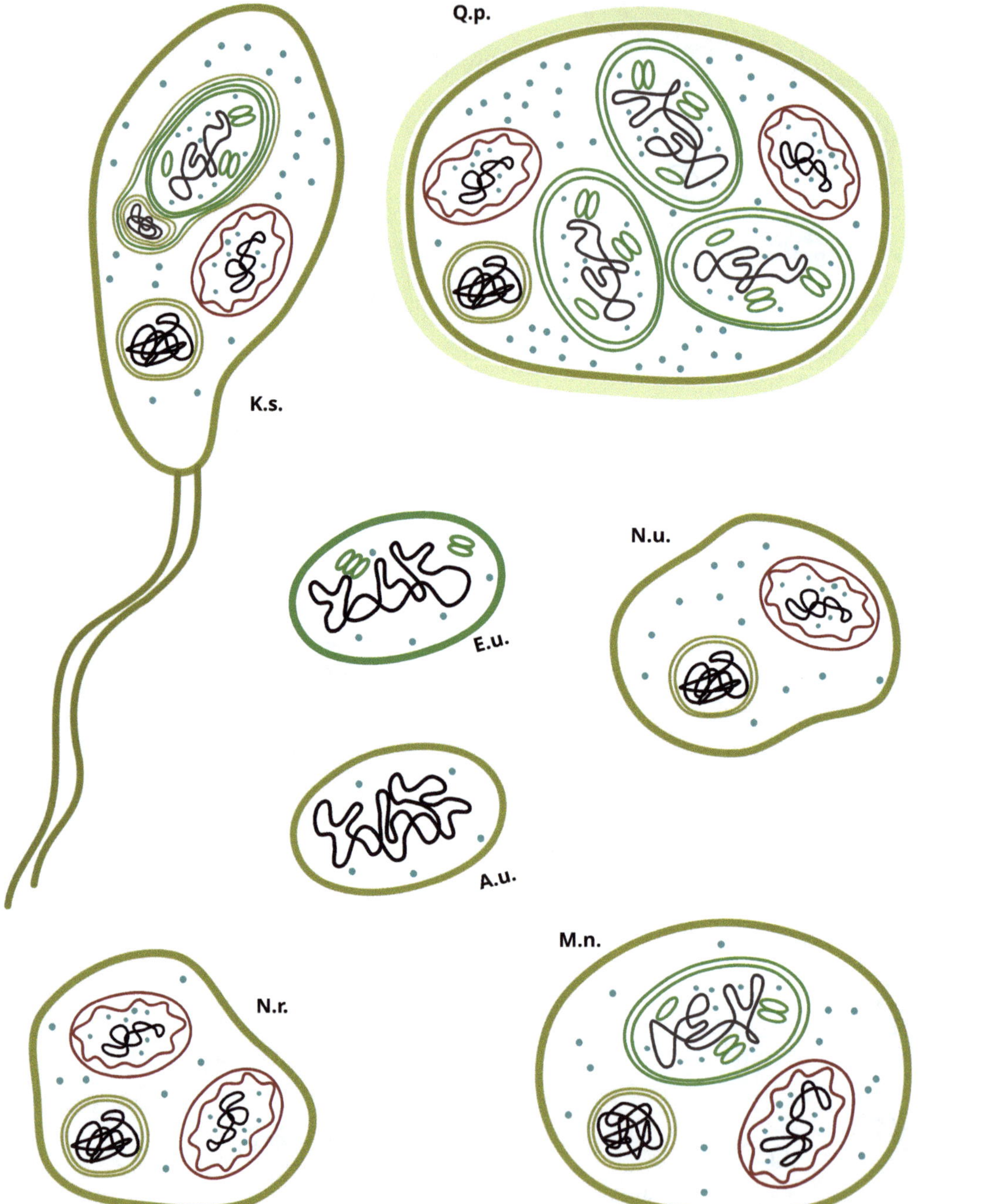
Q.p.
K.s.
N.u.
E.u.
A.u.
M.n.
N.r.

Part IV

Genetics: A Molecular Perspective

Studying Genomics of Natural Populations

16

Daria Shipilina

▶ Genomics is an essential tool for evolutionary biologists and many other scientists. For example, in the study of speciation (how species form), genomics allows us to examine the fundamental differences between groups of individuals and determine if they are distinct enough to be considered separate species. This is especially important when potential species look and behave similarly to the human eye — something that happens more often than one might think!

One such mystery led me, Dasha, a young researcher at the time, to investigate an unusual population of birds in the Ural Mountains. These birds closely resembled two different bird species (Fig. 16.1) — they shared some physical characteristics with each one of them and even sang the songs from each species! So what were these birds-some kind of intermediate or an entirely new species? I will explain how I used genetic data to solve the mystery of these birds, as well as how I gathered and interpreted this data. Along the way, we will trace the journey of a DNA sample from a mountain forest to the computer screen and see how modern genetic approaches help us to answer evolutionary questions.

D. Shipilina (✉)
Department of Botany and Biodiversity Research, University of Vienna, Vienna, Austria
e-mail: daria.shipilina@univie.ac.at

16.1 The Chiffchaff Mystery

An interesting discovery was made by the Russian zoologist Petr Sushkin in 1897. In the Southern Ural Mountains, near the city of Ufa, he found birds with a confusing appearance (Fig. 16.2). They looked like the Siberian chiffchaff (*Phylloscopus tristis*), a well-known species of small warblers that is uniformly olive — brown on its back and head. However, many of the feathers on their chests were yellow instead of white-a trait of another closely related species known as the European chiffchaff (*Phylloscopus collybita abietinus*).

This discovery initially went unnoticed until more and more researchers began to report birds with characteristics of both species from this region, as well as from other locations in the Urals. Moreover, some ornithologists were able to hear and, as recording equipment became available, record these unusual birds singing the songs of both species (Fig. 16.3).

But what are these individuals? Do these unusual birds represent a new species that is genetically *dissimilar* to both species known to exist in the Urals? Or is it possible that the two species

B. Trubenová et al. (eds.), *Everyday Evolution: Unveiling Life's Hidden Patterns*,
https://doi.org/10.1007/978-3-032-07050-0_16

Fig. 16.1 **Siberian (left) and European (right) chiffchaff**

Fig. 16.2 **Distribution map of European (green shading) and Siberian (brown shading) chiffchaffs.** The red area indicates the region where chiffchaffs were sampled. Red dots indicate locations where unusual birds were reported

in this region actually interbreed and produce genetically *intermediate* hybrids?

To answer these questions, I need to take a closer look at these unusual individuals. Part of my research will involve recording the birds' songs and taking their photographs, but the main analysis will focus on their DNA.

The working hypothesis for my project is that birds with mixed phenotypic and behavioural traits are the result of hybridisation between species. Can I find data to support my hypothesis or, alternatively, find a way to disprove it?

I invite you to accompany me on my scientific journey to discover the origin of these unusual chiffchaffs. Let's begin!

16.2 In the Forests of the Ural Mountains

Before any genetic analysis can be performed, I first need to collect a **sample** (or samples) from which I can extract DNA.

We don't usually think of researchers as managers or administrators, but for a project involving fieldwork (what we call trips to collect samples), good organisational skills are a must. Below, I describe the key questions I need to consider when collecting and preparing DNA samples for genomic analysis.

1. **Where should I go to collect samples?**
   After checking the existing literature, I have a list of locations where - the mixed birds have been spotted. One of them — the South Ural Nature Reserve — is particularly suitable, as our colleagues there can provide housing and advice. As part of the state reserve's goals, they regularly monitor all biodiversity on-site. The ornithology team noticed that some chiffchaffs sing mixed songs, making it very difficult to assign them to either the Siberian or European species. My colleagues are as intrigued as I am to learn more about the origins of these "mixed" birds. We decided to work together on this topic, combining their excellent local knowledge with my expertise in genomics.

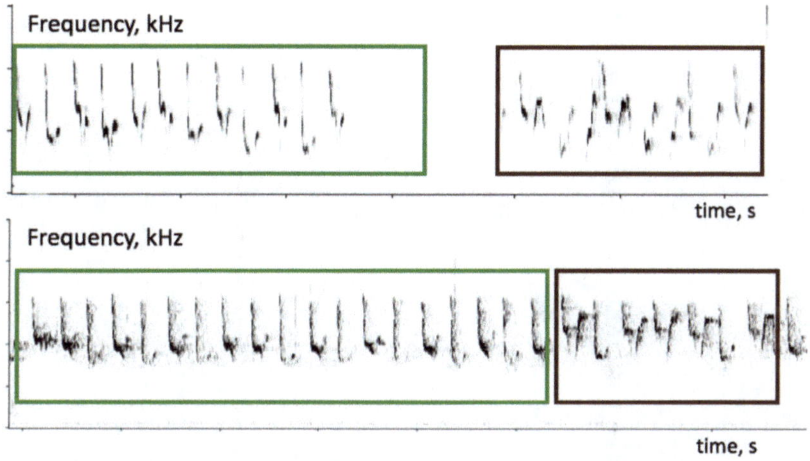

Fig. 16.3 **Sonograms of "mixed" chiffchaff songs.** A sonogram is a visual representation of sound, in this case, of bird song. Each line is a fragment of the song of one bird. Green boxes show European songs, and brown boxes show Siberian songs. The first bird sings the two songs interchangeably; the second one combines them into one song

2. **What kind of samples do I need?**
   To extract DNA, we need multiple cells from the individual. As bird red blood cells (unlike those of humans) contain nuclei, the most common way to collect DNA from birds is by taking a blood sample. The procedure is similar to visiting a doctor and having blood drawn from a vein. It is quick and safe for the bird — just a tiny prick, and it takes only a minute.
3. **How do I get these samples?**
   Of course, chiffchaffs won't willingly line up to donate their blood samples for science! So, how do I catch a bird to take a sample in the first place? The sound trap, a well-known catching technique, appears suitable for this task, and I have already tested it on chiffchaffs living near my department. I install a thin mist net, similar to a fisherman's net, next to the tree where a bird is singing. Then, I switch on the speaker and play the song of the chiffchaff. My "subject" — the resident bird — thinks an intruder has entered its territory. Despite their small size, chiffchaffs can react very aggressively to intruders! The male tries to attack my speaker and is trapped in the almost invisible net (Fig. 16.4).

   I untangle the bird from the net, take a photo, measurements and a blood sample — all very carefully. After making sure the bird has not been harmed, I release it.
4. **How many samples do I need?**
   To support the hypothesis of ongoing gene exchange between the two species, I need to collect DNA from as many "unusual" individuals as possible. I suspect that even birds that look like one parental species may have some genes from both parental forms, so I catch them as well. Of course, I also need to know the exact sequence of nucleotides in the genes of each of the parental forms. I will find this information in a gene database.

After everything has been prepared and the chiffchaffs have arrived in the Ural region from their wintering grounds, I take a train to the Southern Urals with a huge backpack full of equipment. I will soon memorise nearly all the trails in the State Reserve because I will be wandering around looking for singing chiffchaffs. These birds are territorial, and sometimes it takes a long time to move between their territories.

My catching technique is a success — most of the time, the birds come to the mist net very quickly (and very angrily). My handling of the birds is quite gentle, which is not always ideal: A couple of birds managed to escape. In most cases, however, I successfully take measurements, pic-

Fig. 16.4 A chiffchaff trapped in a chiffchaff trap. A researcher points at a chiffchaff trapped in the nearly invisible mist net. A speaker, placed on a nearby branch, was used to attract the bird

Where to Get DNA?

To collect human DNA, we usually swab an individual's mouth or collect hair follicles (an older method). However, we can extract DNA from almost any part of the organism or any of its secretions. We are able to extract DNA from mummies that are thousands of years old, from ancient bones and even from molluscs and eggshells. Scientists have been able to isolate DNA from old parchment to determine that it was made of sheepskin and from Ötzi's clothes to find out that they were made of bear, sheep and deer hide. Even DNA from coprolites (fossilised faeces) can be used to determine the diet of extinct animals.

Depending on the organism and sample type, DNA isolation can be quite laborious. For example, muscle tissue needs to be cut into small pieces with scissors, and dried leaves may need some work with a mortar and pestle. Furthermore, there is a significant challenge in DNA extraction-biological contamination. All the treatments during the DNA isolation process are designed to preserve as much DNA as possible, but what if there is DNA from another organism in the reaction tube? It can seriously confuse the results of our research. For example, even eating a turkey sandwich at the lab table can result in my extracting turkey instead of chiffchaff DNA! To avoid this, most of the isolation and amplification must be done in a sterile lab.

tures and a blood sample. My colleagues showed me a useful trick: After collecting blood droplets on a filter paper, I can dry it out and put it into a new sterile envelope. It then takes up virtually no space in my backpack!

Finally, I have gathered a sufficient number of samples. What a journey this has been! Walking through the forest and climbing the hills, I feel truly united with nature. Holding these little birds and then releasing them and watching them return happily to their homes is a unique experience. It's a powerful reminder of nature's fragility and our responsibility to study and protect it.

16.3 DNA Extraction

After two months of hard work, the DNA samples travelled back safely to the lab in my backpack.

A lot of hard work has been done: 74 birds were sampled this year. One-third of the birds looked unusual (27 of them, to be exact)! Additionally, I obtained samples from 28 to 29 birds that looked like the very typical European and Siberian chiffchaffs, respectively.

However, most of the genetic analysis is still ahead of me. The next step is **DNA extraction**. Extraction simply means separating DNA molecules from all the other molecules found in a cell.

It's a multi-step procedure that must be performed in a research laboratory. I will be adding small quantities (sometimes just microlitres) of various reagents to the test tubes containing DNA. But it will almost always look like a clear liquid, making it difficult to control the process based on visual cues. Therefore, before starting work in a lab, I like to gain a clear conceptual understanding of what will happen during the DNA extraction process.

I picture my sample at the microscopic level: It consists of millions of blood cells. Inside each cell, long strands of DNA are wrapped around proteins, forming chromatin. However, isolating the chromatin is not enough — I need strands of pure DNA. To extract these, I must unwind the chromatin and remove the proteins (such as histones) that keep the DNA compact.

First, I need to break up the sample and, ideally, access the individual cells. My samples are droplets of blood on the piece of filter paper. In order to reveal single cells, I can simply cut the paper into small pieces and soak it in a saline solution.

Then, the sample is treated with a solution containing several reagents. Some of them work similarly to dish soap, destroying the lipids in the cell membrane and releasing the cell's contents. Others break the proteins that fold the DNA compactly in the cell. This is usually done by a special enzyme-proteinase K.

However, broken molecules still remain in the tube. The next step is to discard this debris and clean the DNA. This can be done in many different ways. I decided to centrifuge (spin at high speed) my sample. Hopefully, the DNA molecules will stay intact. As whole DNA molecules are heavy, they will settle at the bottom of the tube during the centrifugation process. All the debris (broken molecules and cell structures) will float to the top. Then, I can simply get rid of the top layers of the liquid. (For greater precision, researchers use special reaction tubes with a microfilter.)

Finally, once the DNA is isolated and purified, it must be stored in a saline solution (the buffer), and only then can it be visualised. In the presence of alcohol and salt, DNA molecules become "hard" (insoluble), and if there is a lot of DNA, we may even see a cloudy or stringy, white structure in the reaction tube-the DNA precipitate.

I repeat the DNA extraction individually for all my samples. For many of them, I could see the DNA in the reaction tubes — it looked like a tiny ball of cotton. For others, I used more advanced techniques to check the quantity of DNA.

16.4 Searching for Suitable Genes

In the third step, I need to decide which gene (or genes) to study in order to determine the origin of the unusual chiffchaffs. I also need to know the sequence of the chosen gene in both parental species. Furthermore, the chosen gene must have sufficiently many differences between the two parental chiffchaff species so that I can tell the species apart by looking at this gene.

Information about all the genes currently known to biologists is collected in online gene or genome databases. One of the most popular is the GeneBank, which is run by the National Center of Biotechnology Information (NCBI). As with other search engines, I simply type the scientific name of the European chiffchaff-*Phylloscopus collybita*-into the search bar, and it returns a list of the known genes and their sequences. Then, I do the same for the Siberian chiffchaff, *Phylloscopus tristis*.

The search is successful: One of the mitochondrial genes is known for both species. Its name is cytochrome b, and it's responsible for cellular respiration. I download the sequence for this gene (from NCBI) — a string of letters A, C, G and T representing the nucleotides, and note which nucleotides differ between the species. It's time to return to the lab.

16.5 PCR and the Principle of DNA Multiplication

Now, the DNA has been extracted from multiple samples. However, another problem remains: I only have tiny amounts of DNA from each bird. I do not want to use it all up in a single analysis because the birds are far away, and in all likelihood, I would not be able to get more samples from the same birds again. Fortunately, **PCR (polymerase chain reaction)** technology can solve this problem. It can be used to amplify DNA fragments by a factor of millions and billions.

PCR is widely used in evolutionary biology and in many other areas of biology and medicine, including molecular biology and medical diagnostics. The core principle of PCR is based on the natural ability of the cell to duplicate DNA. PCR mimics this process in the reaction tube by unfolding the DNA, making a copy of the gene of interest and then repeating this process many times. Despite the complexity of the process described below, it doesn't require a lot of manual work on my part. All I need to do is put the reagents in my reaction tube. After placing the reaction tubes into the machine and adjusting the settings, I can have a cup of coffee and imagine what is happening inside the machine.

PCR

As the name would suggest, polymerase chain reaction (PCR) is a chemical reaction in which each newly copied DNA fragment serves as a template for the next round of copying, resulting in a chain reaction of copying events that exponentially increases the amount of DNA. It requires different temperatures for different steps and is therefore performed in a special thermostat known as the thermocycler.

The ingredients for the reaction are:

- DNA extracted from samples
- Free nucleotides to be used as "building blocks" for making new copies
- Short template sequences (primers) that mark the beginning and end of the gene to be copied (my primers are specially designed to connect to the selected diagnostic gene, cytochrome b)
- Enzymes to catalyse the copying process

After all the reagents are added to the tube, the machine undergoes numerous cycles of heating and cooling, resulting in multiple cycles of DNA replication. Each cycle consists of the following steps (as shown in Fig. 16.5):

1. First, the two strands of the DNA molecule are separated by applying very high temperatures. Now, each of the separated strands will serve as a template for the next step.
2. Upon cooling, primers attach to the DNA strands and indicate which part of the molecule needs to be copied. Primers are specially designed for each experiment (I designed mine for this project): One binds to the forward strand at the beginning of the gene, the other to the reverse strand at its end. This allows the two separated strands from the previous step to be used simultaneously; DNA copying will go in opposite directions along each strand.
3. Then, a process similar to cellular DNA replication begins. The major protagonist here is

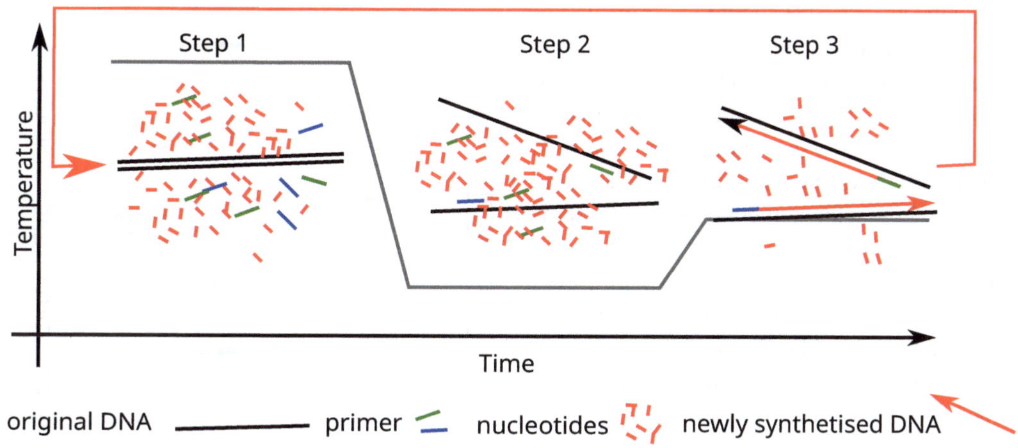

Fig. 16.5 Schematic of the polymerase chain reaction (PCR). DNA strands are separated by applying high temperature (step 1). When the temperature is lowered, primers attach to the complementary sections (step 2), and DNA copying starts, doubling the number of gene copies (step 3). This cycle repeats 25–35 times and takes 2–4 hours, depending on the length of the DNA region being copied. During this time, the target region can be amplified from just one or a few molecules to billions of copies

the Taq polymerase — a DNA polymerase that copies DNA strands by using free nucleotides to synthesise the complementary strand.

4. The copying starts from the point indicated by one of the primers (say, at the beginning of the gene) and stops at the end of the strand. This means that the sequence after the gene — which I don't need — is also copied. However, in the next cycle of PCR, the newly synthesised strand will attract a complementary primer, which attaches at the end of the gene. The copying repeats, but in the reverse direction- this time ending at the beginning of the gene- resulting in a copy of just the desired region! The number of copies of my gene has doubled, and the cycle starts again.
5. Numerous repetitions of the cycle will result in many DNA fragments — some of them longer, but most containing only the desired region. We can simply separate them by length and use only the fragment of interest for sequencing.

Interesting Fact

Primers are short sequences of RNA nucleotides that initiate the synthesis of DNA in living cells. DNA polymerase recognises the RNA primers and starts adding DNA nucleotides complementary to the template DNA strand, working in a specific direction. In PCR, the primers are designed artificially by the researcher to start the copying process at precisely their desired location. Primers are approximately 20 nucleotides long to ensure specificity. Using two primers — one complementary to each strand at each end of the desired gene-allows us to copy the section between the primers in every two cycles.

At the end of PCR-induced DNA multiplication, I have reaction tubes (one per individual) full of copies of my diagnostic gene. Interestingly, if I ever want to study another gene or even all the genes from my organism, I can use the same extracted DNA and amplify another region. However, I still don't know the sequence of the gene. The next step, **DNA sequencing**, will reveal this information.

16.6 From the Reaction Tube to the Computer Screen

I want to go from the contents of the reaction tube (which, again, look like a clear liquid) to a simple text file containing a long sequence made up of just four letters — A, T, C, and G — repeated in various combinations. These letters represent the sequence of nucleotides in the DNA fragment that I have for each bird. DNA sequencing is the process of determining the sequence of nucleotides in a fragment of DNA. Even now, after I understand the details of how sequencing works, it still seems a bit like magic.

There is currently a great variety of sequencing technologies. For my project, the best approach is Sanger sequencing, the oldest and best-known method. It is relatively inexpensive and allows me to sequence a gene of the right length.

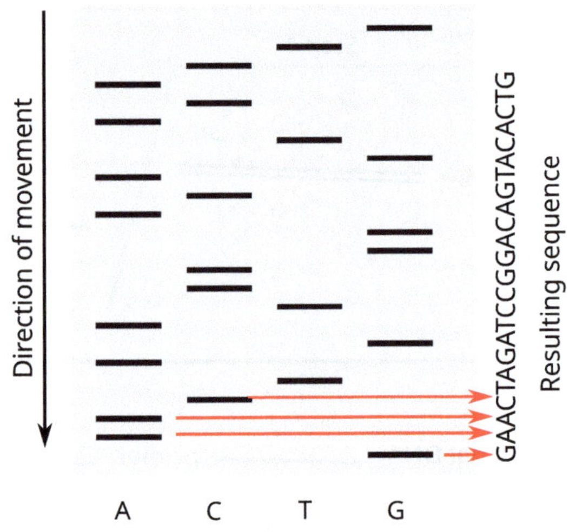

Fig. 16.6 Schematic of gel electrophoresis. Fragments are separated according to size — shortest at the bottom and longest on top. Each column contains fragments from a different tube

Interesting Fact

Taq polymerase is named after the heat-tolerant bacterium *(Thermus aquaticus)* from which it was originally isolated. *T. aquaticus* lives in hot springs and hydrothermal vents. Its DNA polymerase is very heat stable and is most active at around 70 °C (a temperature at which a human or *E. coli* DNA polymerase would not work at all). This heat stability makes Taq polymerase ideal for PCR.

Sanger Sequencing

Similar to PCR, Sanger sequencing also involves making many copies of DNA, but this time, the result of the process is visualised. Everything needed for PCR will also be needed here: DNA polymerase (Taq polymerase), all four types of DNA nucleotides, primers, and the template DNA to be sequenced. However, the sequencing reaction requires additional ingredients: special versions of all four nucleotides, known as dideoxynucleotides. Each of these special nucleotides is labelled with a uniquely coloured dye, and due to its chemical properties, it stops the DNA elongation. Furthermore, only one primer is used to ensure that only one strand of DNA is sequenced — the other one can be easily inferred using the complementarity principle. I combine all the necessary reagents in the tubes and bring them to the research facility that houses the sequencing machine.

During Sanger sequencing, the DNA sample is divided into four separate tubes for four different sequencing reactions. Each tube contains all four of the standard deoxynucleotides and one type of the special, elongation-stopping dideoxynucleotide. Let's look closely at one of the tubes, for example, the one containing the dideoxynucleotide G. During DNA synthesis, DNA polymerase adds nucleotides one after the other, building the complementary strand. However, when it encounters the dideoxynucleotide G, the elongation stops. As in PCR, the strands are then separated, and the process starts again. This results in many complementary strands of different lengths, each of which ends with G. Then we place the fragments in the gel, where they start to move at various speeds-the shorter fragments move faster, so the fragments separate according to their length. By comparing the lengths of the fragments from all four tubes containing different dideoxynucleotides, we can infer the order of nucleotides in the sequence (Fig. 16.6).

As we can see, sequencing is a very complicated process — one that combines several different approaches from molecular biology, physics, and chemistry. However, the process has been automated and can be completed quickly using modern machines. The sequence of my chosen gene in my DNA samples will be written down for me in a very simple text format. I return to the research facility in a few hours with a flash drive, and at this point, my lab work is finished.

16.7 Mystery Solved

Now that I have the long-awaited sequences on my computer, I can start investigating whether they belong to the European or Siberian species or something in between.

In order to answer this question, I will need to use a bioinformatic algorithm known as sequence alignment. Aligning the sequences simply means placing them side by side, one below the other, so that identical nucleotides align. I already know the sequences for the diagnostic gene of the parental species: European and Siberian chiffchaffs. Now, I will align the sequences of all the individuals from the Ural to these reference sequences.

At first glance, I can see that the genes of the unusual chiffchaffs are as unusual as their appearance and song! Nearly half of my birds have some nucleotides from one species and some from the other. And an even bigger surprise is waiting for me when I examine the sequences of (what I thought were) quite normal individuals from the Southern Ural. Among the birds I identified as Siberian chiffchaffs through careful in-hand examination, two turned out to have genes of the European chiffchaff. The results for the birds identified by phenotype as European chiffchaffs are even more shocking: Only 25% of them are true European chiffchaffs.

The results of my analysis very strongly support my initial hypothesis. The number of birds that have mixed genetic origin is so large (17 birds with "hybrid" genes plus nine with a mismatch between phenotype and genotype) that I conclude that hybridisation (consistent mating between different species) takes place in the Southern Urals. My project is finished. Not only have I solved the mystery of the strange chiffchaffs, I have also discovered the place where Siberian and European chiffchaffs regularly interbreed — a hybrid zone.

Hybrid Zones

A hybrid zone is a region where genetically distinct populations of the same species or closely related species meet and interbreed. This interbreeding results in the production of hybrids-individuals that are a genetic mixture of the parental forms. Hybrids usually vary in the proportion of genes they share with each of the parental forms. In particular, hybrids at different ends of the hybrid zone tend to share more genes with the nearer parental form.

Some of the best-studied examples of animal hybrid zones are in central Europe: The hybrid zone between carrion crows (*Corvus corone*) and hooded crows (*Corvus cornix*) extends through Europe from north to south, with hybrids showing various mixes of parental plumage colouration.

The hybrid zone between the fire-bellied toad (*Bombina bombina*) and the yellow-bellied toad (*Bombina variegata*) is located throughout Europe in areas where lowland habitats, typically inhabited by the fire-bellied toad, gradually change into higher-altitude habitats, typically inhabited by the yellow-bellied toad.

16.8 What's Next?

Hybrid zones can persist in their current form for a long time, maintained by the lower fitness of hybrids and other reproductive barriers. But what happens to these hybrid zones over *very* long times? Will the two parental forms eventually merge together, forming a single species? Or will the hybrids become reproductively isolated from the parental forms and eventually produce a third, separate species? Or will the two parental forms become reproductively isolated from one another, forming two distinct species with no hybridisation at all?

All of these scenarios are possible, and the existence of such different possible outcomes is one of the reasons why hybrid zones are so interesting to evolutionary biologists, allowing us to directly observe evolution in action. In so doing, we hope to shed some light on big biological questions such as the origin of species...

DEFINITIONS

Polymerase chain reaction (PCR): A method used to make millions of copies of a specific DNA segment

Primer: A short ($\sim 20bp$), single-stranded piece of DNA that serves as a starting point for DNA replication or PCR

Hybrids: Offspring of individuals from two geographically or genetically distinct populations

SUGGESTED READING

- Thörn F et al. Frequent Hybridisation Between Parapatric Lekking Bird-of-Paradise Species. Mol Ecol. 2025 Jun;34(11):e17780.
- Shipilina D et al. Patterns of genetic, phenotypic, and acoustic variation across a chiffchaff (Phylloscopus collybita abietinus/tristis) hybrid zone. Ecol Evol. 2017 Mar 2;7(7):2169–2180.
- Saiki RK et al. Primer-directed enzymatic amplification of DNA with a thermostable DNA polymerase. Science. 1988 Jan 29;239(4839):487–91.

16.9 Questions

1. **Part of the genome in frog species A reads as follows:**
   - TTATTCTTAGCAGTGTGCCTAATTCGACAGAT TCAGTTCCATTTTACTTCCAAACACCATTTT GGGTTTGAGGCAGC

   while in frog species B, it is
   - TTGTTTTTAGCAGTATGCCTAATTCGACAAAT TCAGTTCCATTTCACCTCCAAACACCAT TTTGGGTTTGAGGCGGC

   Which of the following offspring sequences would indicate a hybrid?
   (A) TTATTCTTAGCAGTGTGCCTAATTCGACAG ATTCAGTTCCATTTCACCTCCAAACACCA TTTTGGGTTTGAGGCGGC
   (B) TTTTTGTTAGCAGTTTGCCTAATTCGACAT ATTCAGTTCCATTTTACTTCCAAACACCA TTTTGGGTTTGAGGCAGC
   (C) TTATTCTTAGCAGTGTGCCTAATTCGACAG ATTCAGTTCCATTTTACTTCCAAACACCA TTTTGGGTTTGAGGCAGC
   (D) TTGTTTTTAGCAGTATGCCTAATTCGACAA ATTCAGTTCCATTTCACCTCCAAACACCA TTTTGGGTTTGAGGCAGC
   (E) TTGTTTTTAGCAGTGTGCCTAATTCGACAG ATTCAGTTCCATTTCACCTCCAAACACCA TTTTGGGTTTGAGGCGGC
2. **The highlighted area in the sequence below shows the gene that we want to study. Which pair of primers can be, in principle, used to amplify (make multiple copies of) the gene? [Note that in the lab, the primers we use are 20-40 bases long to achieve specificity. With primers four bases long, many unspecific regions would be amplified. Real genes are also much longer, typically between several hundred and several thousand nucleotides.]**
   AGG CAATCAGCTAGCATATAGAATGTCTCTGG TCGCTACAC ACGCT
   (A) CAAT and GTGT
   (B) AGGC and TATA
   (C) CAGA and GTTA
   (D) ATAT and GAGA
   (E) CAAT and ATTG
3. **When you returned from a field trip to Ecuador, you found a weird-looking spider inside one of your shoes. You take a detailed picture of the spider and keep it in a jar. First, you try to identify the spider using books and the Internet, but with no success. Then, you go to your newly equipped high-tech lab. Which of the following procedures will you perform in order to identify the spider?**
   (A) Protein extraction
   (B) DNA extraction
   (C) PCR
   (D) Sequencing
   (E) Cloning
4. **When you get the results of your lab analysis (using the procedure identified above), it may lead you to conclude which of the following?**
   (A) The spider is a specimen from an entirely new species.
   (B) The spider is a hybrid between two related species of spiders.
   (C) The spider belongs to a well-known species, but this specimen has a rare mutation causing it to lose all its hair, a characteristic of the species.
   (D) It is not a spider at all!
   (E) The spider was a juvenile specimen of a rare species.
5. **During the isolation of spider DNA, you were not careful enough, and you did not use gloves. You've probably contaminated the sample with your own DNA. What do you do next?**
   (A) The experiment is ruined. I can't do anything.
   (B) Human and spider DNA are completely different, so I don't have to worry. I continue with the analysis.
   (C) Not everything is ruined. As long as the gene I'm amplifying is not present in humans, I should be able to amplify it.
   (D) I put the sample in a $-80^{\circ}C$ freezer, which will degrade the human DNA but not the spider DNA.
   (E) I adjust the design of the primers so that they amplify the part of the spider gene I need, but not the part of the homologous human gene.

6. **The rate at which Taq polymerase enzyme synthesises DNA is approximately 1 kb (kilobase) per minute. Your gene of interest is 2385 base pairs long. Which time setting(s) (length of the cycle) could be used for the amplification step when setting up the thermocycler?**
   (A) 60 seconds
   (B) 120 seconds
   (C) 1 min 30 seconds
   (D) 2 min 30 seconds
   (E) 3 minutes
7. **You have a single copy of the DNA with your gene of interest. After ten cycles of PCR, approximately how many copies of your gene do you expect to find in the tube?**
   (A) 11
   (B) 112
   (C) 1024
   (D) 10,048
   (E) More than a million
8. **Which of the following statements about Taq polymerase are correct?**
   (A) It is stable at high temperatures.
   (B) It was isolated from heat-resistant bacteria.
   (C) It is a DNA polymerase-a crucial enzyme for DNA synthesis.
   (D) It is present in all organisms that carry DNA.
   (E) It has a sequence complementary to the gene of interest.

16.10 Project: Control in a Restaurant

During a routine hygiene control at a restaurant, the hygienists found five unlabeled packages of frozen meat. As the restaurant owner was not able to provide any documentation explaining the origin of the packages, samples were taken from each package and sent to the laboratory for DNA analysis.

You are the scientist in charge of investigating the mystery meat!

Part 1: Lab Work

After you successfully isolate DNA from all of the samples, you use PCR to increase the quantity of the DNA. In particular, you want to focus on one particular region. You may want to download the sequence from our website here: https://en.edub.sk/everyday-evolution.

```
ATGTTCATTAACCGATGACTATTCTCTACCAATCACAAAGACATTGGTACCCTGTATTTACTATTTGGCG
CCTGGGCAGGAATAGTAGGTACCGGTCTAAGTTTGTTGATTCGTGCTGAATTAGGTCAACCTGGCACACT
TATCGGAGACGACCAGCTTTATAATGTTCTAGTGACAGCTCATGCCTTCGTAATAATTTTCTTTATAGTT
ATACCTATTATAATTGGAGGTTTTXXXXXXXXXXXXXXXXXXXXAATAATTGGAGCTCCTGACATAGCAT
CCCGTCTAAACAACATAAGCTTCTGACTACTCCCCCCTTCCTTTCTACTACTAATAGCATCCTCTTCAAT
GAGGCCGGCGCAGGTACAGGCTGAACTGTTTACCCTCCTCTAGCCGGAAATCTGGCACAAATTTGCAGGA
GCCTCAGTAGACCTTACCATTTTCTCTCTACATTTAGCCGGTGTATCTTCAATCCTTGGAGCTATTAACT
TCATCACAACTATTATCAATATAAAACCACCCGCTATAACTCAATACCAAACACCCCTCTTCGTCTGATC
```

You design two primers, shown in Table 16.1, that will amplify the desired region.

Remember that even though only one strand of DNA is shown above, primers attach to different strands and in opposite directions!

Questions

You can highlight the requested sequences in the sequence above using different colors.

1. Where will the primers attach upon denaturation?
2. Which region will be the most amplified?
3. You start with a single copy of DNA. Using the primers above, write down the strands that you will have in your tube after a single cycle of PCR.
4. Write down the strands that you will have in your tube after the second cycle of PCR.
5. How many "pure" genes (sequences that you highlighted in question 2) will you have after five cycles?

Table 16.1 Primer design

Primer 1	AAGTTTGTTGATTCGTGCT
Primer 2	TGTGCCAGATTTCCGGCTA

You have successfully separated out the shortest fragments containing the gene that you want. You notice that part of the desired sequence is unknown to you (even though your PCR machine reproduced it without any errors!). You need to sequence this part.

In the sequencing machine, the sample is separated into four tubes, and the necessary reagents are added.

6. Write down what is in the tubes:
   Tube 1:
   Tube 2:

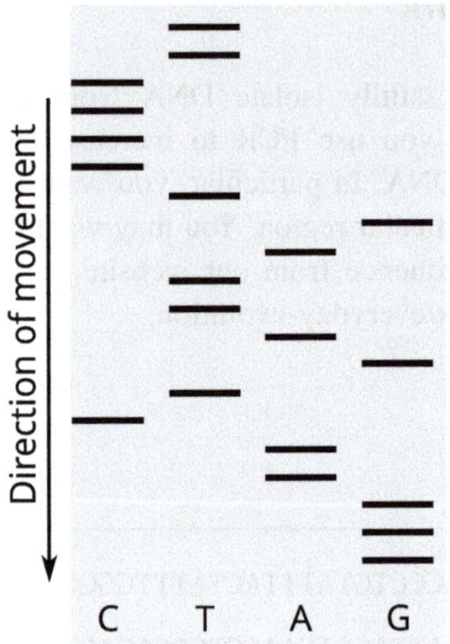

Fig. 16.7 Gel with DNA fragments separated by length

Tube 3:
Tube 4:

After you have had your coffee, it is time to read the data! Unfortunately, the machine is broken, and you have to decipher the data yourself. In Fig. 16.7, the DNA fragments created during the sequencing process are separated by their length, with the shortest at the bottom. Use the DNA fragments in the gel to decipher the pattern and write down the sequence. If you need help, watch the video about Sanger sequencing linked at the end of the article.

7. What is the unknown sequence?
8. Replace the XXXs in the region with the now-identified sequence. Search for the origin of this DNA in the database using BLAST according to the instructions in the box below. Can you tell what gene it is?
9. What animal does it come from?

Part 2: What is in the Packages?

The origin of the meat is still a mystery-it could be from any animal. As you cannot be sure which diagnostic gene would provide the best results, you have decided to amplify two genes for each sample: cytochrome c and insulin. You have isolated the DNA successfully from all five packages, and you have run PCR twice for each sample, each time with different primers, amplifying different genes. You now have ten test tubes that you can sequence. You can download the results here in a simple text format: https://en.edub.sk/everyday-evolution.

Try to identify the species using **BLAST**. See the detailed instructions on how to use BLAST below. Write the possible animal origins of the meat in Table 16.2. For each possible species, write the percentage of query cover and the percentage of identity.

Questions

1. Which samples can you identify using only the gene for insulin?
2. Which samples can you *not* identify using only the insulin gene? Why can't you identify them?
3. Which samples can you identify using the gene for cytochrome c alone?
4. Which samples can you *not* identify using only the cytochrome c gene? Why can't you identify them?
5. After combining the results of comparing both genes, can you identify all the samples? Explain.
6. Suggest possible reasons why it was difficult to identify some samples.
7. The content of which package(s) is a serious problem for the restaurant?

You may find the outcome of this project a little bit gross. Unfortunately, the scenario is quite realistic. :(

Table 16.2 Sample identification

	Specimen 1	Specimen 2	Specimen 3	Specimen 4	Specimen 5
Insulin gene					
Cytochrome c gene					

Instructions for Using the BLAST

1. Go to the BLAST search page: https://blast.ncbi.nlm.nih.gov/Blast.cgi.
2. Choose the "Nucleotide BLAST" and copy the entire nucleotide sequence of your sample and paste it into the "Enter Query Sequence" box (top left).
3. Do not change any settings on the page, just scroll down and click the BLAST button.
4. Wait for the result. It may take a while.
5. Scroll the results page until you see the "Sequences producing significant alignments" section and look at the species listed there.
6. For each species, there are several columns. As we are only interested in species-level determination, we will focus on just two of them: the "Query Cover" column and the "Per. Ident" column.
7. Query cover is the percentage of the query sequence (your specimen) that overlaps with the reference sequence. The percentage shown in this column means how much of the sequence you put into the BLAST can be aligned to the DNA of the species suggested by the database.
8. The Per. Ident column shows the percentage of base pairs that are the same in the query sequence and in the reference sequence. This column shows how much of the aligned part of your gene is exactly the same as the corresponding part of the species suggested by the database.
9. To be able to evaluate the identity of your specimen, you need to consider both values. (For instance, it is possible to have a 99% identity match, but only across 35% of your sequence. In this case, you would have no information on how closely the other 65% of your sequence matches up, and therefore, you cannot be certain of the identity of your specimen.)

Viruses in the World of Bacteria

17

Hildegard Uecker

▸ When we think of viruses, we usually think of diseases that affect us and other people around us: influenza, measles, rubella and many more. However, humans are not the only victims of viruses. In fact, the first virus ever discovered did not cause a disease in humans but in tobacco plants. Today, we know that viruses are ubiquitous and infect all forms of life—humans, birds, plants, insects, etc. There is no living species that is not affected by viruses; not even bacteria are safe from them! Some viruses kill bacteria, while others bring benefits to their bacterial hosts. Let's take a closer look at this miniature world of attack and defence!

Bacteria have populated essentially every environment on Earth. They live in soil, in water, and in and on plants, animals and humans. At any given time, any individual has about as many bacterial cells as human cells! They play innumerable roles in our planet's ecosystem and in human health and disease. By infecting bacteria, viruses enter the scene wherever bacteria are. Viruses that infect bacteria are called bacteriophages or phages, which is the Greek word for "bacteria eaters".

H. Uecker (✉)
Institute of Integrative Biology, ETH Zürich, Zürich, Switzerland

Department of Theoretical Biology, Max Planck Institute for Evolutionary Biology, Plön, Germany
e-mail: uecker@evolbio.mpg.de

Bacteria and Viruses

Bacteria and viruses have something in common: They can cause pretty nasty infections. Both can make us cough and sneeze, cause stomach problems or a fever or even—in the most severe cases—lead to death. Infections caused by both viruses and bacteria can be acute or chronic, mild or severe. They even spread in the same ways: through contact with infected people (e.g. via touch, kissing or sex), infected animals (e.g. pets, livestock or insects) or contaminated surfaces, food or water. However, despite all these similarities, bacteria and viruses are fundamentally different.

Bacteria	Viruses
Larger: ca. 200–2000 nm	Smaller: ca. 20–200 nm
Single-celled organisms	DNA or RNA molecules in a protein shell
Reproduce independently by cell division	Need the machinery of the host cell to reproduce
Infections can be treated with antibiotics	Infections cannot be treated with antibiotics

B. Trubenová et al. (eds.), *Everyday Evolution: Unveiling Life's Hidden Patterns*,
https://doi.org/10.1007/978-3-032-07050-0_17

Bacteria are complex, single-celled organisms that can reproduce on their own. They are clearly living organisms. Viruses are pretty simple in comparison, containing only a few DNA or RNA molecules wrapped in a protein shell. They cannot reproduce on their own and are thus usually not considered to be alive. In particular, to reproduce, they must infect a host cell and hijack its reproduction machinery, which often harms the cell.

In the following sections, we will learn about viral infections in bacteria; the role of bacteriophages in experimental evolution, in oceans' ecosystems and in disease treatment; how bacteria defend themselves; and how scientists have turned the bacterial immune system into a revolutionary technology.

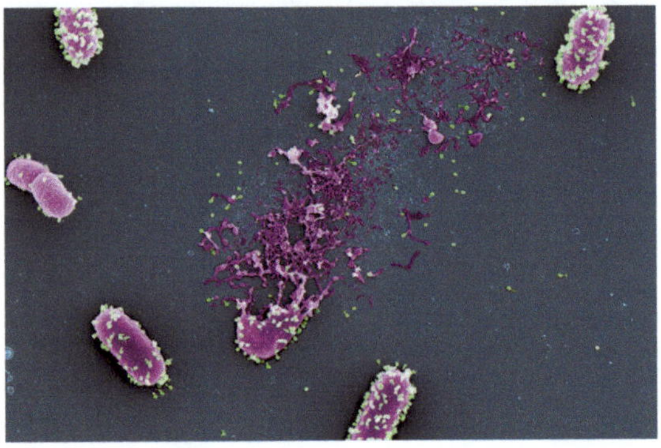

Fig. 17.2 Electron microscopy image of bacteriophages (green) attacking and killing bacteria (purple). Image by courtesy of M. Rohde and C. Rohde (Helmholtz Centre for Infection Research, Braunschweig/Leibniz Institute DSMZ, Braunschweig, Germany) and colorised by Dwayne Roach (Institut Pasteur). ©M. Rohde and C. Rohde. Reproduced with permission

17.1 Viral Attack

How Phages Infect Bacteria

While human or animal viruses enter host cells as a whole, bacteriophages inject their DNA or RNA into the cell. Many phages have tails with which they pierce the bacterial cell wall and membrane. The DNA/RNA is packed under high pressure within the capsid, and for many bacteriophages, it is the release from this pressure that "shoots" the DNA/RNA into the bacterial cell (Fig. 17.1). Sometimes, only part of the DNA enters at first, say 70% of the strand. However, when the bacterial RNA polymerase starts transcribing the viral DNA, the rest is pulled into the cell by various mechanisms. The empty capsid remains outside on the surface of the cell.

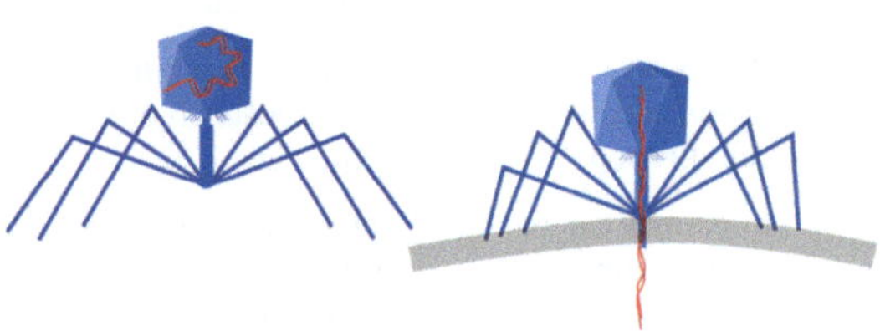

Fig. 17.1 Schematic of a phage. Left: Free virion—the genetic material (in red) is wrapped in the capsid. Right: The phage injects the DNA/RNA into a bacterial cell

Usually, phages have one of two possible life cycles.[1] **Lytic phages** use the host cell machinery to replicate, and once a sufficient number of virions has been produced, they "lyse" (that is, destroy) the cell (impressively shown in Fig. 17.2). In this way, they are released into the environment, where they can infect new cells. **Temperate phages**, on the other hand, integrate into the chromosome of the bacterial cell. Thus, when the bacterial DNA is replicated, the viral DNA is replicated as well and passed on to the daughter cells together with the bacterial chromosome, propagating the phage. This life cycle is called *lysogenic*, and the viral DNA is called a "prophage" during this stage. At some point, triggered by environmental conditions, the phage may switch to a lytic life cycle, causing the bacteria to produce virions, and finally lyse the cell.

Cholera: A Bacterial Disease with a Viral Culprit

Cholera is a diarrhoeal disease that can lead to severe dehydration and even death within a few hours in very severe cases. It is

(continued)

[1]Other life cycles are possible. For instance, some phages cause a chronic infection: They are continuously produced, leaving the cell one by one without killing it.

transmitted via contaminated food or water and is thus particularly prevalent in regions with poor hygienic conditions. It affects an estimated 1.3–4 million people worldwide, causing 21,000–143,000 deaths a year.

Cholera is caused by a bacterium called *Vibrio cholerae*. But here is the twist: The toxin that makes us sick is actually encoded by a virus that was integrated into the bacterial genome. It is passed on from the mother cell to the daughter cells during cell division but can also be horizontally transmitted to neighbouring bacterial cells. Variants of the *Vibrio cholerae* bacterium that do not carry this virus do not cause disease. (On the other hand, carrying the virus is insufficient to make bacteria pathogenic. To be pathogenic, they also need to possess genes that allow them to attach to the intestinal wall.)

Prophages can be useful to bacteria. They often provide the bacterial cell with immunity against infection by certain other types of phages. This can, for example, happen through modification of the cell surface. Temperate phages also sometimes carry genes that prove beneficial to the bacterial host cells. For instance, they might help the bacteria to infect humans or other animals (see Box on Cholera). Prophages can thus provide their carriers with an evolutionary benefit.

(R)Evolutionary Insight from Bacteriophages

When a bacterial population grows in a liquid medium that contains nutrients to feed the bacteria, the liquid will turn cloudy within a few hours due to the increasing number of bacteria. Upon adding phages, the liquid will become clear again very quickly as the phages kill the bacteria. Yet, after some time (hours or days), the liquid will go back to being cloudy. This means that some bacterial variants were not harmed by the phage and were able to multiply. The offspring of these bacteria can withstand the phage as well. This shows that bacteria can somehow become resistant to phages.

By the middle of the twentieth century, the Mendelian ideas of inheritance and mutation were generally accepted, though the structure and function of DNA had not yet been fully understood. However, scientists did not know whether mutations occur randomly or as a reaction to the environment. Thus, is the observed bacterial resistance to the phages due to some sort of mutation that occurred independently of the phage (e.g. even *before* the encounter)? Or does the interaction with the phage *induce* some heritable change in the bacteria? Salvador Luria and Max Delbrück designed an ingenious experiment to find this out.

Salvador Luria let bacteria grow in a liquid in the absence of phages. Then, he spread some of it onto a Petri dish containing a large number of phages. Most of the bacteria died, but the resistant ones formed colonies that he could count. He then repeated this procedure several times. Salvador Luria and Max Delbrück thought about what would happen under both hypotheses—mutations induced vs. occurring independently of phage exposure (Fig. 17.3).

Hypothesis 1: If mutations are induced by the environment, we expect that a certain proportion of bacteria mutate and the others do not. Let's say approximately 1% of the bacteria that encounter the stimulating environment mutate. All these mutations happen (or do not happen) at the time of encounter with the phage. If the experiment is repeated many times, we expect the number of colonies to be roughly similar in each experiment. In other words, the variance in the number of colonies across different experiments would be rather low. (More precisely, it is known that in such situations, the variance is equal to the mean.)

Hypothesis 2: If mutations are independent of the environment, they may arise at any time: a long time before the encounter with the phages, briefly before the encounter or not at all. The early mutations are like "jackpots". The mutated

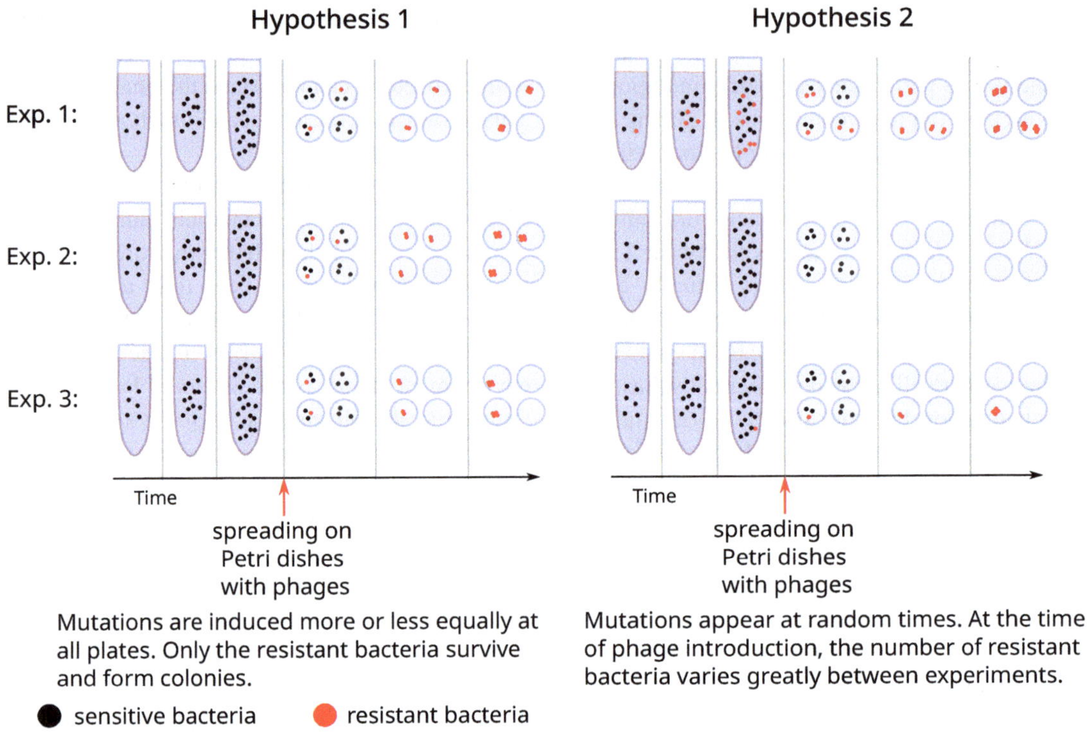

Fig. 17.3 Luria-Delbrück experiment. Illustration of the Luria-Delbrück experiment and the two hypotheses for how mutations occur, as explained in the text. Additionally illustrated: If samples from the same experiment are spread on different plates, the variance in the number of colonies across plates is rather small (equal to the mean) under both hypotheses. (Figure based on Schlegel, S., Genevaux, P. and de Gier, JW. (2017). Isolating *Escherichia coli* strains for recombinant protein production. Cell. Mol. Life Sci. 74, 891-908, https://doi.org/10.1007/s00018-016-2371-2, and Robeva, R. S., and Jungck, J. R. (2023). Fascination with Fluctuation: Luria and Delbrück's Legacy. Axioms, 12(3), 280, https://doi.org/10.3390/axioms12030280)

bacteria will have left many descendants by the time of the phage encounter. Hence, if a "jackpot mutation" occurs, we will count many resistant colonies on the plate. By contrast, if the mutations occur late (or not at all), there would be very few or no colonies. Repeating the experiment many times, we should observe some plates with many colonies and some with none—that is, we expect a high variance in the number of colonies.

Max Delbrück calculated the distribution of colony numbers that one would expect under each hypothesis and compared these theoretical predictions to the experimental results (mathematics was really important to solve the puzzle!).

The experimental observations that Salvador Luria made were consistent with the second hypothesis. Hence, the puzzle was solved: Resistance mutations happen spontaneously and not as a result of some interaction with the phage.[2] The Luria-Delbrück experiment is very important for a second reason. The colony count allows us to estimate the mutation rate of bacteria. For this purpose, similar experiments are still used today.

Viruses in the Ecosystem

Viruses play a major role in the ecosystems of the Earth. As an example, let's look at the world's oceans. Overall, the world's oceans contain on the order of 10^{30} viruses. Researchers calculated that,

[2]Much later, researchers found that stress increases the mutation rate in bacteria. But as far as we know, it increases the rate of *all* mutations, not specifically those mutations that confer resistance.

if we put all these viruses into a line, it would stretch 100 times the distance across our galaxy. If we take one litre of seawater from the surface of the ocean, we will find more viruses than there are humans on Earth (10^{10} versus 8.2×10^9). Every single second, approximately 10^{23} viral infections happen in the oceans of this world. All ocean life, be it whales, fish, plants, algae, bacteria or other microorganisms, is susceptible to infection by viruses. Clearly, this has an immense effect on life in the ocean. Let us look at the viruses in the ocean that infect bacteria.

To begin, phages play a big role in nutrient cycling. About 20–40% of bacteria living in the ocean are lysed by viruses every day. Through the lysis of bacteria and other microbes, nutrients such as carbon or phosphorus are released. This is called **viral shunt** because it makes these nutrients directly available to other microorganisms to uptake. If the microbes were eaten by other, larger organisms rather than being lysed, these nutrients would be transported to higher levels in the food web.

But phages also have less obvious effects on the ecosystem. A striking example relates to photosynthesis. Photosynthesis is best known from plants, which convert water and carbon dioxide into oxygen and sugar using energy provided by sunlight. The sugar can then be used as an energy source for other organisms, and of course, we rely on oxygen for breathing. In fact, about half of all oxygen in the atmosphere is produced not on land but in the ocean, mainly by algae and bacteria (the so-called cyanobacteria). Some of the viruses that infect such cyanobacteria also carry photosynthetic genes, and this can help stabilise photosynthesis in the bacteria. In particular, without prophage infections, photosynthesis in the bacteria is inhibited at high light intensities; the system breaks down. However, with the help of the phage, photosynthesis continues.

Interesting Fact

Sometimes, when the viral DNA is packed into the capsid to be released into the environment, some bacterial DNA is packed in as well. In this way, viruses can acquire new genes. This happened, for example, with the photosynthetic genes in cyanobacteria: Phages actually picked up these genes from the bacteria, and now, both viruses and bacteria carry them. Moreover, viruses can transfer the acquired genes to their new host, a process called **transduction**. Researchers estimate that in the ocean, about 10^{24} genes are moved from viruses to hosts each year via transduction—yet another way viruses affect life and evolution in the sea.

Phage Therapy

As we mentioned before, bacteria can also be pretty nasty. Many diseases—tuberculosis, for example—are caused by bacteria. Bacteria are also responsible for most hospital-acquired infections. We luckily have antibiotics to treat bacterial infections, but antibiotic resistance in bacteria is a huge threat to effective healthcare. A total of 35,000 patients die every year in the European Union (+ Iceland, Liechtenstein and Norway) due to resistant bacteria[3] (see Chap. 6). Now, as we have read, phages kill bacteria. Why not use phages instead of (or in combination with) antibiotics to treat bacterial infections? The foe of our foe is our friend, right? Félix d'Hérelle, a doctor, had this precise idea around 100 years ago. First, he swallowed phages himself to see whether it was safe to swallow phages, and then, he gave them to patients suffering from dysentery, cholera and plague. According to his reports, the phage treatment helped. With the discovery of antibiotics, however, the idea of phage therapy essentially disappeared. Antibiotics worked more reliably. Moreover, doctors felt more comfortable treating their patients with antibiotics than with viruses. However, a scientific institute founded in 1923 in Tbilisi, which specialises in phage therapy, has persisted and continued research and clinical trials. The institute still exists today, and

[3]European Centre for Disease Prevention and Control. Assessing the health burden of infections with antibiotic-resistant bacteria in the EU/EEA, 2016–2020. Stockholm: ECDC; 2022.

with the rise of antibiotic resistance, interest in phage therapy is increasing again. While phage therapy is still experimental, it is sometimes successfully used as a last-resort treatment.

Of course, bacterial resistance to phage attack (as observed by Salvador Luria and Max Delbrück) is a threat to successful phage therapy. Spontaneous mutations can, for example, alter the receptors at the bacterial cell surface to which the phages attach. And resistance through spontaneous mutation is not the only way in which bacteria can escape viral attacks.

17.2 Bacterial Defence

One might wonder if bacteria also have defences against viruses, as humans have immune systems. Indeed, bacteria also possess immune systems (though they are, of course, extremely different from ours). Let's have a closer look.

Innate Immunity

One component of the bacterial immune response are the so-called restriction-modification systems. The main building blocks of these systems are a restriction enzyme and a methyltransferase. Let's see what these two building blocks do and how the system works. The restriction enzyme cuts viral DNA at specific sites, thus destroying it. In this way, it protects the cell from incoming foreign DNA, such as viral DNA. Different restriction enzymes recognise different sequences, and often, the bacterial cell contains more than one type. But how does the bacterium prevent the enzyme from cutting its own DNA? After all, it is quite likely that the bacterial genome contains the same (usually short) sequences that the restriction enzymes bind to. This is where methyltransferase becomes important. The methyltransferase modifies these exact sequences by binding a molecule (a methyl group) to the bacterial DNA. With this molecule in place, the restriction enzyme is unable to bind to the DNA and cannot take action. It is thus a question of what acts first: the restriction enzyme or the methyltransferase. Bacterial DNA is immediately modified by the methyltransferase after replication and is therefore protected. In contrast, incoming viral DNA is most likely cut and degraded as the restriction enzyme is faster than the methyltransferase. Restriction-modification systems are part of the innate immune response of bacteria to phage infection: They are readily available but not flexible—they either fit a particular virus or they don't.[4]

Adaptive Immunity

However, many bacteria also possess a second kind of defence system, which is less specific and more flexible: the so-called CRISPR-Cas system. These constitute the adaptive immune system of bacteria (though not all bacteria have it). The CRISPR locus contains short repetitive sequences interspaced by so-called spacers that are themselves short DNA sequences. The repetitive sequences are (partially) palindromic, i.e. they read the same forwards and backwards. This structure gave the locus its name: CRISPR stands for <u>c</u>lustered <u>r</u>egularly <u>i</u>nterspaced <u>s</u>hort <u>p</u>alindromic <u>r</u>epeats. When this locus was first discovered in a bacterial genome, its purpose was unclear. Later, researchers noticed that the spacers (the DNA sequences between the repeats) match the DNA sequences of phages (and plasmids). Today, we know why they are there: They serve as a memory of prior phage infection and help the cell fight the same virus in the future. There are many different CRISPR-Cas systems acting against phages, but the general principle is the same for all of them. Let us look at the entire process (see also Fig. 17.4).

The build-up of immunological memory requires the acquisition of phage DNA; this step is called "adaptation". It is accomplished by Cas

[4]Another response to phage infection is an abortive infection, where phage-infected bacterial cells commit suicide to prevent phages from replicating and spreading to sister cells. This does not save the cell itself but may save the population.

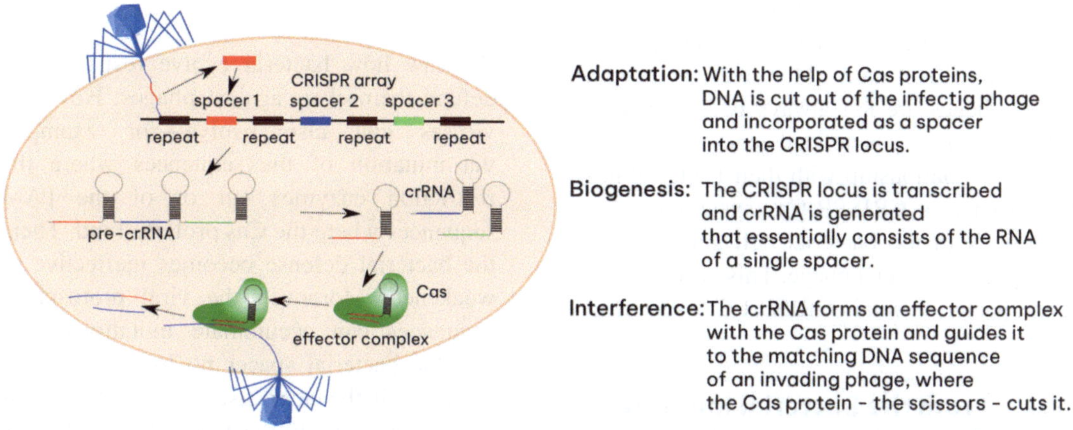

Fig. 17.4 The stages of CRISPR-Cas immunity. The figure illustrates the three steps—adaptation, biogenesis and interference—of the "adaptive immune system" of bacteria. (Figure based on Loureiro A. and da Silva G.J. (2019) CRISPR-Cas: Converting a bacterial defence mechanism into a state-of-the-art genetic manipulation tool. Antibiotics 8(1), 18. License: CC BY, http://creativecommons.org/licenses/by/4.0/)

(CRISPR associated) proteins, which bind to specific short nucleotide sequences in the phage DNA known as *PAM sequences* (protospacer adjacent motif). The Cas proteins then cut out a DNA segment from the phage genome—the *protospacer*—and incorporate it into the CRISPR locus as a new spacer. (Of course, this process is only possible if the phage does not kill the cell, e.g. because it has already been neutralised by restriction enzymes.)

Next, the cell needs to get ready to fight the phage upon reinfection. For this, the CRISPR locus is transcribed, and crRNA (CRISPR RNA) is generated in a step called "biogenesis". The crRNA consists of a single spacer flanked by partial repeat sequences. Together with a Cas protein, the crRNA forms a complex that patrols the cell in search of matching foreign DNA. The role of the crRNA is to guide the Cas protein to the DNA sequence in the phage genome that matches the spacer. If this sequence is next to a PAM sequence, the complex binds, and the Cas protein (the "scissor") cuts and destroys it. This final step is called "interference".

CRISPR-Cas systems thus provide a form of acquired immunity. In this respect, they resemble our own adaptive immune systems that memorise pathogens with the help of memory B cells (see Chap. 7). But there is a fundamental difference: Since the bacterial genome is modified as a result of spacer acquisition, immunity is passed on to the daughter cells and is hence heritable.

New spacers are usually added to the same side of the CRISPR locus. This is exciting as the sequence of the CRISPR locus, therefore, reveals the history of phage infections in the cell lineage. One might then wonder if the bacterial genome could become intolerably large as ever more spacers are added. Yes, it would indeed, but spacers are not only acquired. They can also be lost. Normally, bacteria contain fewer than 50 spacers, but in exceptional cases, there can be several hundred of them.

Interesting Fact

It is very important that the bacterium does not use its CRISPR-Cas system against itself. Otherwise, it would harm itself. First, to avoid autoimmunity, the cell should not use its own DNA as a protospacer, and although this sometimes happens, bacteria have evolved ways to minimise the risk. Second, the cell needs to be able to distinguish the spacer in the CRISPR locus from foreign DNA. This works because the CRISPR locus does not contain the PAM sequence to which the Cas protein binds.

While CRISPR-Cas systems are a great way to fight viruses, they also have disadvantages. One is the risk of self-harm. Moreover, as discussed above, phages are not always bad for bacteria; they can also carry beneficial genes and develop a symbiotic relationship with their bacterial hosts. In that case, a CRISPR-Cas system would be deleterious to the bacterial cell since it would destroy the beneficial phage. This is perhaps why not all bacteria contain CRISPR-Cas systems.

17.3 From the Bacterial Immune System to Genome Editing

In the last 10 years, there has been much talk of CRISPR-Cas9, which is a new, extremely powerful tool for gene editing that allows scientists to edit the genomes of almost any organism by inserting, deleting and modifying genes. As one might guess, there is a connection to the bacterial immune system. Indeed, one of the bacterial CRISPR-Cas immune systems (that uses a Cas protein called Cas9) is the basis of the CRISPR-Cas9 technology. In this specific system, a second RNA molecule, known as trans-activating crRNA (transcrRNA), is needed in addition to the crRNA. TranscrRNA contains the sequence of the repeats of the CRISPR locus. Together, crRNA, transcrRNA and Cas9 form the complexes that find and cleave the target DNA. This structure and mechanism were discovered by Emmanuelle Charpentier, Jennifer Doudna and their collaborators, who published their findings in a seminal article in 2012 and also recognised the potential to turn this into a gene-editing technology. Although the technology is derived from bacteria, it also works in almost all other cells, particularly in eukaryotic (including human) cells, which is quite amazing. In 2020, Emanuelle Charpentier and Jennifer Doudna were awarded the Nobel Prize in Chemistry for their groundbreaking discoveries. How are genes edited with CRISPR-Cas9?

Interesting Fact

We saw how bacteria evolve resistance or defend themselves against phages. However, viruses can also evolve—for example, via mutation of the sequences where the restriction enzymes cut or of the PAM sequences where the Cas proteins bind. Then, the bacterial defense becomes ineffective or weakened. Moreover, the viral protospacer sequences can accumulate mutations such that the bacterial spacer no longer matches. However, in the latter case, some CRISPR-Cas systems are able to react quickly. While they do not provide good immunity to the mutated phage, the old crRNA-Cas complexes with the almost-matching spacer are used to speed up the acquisition of new, perfectly matching spacers. This is termed "primed adaptation" in contrast to the "naive adaptation" that occurs when the bacterial cell has not encountered any version of the phage before.

Instead of the crRNA of the natural system, scientists synthesise an RNA fragment of about 20 nucleotides—the gRNA ("guide RNA")—that corresponds to the fragment in the DNA that they want to edit. They combine it with the transcrRNA to obtain the sgRNA ("single guide RNA"). Last, the sgRNA forms a complex with the Cas9 protein that is delivered into the cell whose genome is supposed to be edited. The "g" stands for "guide" since the RNA molecule guides the complex to the corresponding gene in

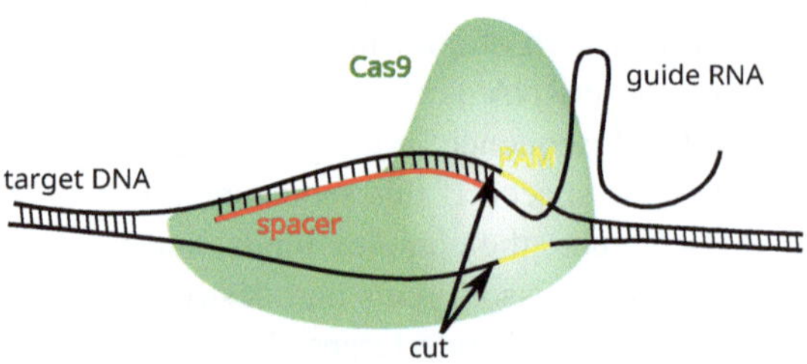

Fig. 17.5 Schematic of the CRISPR/Cas9 complex targeting DNA. The guide RNA directs the Cas9 protein to the target DNA, matching the spacer. The Cas9 protein binds to the adjacent PAM sequence and cuts the DNA

the genome, where it cuts the DNA (Fig. 17.5). As we saw above, for this to work in a bacterial cell, a PAM sequence must be adjacent to the protospacer. This is the same here: Scientists need to choose DNA sequences that are close to a PAM sequence for editing. This restricts the possible sequences to some extent. Now that the DNA is cut, the cell will repair it. Normally, this will not work perfectly, and the gene is broken. In this way, CRISPR-Cas9 can be used to knock out genes. However, scientists can also provide the cell with a DNA repair template that contains additional or altered DNA and matches the broken genome pieces at the ends. In this way, scientists can introduce mutations or add entire DNA sequences into the genome.

The CRISPR-Cas9 system was the first one to be turned into a genome-editing tool. Since then, additional technologies that are based on other bacterial CRISPR systems with different Cas proteins have been developed. For instance, a close cousin of the Cas9 protein, Cas13, is able to edit RNA instead of DNA. During protein production, DNA is transcribed to RNA, which is then translated into a protein. By editing RNA, CRISPR-Cas13 can be used to modify the expression of genes without the permanence of a DNA edit.

CRISPR-Cas genome editing is a revolutionary technology that not only lays the ground for future scientific discoveries but also for innumerable applications, for example, in agriculture and disease treatment. One of its advantages is that it is cheap, quick and easy (compared to other gene editing techniques).

Applications of CRISPR-Cas Genome Editing in Agriculture and Medicine

Following the discovery of the CRISPR-Cas mechanism and the development of the technology, scientists have quickly started to perform numerous studies to push its application in agriculture and disease treatment. Let us have a glimpse at some of the possible applications. The CRISPR-Cas technology can, for instance, be used to develop crops that are more nutritious, delicious, drought-tolerant and/or resistant to pests, and numerous successes have already been achieved. For example, researchers successfully applied CRISPR-Cas to generate wheat that is resistant to powdery mildew, a fungal disease. Other researchers are trying to use it to make bananas resistant to the Panama disease, a fungal disease that is threatening banana production all over the world. Moreover, CRISPR-Cas has great potential to treat genetic diseases such as Huntington's disease, cystic fibrosis and sickle cell anaemia. The principle behind this is to edit the genes in the concerned tissues. E.g. to treat sickle cell anaemia, the genomes of stem cells in the bone marrow (from which the red blood cells are continuously generated) are modified to produce healthy haemoglobin. In 2023—a mere decade after the seminal paper by Emanuelle Charpentier and Jennifer Doudna—CRISPR-Cas treatment for sickle cell anaemia and for transfusion-dependent β-thalassemia has been approved as the very first CRISPR-Cas medicines in the UK, the EU and the USA. Furthermore, in 2025, it was reported that CRISPR-Cas has been successfully applied as personalized medicine to treat a baby with a rare genetic disease—the first time of such a personalised application. Importantly, these treatments do not involve modifications of DNA in the germline, meaning that the modifications are not inheritable. Most researchers consider editing genomes in the germline to be unethical. Another potential medical application is the treatment of infections caused by human viruses that incorporate their DNA in the human genome—such as HIV. In these cases, CRISPR-Cas could be used to cut the viral DNA out of the human genome. In *in vitro* experiments, scientists have managed to remove the HIV genome from T-cells obtained from infected patients. Tests in animals and in-human trials are on the way.

Questions of Safety and of Ethics

As with all new technology, safety is a big concern. Moreover, ethical questions may arise—how should we be allowed to use gene editing?

Crops: While CRISPR-Cas can be used to insert entirely new genes, other applications rely on the modification of existing genes that could also occur naturally, as in traditional breeding. In traditional breeding, mutations occur randomly, and irradiation is often used to increase the number of new mutations and, hence, the genetic variation. In contrast, with CRISPR-Cas, specific mutations can be purposefully introduced, tremendously speeding up the development of new crop variants. Although these crop variants are no different from those that could arise naturally, the Court of Justice of the European Union decided in July 2018 that *all* CRISPR-Cas-edited crops fall under the regulations for genetically modified organisms. This means that they must undergo a long approval process before they can be cultivated.

Humans: While the CRISPR-Cas technology is normally precise, some small fraction of edits happen at the wrong site. Moreover, some studies have shown that CRISPR-Cas editing may select for cancer-prone cells. Another problem is potential immune reactions to the editing tool. All these risks need to be assessed and minimised. Furthermore, especially when it comes to editing cells in the germline or of human embryos, where the future individual will carry the edits in many or all the cells of its body, one huge risk is that we do not understand the function of most genes well enough. Therefore, the edit might well have deleterious side effects that were unforeseeable.

Apart from these risks, there are also concerns that editing embryonic cells may not only be used for disease prevention but also to "design" babies (although we are still far from being able to do this). This leads to very fundamental ethical questions: What are we allowed to do? Will we value some human life (with the desired characteristics) more than others? What does it imply for the humans that are born from gene-edited embryos?

These risks and ethical concerns are why most scientists are not (yet) using CRISPR to edit human DNA in the embryo or germline. However, in November 2018, a scientist in China announced the birth of twin girls whose genes he had edited with CRISPR-Cas. The modification was supposed to make them resistant to HIV. The girls had been born in secrecy in October 2018. This experiment has been widely condemned as immoral and irresponsible. The scientist, He Jiankui, was sentenced to 3 years in prison on 30 December 2019.

Tiny as they are and not even properly alive, phages have a huge influence on our world. By infecting bacteria in natural environments, they contribute to shaping the ecosystems of the Earth. By infecting human pathogens, they can be harmful or beneficial to human health. Thanks to their small sizes and their fast replication, bacteria and phages are great study systems in the lab, in addition to their roles in the environment. In this way, they have contributed to fundamental scientific discoveries and the development of new technologies. We don't know what new breakthroughs the future will bring, but they will certainly be exciting!

DEFINITIONS

CRISPR: locus in the genome of many bacteria that contains short repeat sequences and spacer sequences and is part of the adaptive immune system

Spacer: sequence in the CRISPR locus that corresponds to a DNA fragment of a previously encountered phage (or plasmid)

Protospacer: DNA fragment in the viral genome that is incorporated as a spacer

Cas: proteins that are involved in spacer acquisition and in the cutting of invading phage DNA

Cas9: specific Cas protein

crRNA: RNA sequence of a spacer that is used to guide the Cas proteins to the corresponding position in the viral genome

transcrRNA: additional RNA molecule in the CRISPR-Cas9 system needed to build the CRISPR-Cas complex

PAM: short DNA sequence that is required for attachment of the Cas proteins to the viral genome

gRNA: RNA fragment in the synthetic system that corresponds to the crRNA of the natural system

sgRNA: gRNA + transcrRNA

SUGGESTED READING

- Luria SE, Delbrück M. Mutations of Bacteria from Virus Sensitivity to Virus Resistance. Genetics. 1943;28(6):491–511.
- Berryhill BA, Gil-Gil T, Smith AP, Levin BR. The future of phage therapy in the USA. Trends Mol Med. 2025;31(11):982–991.
- Suttle, C. Viruses in the sea. Nature. 2005; 437:356–361.

17.4 Questions

1. **Which of the following statements is/are correct?**
   (A) All bacteria cause diseases in humans.
   (B) All viruses cause diseases in humans.
   (C) All viruses need bacterial cells to reproduce.
   (D) Most viral infections can be treated with antibiotics.
   (E) Suitable viruses can sometimes be used to treat bacterial infections.
2. **What is necessary for viral reproduction?**
   (A) Viral DNA/RNA
   (B) Cell enzymes
   (C) Bacterial cell
   (D) Free nucleotides
   (E) Ribosomes
3. **Viruses are capable of which of the following?**
   (A) They can metabolise organic compounds.
   (B) They can use the host cell to make copies of themselves.
   (C) They can have their proteins made by the host cell.
   (D) They can mutate and evolve.
   (E) Some viruses are able to reproduce on their own, given the right humidity and temperature and the availability of suitable nutrients.
4. **Diphtheria is a bacterial infection of the upper respiratory tract. If it remains untreated, the mortality rate is high. The bacterium that causes diphtheria is called *Corynebacterium diphtheriae*. The deleterious effect on our health is caused by a toxin—the diphtheria toxin—that kills human cells. The *tox* gene encoding this toxin is located on a prophage. Which of the following statements is/are true?**
   (A) Strains of *Corynebacterium diphtheriae* that do not carry the prophage cause diphtheria as well.
   (B) Since the toxin is produced by a virus, diphtheria is a viral disease. It thus cannot be treated with antibiotics.
   (C) Vaccination protects against diphtheria.
   (D) Antitoxins can be used for treatment.
   (E) Mutations in the *tox* gene can lead to the production of an altered nontoxic protein.
5. **The virulence of the *E. coli* strain reported to cause food poisoning is mainly caused by genes from a virus that infects bacteria. Which of the following statements correctly describes why this *E. coli* strain is harmful to humans or describes the events that result in this *E. coli* strain being harmful to humans?**
   (A) The virus infected a bacterial cell that immediately started to produce new viruses.
   (B) The virus entered the cell and accidentally took some bacterial genes, resulting in the next generation of viruses being more harmful to humans.
   (C) The virus infected the bacterial cell, the viral genome became part of the bacterial DNA, and so the copy of the viral genome was present in the whole population of bacterial cells originating from it.
   (D) The virus infected the bacterial cell, which burst and released thousands of viruses within a few hours.
6. **Which of the following statements about the CRISPR-Cas9 technology are correct? CRISPR-Cas9...**
   (A) is a new tool for DNA editing.
   (B) is based on a defence mechanism in bacteria.
   (C) is based on the viral immune system against bacteria.
   (D) does not work in cells of the germline.
   (E) does not work in eukaryotic cells.
7. **Consider a population of bacteria that do not possess any CRISPR-Cas system. Which of the following statements is/are correct?**
   (A) When phages are added to the bacterial culture, beneficial mutations are induced that allow the bacteria to survive.
   (B) When phages are added to the bacterial culture, some "lucky" bacteria may contain beneficial mutations that allow them to survive.
   (C) If bacterial mutations were induced by the addition of phages, we would expect all bacteria to survive.

(D) If bacterial mutations were induced by the addition of phages, we would expect more or less similar numbers of bacteria to survive and produce a colony every time we repeat the experiment.

(E) If bacterial mutations were independent of the environment, we would expect more or less similar numbers of bacteria to survive and produce a colony every time we repeat the experiment.

17.5 Project: Your Own Title

CRISPR is a hot, important topic that is not only exciting for biologists—it could change mankind and nature forever, and it should be of great interest to every human. This time, you won't need any colourful paper, scissors or dice. Instead, you will need to look for reliable information, read, think, think even more and write!

Write a short essay (about 700 words) on the CRISPR technology. Follow the instructions below to write a good essay. You can use AI tools, but remember that you are responsible for the content and the accuracy of information.

1. Give an example of a specific hereditary disease that could potentially be treated with this technology. Briefly describe the disease.
2. Describe in detail one possibility of how CRISPR could be used to treat this disease: which part of the genome would need to be edited, which cells would be treated, at what age or life stage should the patient be treated, etc.
3. Describe how the CRISPR editing procedure could change the condition caused by the deficient genes or what in the human body would work differently following the procedure.
4. Describe the possible drawbacks and ethical considerations that would arise if the CRISPR method of gene editing became widely used.
5. Suggest some government regulations and guidelines that should be followed by scientists and medical workers using CRISPR to edit the human genome.
6. Cite your references and provide a bibliography.

Express yourself clearly and concisely. When rereading your essay, ask yourself:

7. Did you argue well why the chosen disease would be suitable to be treated by CRISPR?
8. Did you explain clearly how your proposed procedure would work?
9. Do your statements and reasoning make sense?
10. Are your ideas based on clearly stated facts?
11. Is it easy to understand your thoughts?
12. Do sentences follow a logical order and make sense together?

This project was developed by Barbora Trubenová and Kristína Hudáková.

18 Genetically Modified Organisms: History, Applications and Social Concerns

Jarmila Šimková

▶ Most of us have seen the label "GMO-free" at the supermarket. What does it mean? Why does society raise concerns about genetically modified organisms (GMOs), and why do many countries impose strict regulations on GMOs? In this chapter, we will delve deeper into this fascinating topic and address these questions.

The debate over GMOs has been going on for almost 40 years. The first GMO was created in 1973 by biochemists Herbert Boyer and Stanley Cohen. They constructed a **recombinant DNA**—DNA that combines genetic material from multiple sources—consisting of a plasmid into which they artificially inserted a kanamycin antibiotic resistance gene.[1] This recombinant DNA was then inserted into *E. coli* cells, providing them with a trait not previously found in that organism: kanamycin resistance!

But what are GMOs? Our ancestors have been bringing about changes in the genetic composition of plant and animal populations for thousands of years with artificial selection and crossbreeding. Individuals with desired traits would be selected to produce offspring that would inherit those traits. Over many generations, this resulted in dramatic genetic changes. Many of the plants and animals we know today did not look the same thousands of years ago—they changed drastically in the process of domestication (see Chap. 1). However, these processes are NOT considered **genetic modification** because selection, mating and recombination occur naturally. A GMO is defined as an organism whose genetic material (DNA) has been modified in a way that does not occur naturally by mating or natural selection. Genetic material can be modified by random mutagenesis and by directed mutagenesis. Random mutagenesis relies on chemicals or radiation (e.g. UV) that induce random mutations. Scientists or breeders must then select desirable mutants. Directed mutagenesis, also called **genetic engineering**, employs techniques that allow site-specific changes in DNA. In this chapter, we will focus on the techniques of genetic engineering.

Many countries strictly regulate or even ban GMOs, especially for crops, because of concerns about human and animal welfare as well as the environmental impact of releasing GMOs into nature. Despite these concerns, genetic modification techniques—such as the insertion of new genes, deletion of existing ones or modification of their

[1] We will talk more about plasmids and their function later in this chapter, as well as discuss the techniques used by the two scientists.

J. Šimková (✉)
eduB, Bratislava, Slovakia

KBVE FEE Technical University in Zvolen, Slovakia
e-mail: jarmila.simkova@edub.sk

B. Trubenová et al. (eds.), *Everyday Evolution: Unveiling Life's Hidden Patterns*,
https://doi.org/10.1007/978-3-032-07050-0_18

expression—have proven useful in fields ranging from medicine to environmental science. Since its invention, applications of recombinant DNA have drastically improved the lives of millions of people worldwide. Let's have a look at a few examples. One example is insulin production for patients with diabetes. Insulin, a hormone produced by the pancreatic islets, lowers blood glucose levels and is insufficient in diabetic patients. Thus, patients suffering from diabetes can be injected with insulin to maintain blood sugar levels. Before 1977, scientists extracted this hormone from pigs and cows, but the amino acid sequence of pigs' and cows' insulin slightly differs from that of humans, and also, the supply of the animals could not keep up with the increasing demand. Since the amino acid sequence for human insulin was known, scientists were able to design and synthesise a nucleotide sequence that was compatible with the molecular machinery of *E. coli*. At the time, researchers already knew that protein chain A and protein chain B make up the structure of insulin. Thus, the gene for each of the chains was separately inserted into two plasmids. The plasmids were then inserted into *E. coli* cells, creating new, transformed organisms. After the *E. coli* expressed protein chains A and B, they were purified and joined *in vitro* to form active insulin. Thus, insulin became the first protein to be produced by recombinant DNA technology.

Another example is the production of enzymes by microorganisms for use in laundry detergents, which improves stain removal. Stains are composed of proteins, lipids and carbohydrates—polymers that can be broken down into smaller oligomers (shorter chains of monomers) by specific enzymes. These oligomers then more easily separate from the textile fibres and are washed away. Some enzymes can be active (able to break down polymers) already at 30°C, which can potentially save energy and household costs. Since enzymes are proteins, microorganisms that naturally produce them can be genetically engineered to produce higher yields, and organisms that do not produce them naturally can be modified to do so. For example, filamentous fungi are used for the large-scale production of enzymes that we can find in many laundry detergents.

But how do we manipulate the genome of a particular organism to create a genetically modified organism? And, more importantly, how do we manipulate the genome of a particular organism? Over the decades, many techniques have been developed. One way to construct a GMO is through recombinant DNA, which involves several crucial ingredients and steps. As we mentioned earlier, recombinant DNA joins DNA from at least two different organisms together. It is created by inserting a specific gene from one organism into the genome of a different host organism using a plasmid vector. The joining of a gene and a plasmid is usually done with the help of various enzymes that interact with DNA.

After creating recombinant DNA, it needs to be introduced into a host cell for replication and expression, a process called **transformation**. This is not a simple task, since the structure of prokaryotes differs from that of eukaryotes, and specific transformation techniques must be used for different organisms. In addition, not all cells would have accepted the new, foreign DNA; therefore, we must select and cultivate only the cells that have accepted it. For this, we can apply **selection techniques**. Overview of the steps can be seen in Fig. 18.1.

18.1 Crucial Ingredients

Now that we have a general idea of the big picture, we will examine the crucial ingredients. The most important ones are (1) the gene of interest we want to insert, (2) plasmids to deliver the gene into the target cell, and (3) restriction enzymes to cut DNA at specific locations, enabling the gene of interest to be integrated into the plasmid. Let's take a closer look at them.

Gene of Interest

In the genome of the fungus *Aspergillus carbonarius*, the *frd* gene codes for the enzyme fumarate reductase. This enzyme is responsible for turning fumarate into succinate inside fungal cells. Succinate is a building block that can be

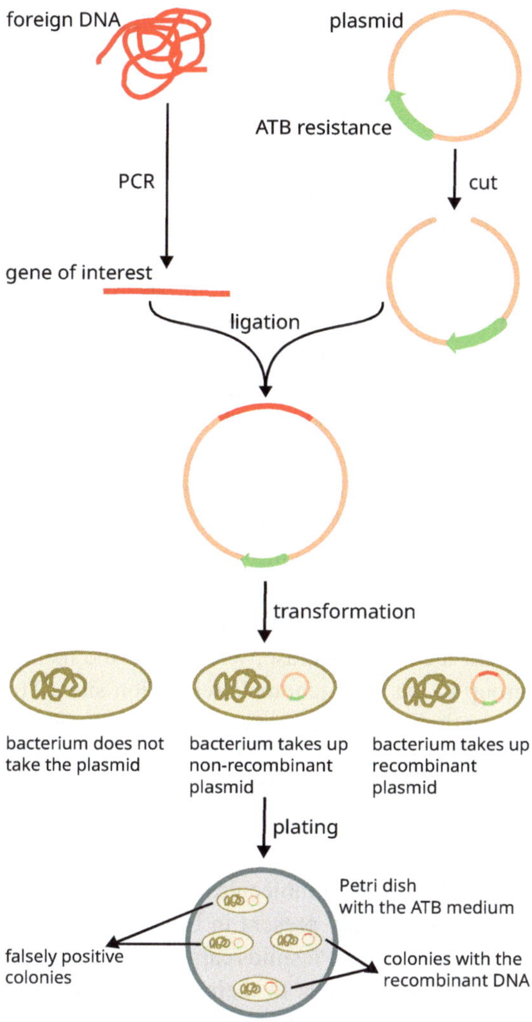

Fig. 18.1 Constructing a GMO. A gene is copied from a foreign DNA by PCR. The gene is then inserted into the cut-open plasmid that already contains an antibiotic resistance gene. The new recombinant DNA is transformed into a host bacterium. The bacterium can either not take up the plasmid, take up the original plasmid without the gene or take up the recombinant plasmid. Only the cells that took up the plasmid will survive on a medium containing the antibiotics. We must screen for the cells with the recombinant plasmid

used in bioplastic production. A closely related fungus, *Aspergillus niger*, lacks this enzyme. However, scientists want to modify *Aspergillus niger* to produce succinate as well because it grows in an environment with a lower pH. Therefore, it can tolerate higher concentrations of succinate, making its production more efficient and thus cheaper. Therefore, they decide to insert the gene *frd* from *A. carbonarius* into *A. niger*. Thus, *frd* is the **gene of interest**.

But how do we obtain our gene of interest? First, we must determine its nucleotide sequence. For genes that are already known and described, we can simply search in a genome database, such as NCBI (National Center for Biotechnology Information) and JGI (Joint Genome Institute). These databases provide us with a list of organisms known to have the gene. If we suspect that a gene might be present in an organism but do not know its exact sequence, we can use bioinformatics tools to identify in which other organisms a similar or even the same gene is present and use this information to infer the nucleotide sequence in our organism. Once we know the nucleotide sequence of our gene of interest, we must obtain many copies of it. Why more than one copy? Later in the process, some copies may not be incorporated into the plasmid, and their number may decrease. Moreover, even if they are incorporated into the plasmid, not all of them will make it into the cells. We have already read about polymerase chain reaction, or PCR, which enables us to obtain millions of copies of a gene from a single copy (see Chap. 16). PCR utilises primers—short sequences that bind to specific locations at the two ends of our gene of interest and initiate copying of the gene by DNA polymerase. Because primers are designed in a specific way, only the fragment between the primers will be amplified, resulting in a sample containing multiple copies of the gene of interest.

Restriction Enzymes

Restriction enzymes are a type of endonucleases that cleave both strands of DNA at specific sequences known as restriction sites and are primarily produced by bacteria. They provide the foundation for recombinant DNA technologies and have improved the development of medical drugs. In 1978, Werner Arber, Dan Nathans and Hamilton Smith received a Nobel Prize for the discovery and description of these enzymes.

Restriction enzymes evolved as a defence mechanism in bacteria and archaea against

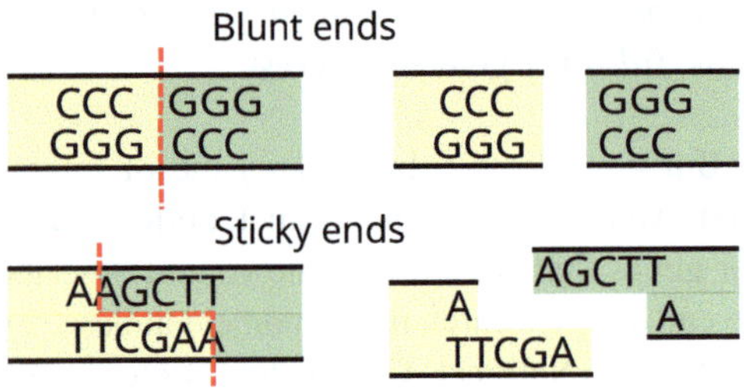

Fig. 18.2 Blunt and sticky ends. Blunt ends result from restriction enzymes cutting the double-stranded DNA on the same spot on both strands (in the middle of the palindromic sequence). There are no overhangs on the resulting DNA fragments. Sticky ends are created by restriction enzymes that cut elsewhere, but after the same base on both strands. Notice that the palindromic sequence is not read on the same strand, but from left to right on one strand and from right to left on the other strand

invading viruses. Scientists first noticed them when studying that different bacterial strains showed varying levels of resistance to specific bacteriophages—viruses that infect bacteria (see Chap. 17). Later, it was discovered that bacteria possess a so-called **restriction modification system** that cleaves invading viral DNA into fragments. Restriction enzymes are classified into five types depending on where they cut the genome. In genetic engineering, enzymes belonging to type II are of special interest because, unlike other types of enzymes, we *know the cleavage site precisely*. These enzymes recognise restriction sites of **palindromic**[2] sequences, usually 4-8 base pairs long, on the viral genome and cut the viral DNA within the restriction site. For example, *HindIII*, isolated from *Haemophilus influenzae*, recognises DNA at 5′AAGCTT 3′ and cuts it after the first adenine on both strands. Notice the overhangs of DNA strands in Fig. 18.2, which we call **sticky ends**. **Blunt ends** are produced by *SmaI* from *Serratia marcescens* by cleaving after the last cytosine in the 5′CCCGGG 3′ sequence. Sticky ends can ligate (seal together) based on complementarity (A-T, C-G) to other DNA fragments, which we will see later in the context of cloning strategies.

[2]Palindromic means that the DNA sequence is the same from 5′ to 3′ on both strands.

We described the restriction part of the restriction-modification system. So, when and why does the modification part happen? Since the restriction sites are relatively short, it is highly probable that they also appear in the genome of the host cell itself. The probability that the restriction site occurs in the host's genome is $1/4^4$ for restriction sites that are four base pairs long and $1/4^8$ for a site that is eight base pairs long. This might seem to be a small number, but in the genome of *Haemophilus influenzae*, which has 1.8 million base pairs, there could be, in theory, approximately $1{,}800{,}000/4^6$, i.e. 440 recognition sites for *HindIII*, meaning its genome would be cut into 441 smaller pieces! So, how does the host cell protect its DNA? The trick lies in a simple modification: the addition of a methyl group (methylation) to the host cell's genetic material. This is achieved using a modification enzyme called methyltransferase, which adds the methyl group to the host genome's restriction sites. The restriction-modification system thus ensures that the cell is protected against bacteriophages while not degrading its own DNA.

Plasmids

Plasmids—circular molecules of double-stranded DNA—are mainly present in bacteria, yeast and some other fungi. Plasmids carry information that can be advantageous to their host, for example, for antibiotic resistance proteins. They can be transferred vertically (during cell division from the mother cell to the daughter cells) or horizontally (from one cell to another, non-daughter cell). See Chap. 6.

A plasmid is an autonomously replicating unit, which means it replicates independently from the chromosomal DNA but only within a suitable host. Plasmids have their own **origin of replication**—a specific location on their DNA where their replication is initiated. It also determines in which organism the plasmid is able to replicate and the number of plasmid copies in a cell. Scientists can now construct artificial plasmids for various purposes, for example, by cutting DNA sequences with restriction enzymes and ligating them together. They can even visit the website of

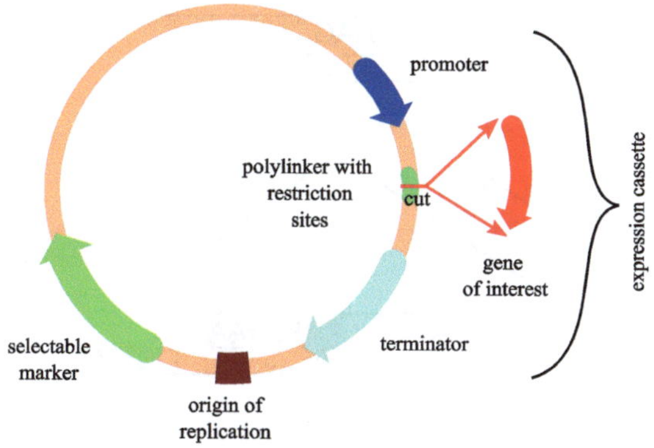

Fig. 18.3 Plasmid map. Plasmids are usually depicted in a plasmid map with all the elements

a company, browse their catalogues and choose a plasmid—just another kind of online shopping! This has been very useful, as it allows us to study the functions of novel genes or express known genes in other organisms by integrating them into plasmids and subsequently into cells. These artificial plasmids contain additional useful elements, for example, an **expression cassette**. It consists of a promoter, terminator and a place between them (**polylinker**) where we can insert our gene of interest. Polylinker contains many restriction sites, so different restriction enzymes can be used to cut it at the desired location. An expression cassette ensures that the gene is expressed (transcribed and translated) in the cell.

Another useful element that can be inserted into a plasmid is a **selectable marker**. Using a selectable marker, we can distinguish between cells that carry the plasmid with the gene of interest from those that do not. Very often, a selectable marker is a gene coding for antibiotic resistance, making the test very simple: only those bacteria that contain the marker (and thus the plasmid) can grow in a medium containing antibiotics (Fig. 18.3).

18.2 Creating GMOs

Now that we have described the ingredients necessary for genetic modification, let's look at the most important steps in this process.

Cloning Techniques

First, we need to integrate our gene of interest into a plasmid backbone to create a circular molecule of recombinant DNA. **Cloning technique** is a general term that refers to the process by which we create a plasmid containing the gene of interest.

Cloning techniques are based on the fact that when two DNA molecules are cut by the same restriction enzyme, they have the same ends and can thus be joined together by a complementary base pairing—this is exactly what we use in cloning techniques. First, we must find a restriction enzyme that cuts our plasmid at the polylinker site. Then, this same restriction enzyme must cut the gene out of its flanking regions that surround it. However, at the same time, the open reading frame of the gene *cannot* contain restriction sites for this particular enzyme, or else, it would disrupt the gene's function.

Now, we will describe the two main strategies used to insert the gene of interest into the plasmid. In the first, known as the **restriction enzyme strategy**, we use the same restriction enzyme to cleave both the plasmid and the gene of interest, thus creating complementary sticky ends (Fig. 18.2). Enzymes that introduce sticky ends are usually used because they are more efficient at constructing recombinant DNA compared to enzymes introducing blunt ends.

For example, let's take a gene of interest that contains the restriction site for *HindIII* in its flanking regions. The plasmid we want to use contains a polylinker that also has the restriction site for *HindIII*. If we mix the gene and the plasmid with *HindIII* (we do this in separate tubes in the lab), we will end up with 5′A(cut)AGCTT 3′ sticky ends on both. After mixing the already cleaved plasmid and the gene in one tube, they bind together based on complementarity. An enzyme, DNA ligase, must also be present, as it glues the ends of the two DNA fragments.

There is one important catch. Since both the plasmid and the gene have two ends, we must ensure both ends are correctly ligated. The promoter on the plasmid always initiates transcription in one direction, so the orientation of the gene plays

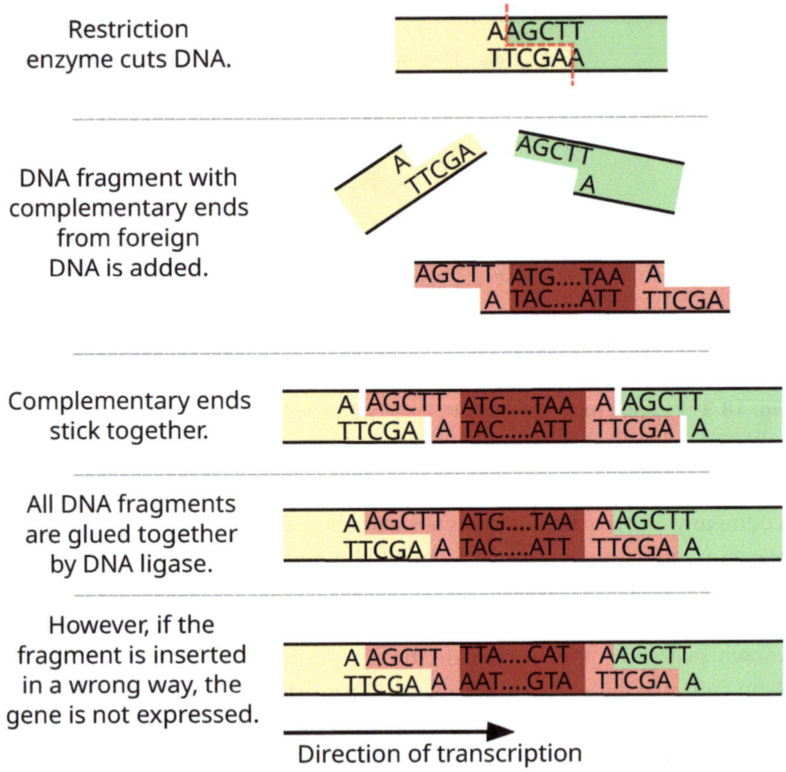

Fig. 18.4 Inserting a gene of interest into a plasmid DNA. If both ends are cut with the same enzyme, the gene can be inserted in two ways, but only one leads to a functioning gene

an important role in the transcription and translation of the gene. For instance, if we used only one type of restriction enzyme, the gene could be flipped around and inserted into the plasmid in the wrong direction, causing the transcription of the reverse gene sequence (Fig. 18.4). Furthermore, the plasmid could close again without the gene being incorporated. To avoid these problems, we need to apply two different restriction enzymes. One enzyme cuts the upstream flanking region of the gene (before the open reading frame) and the corresponding region on the plasmid. The second enzyme cuts the downstream flanking region (after the open reading frame) and the corresponding region on the plasmid.

Gibson assembly is another cloning strategy used for joining a gene of interest and a plasmid. It is mainly applied when we are limited to the use of a single restriction enzyme, for example, when using plasmids containing only one restriction site. As in the previous technique, we open up our plasmid with one chosen restriction enzyme. However, instead of applying the restriction enzyme to the gene, we use PCR to amplify the gene. Normally, we would design the primers that align perfectly at the beginning and the end of the gene. In this technique, however, we design primers with overhangs that match the sequences on the plasmid around its cleaved site. Thus, the amplified gene contains adjacent flanking regions that are *exactly* the same as the parts of the plasmid. Designing the primers is usually carried out by bioinformatics programs that analyse the best nucleotide sequence. Then, we mix the opened plasmid, the amplified DNA of the gene with flanking regions and a mix of enzymes called the Gibson assembly master mix. This merges the plasmid with the gene of interest, and we have our recombinant DNA!

Transformation

After constructing a recombinant DNA, either by restriction enzyme strategy or by Gibson assembly, we still need to get it into the organism that we want to genetically modify. For this, we use so-called **transformation techniques**. Different transformation protocols are used for filamentous fungi, yeast and bacteria as their cell structures vary, but the aim is the same: to make the host cell accept the recombinant DNA. Some bacteria, for example, the species of the *Bacillus* genus, happily accept recombinant plasmids directly from the environment around them. However, most other species require a bit more "persuasion", for instance, by increasing the permeability of their cell membrane by chemical treatment, heat or electric shock.

Selection

After transformation, we must verify that both the cloning part (construction of the recombinant DNA) and the transformation into the host cells were successful.

As we mentioned before, plasmids used in genetic engineering contain selectable markers, for example, a gene for beta-lactamase, which provides resistance to ampicillin. If we plate the transformed host cells, for example, *E. coli*, which is normally sensitive to ampicillin, on a medium containing this antibiotic, cells either grow or die. If the bacteria grow, the transformation must have been successful, and the cells possess a plasmid with the selectable marker gene—the resistance gene. Of course, we must also make sure that the *E. coli* strain we transform in this case does not naturally contain an ampicillin resistance gene. Otherwise, we would not be able to conclude that the resistance was acquired via the plasmid. Even though we selected colonies based on the selectable marker, we cannot exclude the option that the host cells took up only the plasmid without the gene of interest.

As we discussed above, it is also possible that the plasmid closed before taking up the gene of interest, but the transformation was successful. Such bacterial colonies would contain plasmids with the antibiotic resistance gene but without our gene of interest—these are called falsely positive colonies. Therefore, screening of these resistant colonies using PCR and gel electrophoresis is necessary to determine whether or not the gene of interest was really incorporated into the vector (see Box: Colony PCR and Gel Electrophoresis).

After successfully identifying the modified colonies containing the gene of interest, we can select and cultivate them in a corresponding medium and store them for later use. Now, we have constructed our own genetically modified organism!

The restriction enzyme cloning technique was used and developed by Boyer and Cohen, and it launched a new area of molecular biology and cellular process research. Since then, many methods have been developed to produce GMOs, and in 2012, another breakthrough technique was described, the CRISPR/Cas9 system. This technique was honoured with a Nobel Prize in 2020. You can find more about this in Chap. 17.

18.3 Case Studies and Concerns

Let's go back to the concerns about GMOs. There is a big debate over genetically modified crops, their environmental impact and the consequences of incorporating them into the food chain. Agricultural crops are mainly modified to improve resistance to insects and viruses and herbicide tolerance and enhance nutritional content. This allows for more prudent use of insecticides and herbicides, increased yields and lower production costs. Moreover, genetically engineered crops could help alleviate food shortages and malnutrition in regions with limited food access, though GMOs are still not widely adopted there. In other places, GM crops have become ubiquitous; for instance, the majority of corn and soybean crops in the USA are now genetically engineered.

However, many scientists and environmentalists point to the many negatives of GMOs. For instance, one drawback of genetic engineering is that upon insertion of a new protein-coding gene, the resulting protein interacts with other cell compartments and proteins. Transgenic DNA might break down and integrate into other parts of the chromosome, which can, for example, lead to the production of allergenic compounds.

Colony PCR and Gel Electrophoresis

One method that allows us to distinguish the falsely positive colonies from the mutants is the use of PCR. The key idea comes from the length of the amplified fragment. We simply design the primers that anneal on the plasmid backbone before and after the sequence at which the gene of interest should be inserted. Successful insertion of the gene of interest will result in a longer amplified fragment compared to the fragment amplified without the insert. Gel electrophoresis is used to visualise these fragments and determine their size. This technique exploits the natural negative charge of DNA, causing the fragments to migrate towards the positive electrode when an electric field is applied across an agarose gel. The agarose gel matrix acts as a molecular sieve, slowing down the larger fragments while allowing smaller fragments to move more quickly. Samples from different bacterial colonies are loaded into separate wells in the gel, allowing for direct comparison. A DNA ladder containing fragments of known lengths is loaded into a separate well and serves as a standard to estimate the size of the amplified fragments in each sample.

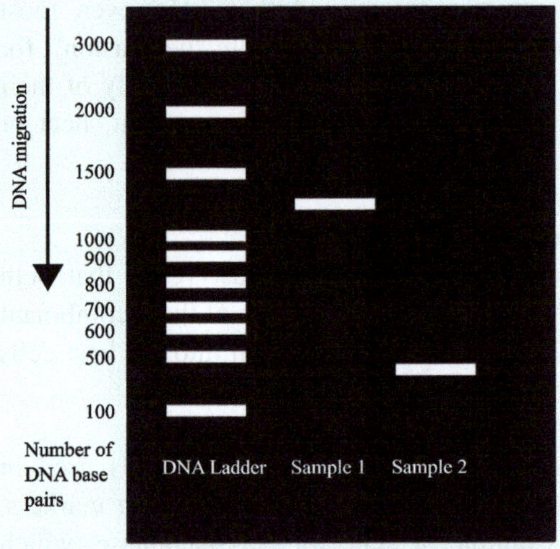

Electrophoresis results. Sample 1 and Sample 2 are visualised in the gel in separate columns. Since Sample 1 is longer, we can deduce that it contains our gene of interest and that the cells are successful mutants. Cells representing Sample 2 are falsely positive colonies

Interesting Fact

Papayas are one of the most exported crops from Hawaii, but in 1997, the industry struggled to survive after the papaya ringspot virus destroyed almost all papaya plants. In response, scientists developed the Rainbow papaya variety, which is resistant to the ringspot virus. Many farmers were saved from shutting down production, and the country's exports were re-established.

Interesting Fact

Mosquitoes are carriers of many pathogens that cause dangerous diseases such as malaria, yellow fever and Zika fever, which result in the deaths of many people every year. Pathogens (viruses or protozoa) carried by mosquitoes infect humans when female mosquitoes bite people to feed on their blood. *Aedes aegypti* mosquitoes, which also transmit some of these diseases, were genetically modified to control the infection spread. In particular, male mosquitoes were modified so that their female offspring do not reach maturity and thus cannot spread diseases. Hundreds of millions of modified males were released in the Florida Keys since 2021, and declines in population numbers of *Aedes aegypti* have been recorded. The project has shown promising results in combating local populations of this invasive species and demonstrated a potential strategy for controlling vector-borne diseases without harming the environment.

A big problem is the emergence of antibiotic resistance in harmful bacteria. As we already know, selectable markers for mutants containing recombinant DNA are often antibiotic resistance genes. Some scientists proposed to avoid antibiotic selectable markers, as gut bacteria could potentially gain antibiotic resistance by taking it up from the transgenic food we consume, lowering the effectiveness of antibiotic treatment when it is needed. On the other hand, there is no evidence that there has been DNA transfer from consumed food to gut microflora.

Another problem is cross-pollination between just GM plants and wild types and the uncontrolled spreading of the genes in the GMOs. Wind, bees and other pollinators do not differentiate between GM and non-GM, plants and can easily transfer pollen between the two. Organic farmers are concerned about contamination from GMO agricultural areas spoiling their crops. All of these intricacies add to the reasons GMOs must be strictly assessed and regulated.

Even though many scientists say genetically modified crops are safe, there are studies questioning the reliability of methods used to assess the safety of these crops and highlighting the fact that the claims were published by biotech companies producing GM crops. Society is polarised when it comes to genetically modified crops and organisms in general.

A good example is genetically modified Bt corn, which has higher grain quality and yield and, moreover, requires less insecticide and pesticide due to a gene inserted from *Bacillus thuringiensis*. This gene codes for a toxin lethal to certain insects, primarily those in the orders Lepidoptera (moths and butterflies), Coleoptera (beetles) and Diptera (flies and mosquitoes), and is considered a natural insecticide. However, in 1999, a study conducted by John Losey and colleagues claimed that nearly half of the monarch butterfly caterpillars died after four days of being fed leaves dusted with Bt maize pollen. Activists marched in front of the US Capitol to protest GM maize. The pressure imposed on the seed companies to stop selling the seeds resulted in six further studies that showed that the concentrations of Bt maize pollen that insects encountered in the fields would not be toxic. Later, it was clarified that Losey had exposed larvae to higher concentrations of the toxin in his studies.

The controversy around Bt maize heated up again in 2007, after a study highlighting the adverse effects of pest-resistant GM maize on streams was published. The authors pointed to the possibility of streams neighbouring Bt corn fields being contaminated with the toxin and demonstrated possible negative effects on nontargeted aquatic organisms. There was a wave of criticism of this study, questioning the laboratory methods used as well as the preliminary assumptions of the authors. But some other scientists who looked into the safety of GMOs confessed that their research was also criticised at the level of hate speech and harassment.

A study from 2020 reflected on the multiple benefits of Bt maize (and other Bt crops) and showed little detectable harmful effects on nontarget insects. Nevertheless, the same study cautions against the widespread application of genetically modified crops. Using GM crops has obvious benefits, such as less need to spray insecticides and higher yields, but GM crops can alter the ecosystem dynamics in multiple ways. For example, when Bt or other toxic genes kill target pests, other pest species that are not susceptible to these toxins overtake the new areas because they do not have competition for food and space.

Another issue is the escape of toxic genes. The inserted genes are transferred to wild-type plants of the same species by cross-pollination. GM crops may also escape outside the planted fields and grow in the wild. If these GM varieties have a selective advantage over wild-type crops, valuable genetic diversity disappears.

Many crops are engineered to be resistant to herbicides, so when herbicides are applied to a field, weeds die but crops survive. Today, most plants are engineered to be resistant to glyphosate, which is a potent herbicide for most weeds. Instead of spraying many herbicides as they would have with wild-type crops, farmers simply grow glyphosate-resistant varieties and spray only glyphosate. However, glyphosate is an endocrine disruptor and potentially toxic to the environment and human life.

Furthermore, the target species can develop resistance to Bt or herbicides. We see this already happening, and strategies are being deployed to slow the evolution of resistant pests.

Farmers often opt for GM varieties because of their benefits mentioned above and grow them in monocultures year after year. In addition to the other risks, this depletes the soil of nutrients. Scientists and agronomists now look towards farming practices that increase yield while preserving biodiversity. These include seeding different crops every year and planting cover crops and flower strips on the edges of the fields to attract pollinators and natural predators of pests. To improve public health and ensure biosafety (prevent accidental release and adverse effects on the environment) and biosecurity (prevent the deliberate release of biological weapons) hazards, regulations regarding GMOs are imposed in many countries. The EU, for instance, publishes extensive guidelines for working with GMOs in laboratories. For food crop cultivation, 64 countries—including all EU members—require GM food to be labelled. In order to be put on the market, the GMO (or derived product) must first be shown to be in compliance with EU requirements. In many countries, for example, Denmark, the Netherlands and Austria, the use of transgenic plants has already been banned. In others, only the import of already approved GM crops is allowed.

And what about other GMOs? There are real concerns about editing the human germline or creating biological weapons from human pathogens that need serious discussion in present-day society.

The possibilities of genome editing continue to expand year by year. Inventions in biotechnology and other related fields give us abilities that have seemed far-fetched in the past decades. Scientists must make careful decisions and actively engage in ethical discussions, as interference with DNA can have unpredictable consequences. The regulations in place vary from country to country, and this leads to many questions about what is ethical. We have seen examples of recombinant DNA saving lives, but it could also have unknown long-term effects on the environment and human health, which have not yet been well studied. And at the end of the day, who is responsible?

DEFINITIONS

Plasmid: A small, extrachromosomal DNA molecule within a cell that is physically separated from chromosomal DNA and can replicate independently

Transformation: The genetic alteration of a cell resulting from the direct uptake and incorporation of exogenous genetic material from its surroundings through the cell membrane

Recombinant DNA: DNA molecules formed by laboratory methods of genetic engineering to bring together genetic material from multiple sources

Transgenic DNA: Foreign DNA (DNA from another organism) that is inserted into an organism by artificial means

SUGGESTED READING

- Lerner A, Benzvi C, Vojdani A. The Potential Harmful Effects of Genetically Engineered Microorganisms (GEMs) on the Intestinal Microbiome and Public Health. Microorganisms. 2024 Jan 23;12(2):238.
- Spinner SAM et al. New self-sexing Aedes aegypti strain eliminates barriers to scalable and sustainable vector control for governments and communities in dengue-prone environments. Front Bioeng Biotechnol. 2022 Oct 25;10:975786.
- Waltz E. GM crops: Battlefield. Nature. 2009 Sep 3;461(7260):27–32.
- Mackelprang R, Lemaux PG. Genetic Engineering and Editing of Plants: An Analysis of New and Persisting Questions. Annu Rev Plant Biol. 2020 Apr 29;71:659–687.

18.4 Questions

1. **Which of the following actions can be considered genetic modification?**
   (A) Addition of a single gene, Sub1A, to the rice genome to allow rice to survive an extended time underwater.
   (B) Bacteria acquiring resistance to antibiotics via conjugation with other bacteria in the gut flora.
   (C) Knock out of a p53 gene (tumor suppressor gene) using CRISPR/Cas9 technique.
   (D) Crossing drought-resistant maize with sweet maize to produce corn that is both sweet and drought resistant.
   (E) Transplantation of bone marrow in a patient suffering from cancer.
2. **Which of the following statements is/are correct?**
   (A) Humans began to genetically modify dogs several thousand years ago.
   (B) It is not possible to genetically modify eukaryotic cells, because they don't have plasmids.
   (C) A modification of an organism's genes cannot be passed on to its offspring.
   (D) Prokaryotes can be genetically modified to produce mammalian proteins.
   (E) Genetic modification can only be done by inserting foreign DNA into the host organism.
3. **A short sequence of DNA shown in Fig. 18.5 has two restriction sites I and II. Applying restriction enzymes would result in three fragments A, B and C. How would the fragments be visualised by electrophoresis, shown in Fig. 18.6?**
4. **Which of the DNA sequences listed below would be cut by a restriction enzyme?**
   (A) 5′AATTCCGC3′
   3′TTAAGGCG5′
   (B) 5′ATTATAAT3′
   3′TAATATTA5′
   (C) 5′AGTAGTAGT3′
   3′TCATCATCA5′
   (D) 5′AAACCTTT3′
   3′TTTGGAAA5′
   (E) 5′AAACGTTT3′
   3′TTTGCAAA5′
5. **Imagine that you want to insert a gene into a plasmid. The relevant sequences are shown below: gene of interest (open reading frame is highlighted):**
   5' AAGCGATCTTAATTAACCGAATGTTCACC GTT AATTAACCGGTTATTGCAAGATCAATTG 3'
   Plasmid polylinker site:
   5' ATCGATTTAATTAAAATTAATTGATCCCAG TCGAATT 3'
   Enzymes and corresponding restriction sites are shown in Table 18.1. Which of the approaches described below will lead to insertion of the gene into the plasmid's polylinker site?
   (A) Mix the plasmid and fragment in one tube.
   (B) Cut the plasmid using BavCI, amplify genes with PCR to match the plasmid, and use the Gibson Assembly Master Mix.
   (C) Cut both the plasmid and flanking regions of the gene using the HacI enzyme.
   (D) Cut both the plasmid and flanking region of the gene using PacI enzyme.
   (E) Mix bacteria with the plasmid and the gene.
6. **How would you verify that your gene of interest is inserted into the plasmid?**
   (A) Grow host bacteria that you assume accepted your plasmid on a medium with antibiotics. Those that survive have the gene of interest.
   (B) Perform colony PCR by designing primers that align to the plasmid before and after the gene of interest. Gel electrophoresis is performed with the PCR product.
   (C) Observe the droplet of recombinant DNA under the microscope.
   (D) Perform colony PCR by designing primers that align to the plasmid before (and after) the gene of interest and the subsequent part of the gene of interest. Gel electrophoresis is performed with the PCR product.

Table 18.1 Enzymes and corresponding restriction sites

BavCI	HacI	PacI
5′ ATCGAT	5′ GATC	5′ TTAATTAA
3′ TAGCTA	3′ CTAG	3′ AATTAATT

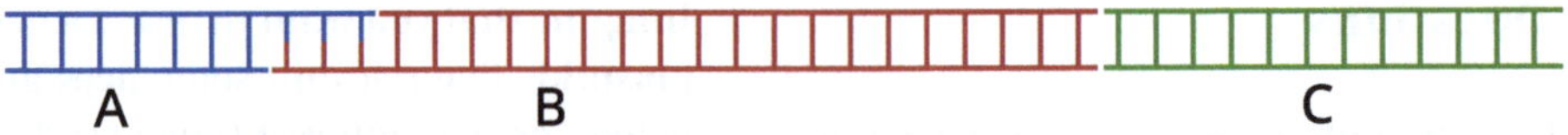

Fig. 18.5 DNA sequence with two restriction sites

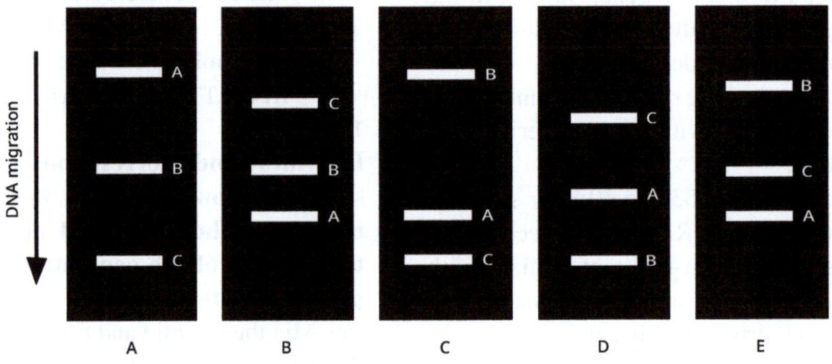

Fig. 18.6 Potential results of electrophoresis

(E) Perform colony PCR by designing primers that align only to the gene of interest. If there is any PCR product, the gene is sure to be inserted into the plasmid.

7. **Which of the following sequences shows one strand of a short palindromic fragment of a DNA molecule?**
   (A) 5′ AAAACGCGCAAAA 3′
   (B) 5′ TTAGCACGTGCTAA 3′
   (C) 5′ CCCGGG 3′
   (D) 5′ ATTGCGTTA 3′
   (E) 5′ CTTAAG 3′

8. **Table 18.2 lists the restriction sites present in the plasmid and in the gene. Which restriction enzymes would you use for the cloning strategy to ligate the gene and the plasmid?**

Table 18.2 Restriction sites

Individual restriction sites present in the multiple cloning site of the plasmid		Restriction sites present in the coding region of the gene	
AsiSI	NotI	NotI	NdeI
ClaI	PacI	PaeI	SacI
PaeI	SacI	HindIII	SalI
HindIII	SalI	NdeI	TaiI
NdeI	KpnI	HapII	KasI

(A) TaiI
(B) KpnI
(C) HapII
(D) ClaI
(E) AsiSi
(F) PacI
(G) PaeI
(H) ScaI
(I) KasI

18.5 Project: Creating a Genetically Modified Organism

You were offered a student job to take over the work of one of your teachers, who has left for a long sick leave. He was in the middle of a project and asked you to finish it, but due to his illness, he was unable to provide you with detailed instructions. The only thing you know is that he was designing recombinant DNAs using specific plasmids and genes to conduct further experiments later on, but he did not manage to complete the design, nor the laboratory work. Luckily, you have found some notes on his desk and computer that show the preliminary designs your teacher had made. The notes show two plasmids and two genes with some highlighted regions and annotations (see Materials: https://en.edub.sk/everyday-evolution).

Part I: The Theory

To create a recombinant DNA, we first need to design the DNA theoretically to avoid unnecessary steps in the laboratory and associated costs.

1. Look at the sequences. What do they show? Do they show one or two strands of DNA? What was the teacher probably trying to do?
2. Do some research and describe the genes. What do they do? Which organism(s) can they be found in?
3. Some parts of both plasmids and genes are annotated. For each annotation, explain what it means. If there is an annotation you are unsure about, make a note of it, and we will review it later.
4. Draw plasmid maps for both plasmids and include the annotations. Keep in mind that plasmids are circular DNA (both ends must meet). You can use the templates in Fig. 18.7.
5. Is there any place in plasmids where you could insert a new gene? Highlight part of the sequence on both plasmids.
6. Now, look at the restriction sites. In Table 18.3, write down the double-stranded DNA sequence that is recognised by each restriction enzyme and indicate the restriction cut.
7. For both plasmids and both genes, indicate the positions of all the restriction sites within them (information given under the respective sequence) and note the respective enzymes that cut these sites. You can use templates (Figs. 18.7 and 18.8) to show the positions of restriction sites on plasmids and genes.
8. Look at the highlighted restriction sites in plasmids. Which restriction enzymes could be used to cut the sequences for later insertion of genes into plasmids?
9. Look at the highlighted restriction sites in genes. Which restriction enzymes could be used to cut the sequences for later insertion of genes into plasmids?

Let's get back to the annotated parts you might not be sure about. In your lecture, you learned that a promoter is responsible for initiating transcription of DNA into mRNA and a terminator is responsible for terminating the transcription. In the plasmids, transcription takes place from left to right with respect to their linear notations.

10. Look at the sequence on both plasmids between the promoter and the terminator. Think of the directionality of DNA. How should the gene be inserted in order to be transcribed correctly?

Transcription is only the first part of successful gene expression. The second part is translation. To translate mRNA into an amino acid sequence, a ribosome binds to the *ribosome binding site* on the mRNA transcript and then initiates translation of the first codon (methionine). Therefore, the ribosome binding site has to be located before the gene coding sequence.

11. After you learned about ribosome binding sites, is your answer to point 8 still the same?
12. Into which plasmid can you insert gene 1 by using only the restriction enzyme strategy?
13. Into which plasmid can you insert gene 2 by using only the restriction enzyme strategy?

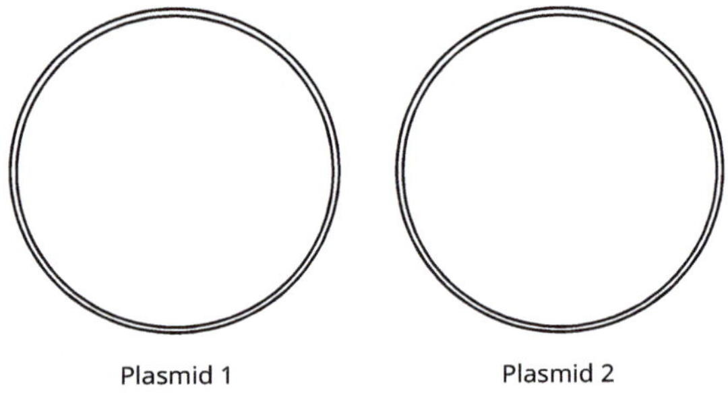

Fig. 18.7 Plasmid templates

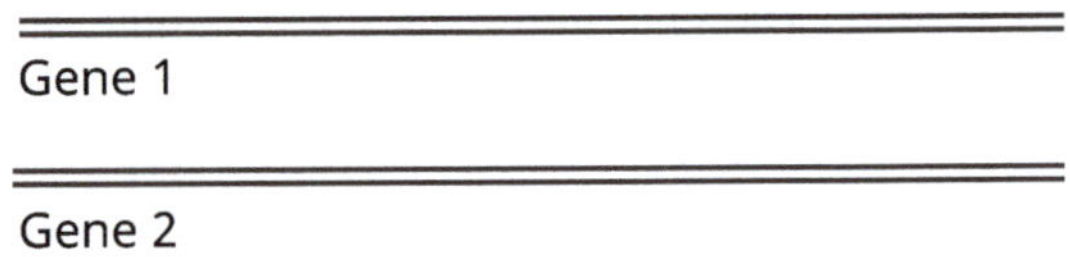

Fig. 18.8 Gene templates

Table 18.3 Enzymes, their corresponding restriction sites and the cuts

BssHII	PvuI	PstI	AvaI	NotI	HindIII

14. Will you face any issues with the directionality of DNA? Explain your answer.
15. Could you use any other genetic engineering technique to avoid directionality issues?
16. Decide on your final design of recombinant DNAs. Describe which gene you will insert into which plasmid and which restriction enzymes you will use for each of them. If you use other genetic engineering methods, briefly describe what is needed for that, too.

Now that we have completed the theoretical part of designing our recombinant DNA, we ideally designed it in a way that enzymes will not cut any undesired sites, our genes and plasmids join together, and the gene is translated correctly (without directionality issues). However, we never know whether our theoretical design works until we go into the laboratory and test it.

Part II: Lab Work

Your recombinant DNA design is ready. Now, it is time to go into the laboratory and do practical work. You have found both the plasmids and genes in the laboratory freezer, and your colleagues confirmed that their sequences match the teacher's notes. They are in four separate tubes. It is very important to avoid any contamination during all practical steps. To complete the laboratory part of the project, label the steps in the Fig. 18.9 and answer the questions.

1. First, you must apply restriction enzymes to the plasmids and genes. This is done in separate tubes. Label step 1: Write down the restriction enzymes you will add to each of the tubes. If you do not need to add any restriction enzymes, describe what you have to do instead.
2. Now, your plasmids and genes are ready to be joined. Label step 2: Prepare two fresh tubes for this step. For each of the fresh tubes, write down which plasmid and gene you will add to it. Which other enzyme needs to be added to make the joining successful?
3. Hopefully, your gene is now part of the plasmid. Now, you need to transform it into a suitable host. Label step 3: Which organism would you choose and why?
4. Step 4: Mix your organism and recombinant DNA. Make sure you do this in separate tubes for each recombinant DNA.
5. Now, you need to verify whether your organism took up the recombinant DNA. This is done by plating the organism on agar in a Petri dish. Label step 5: What chemical do you need to add to the Petri dish to select only

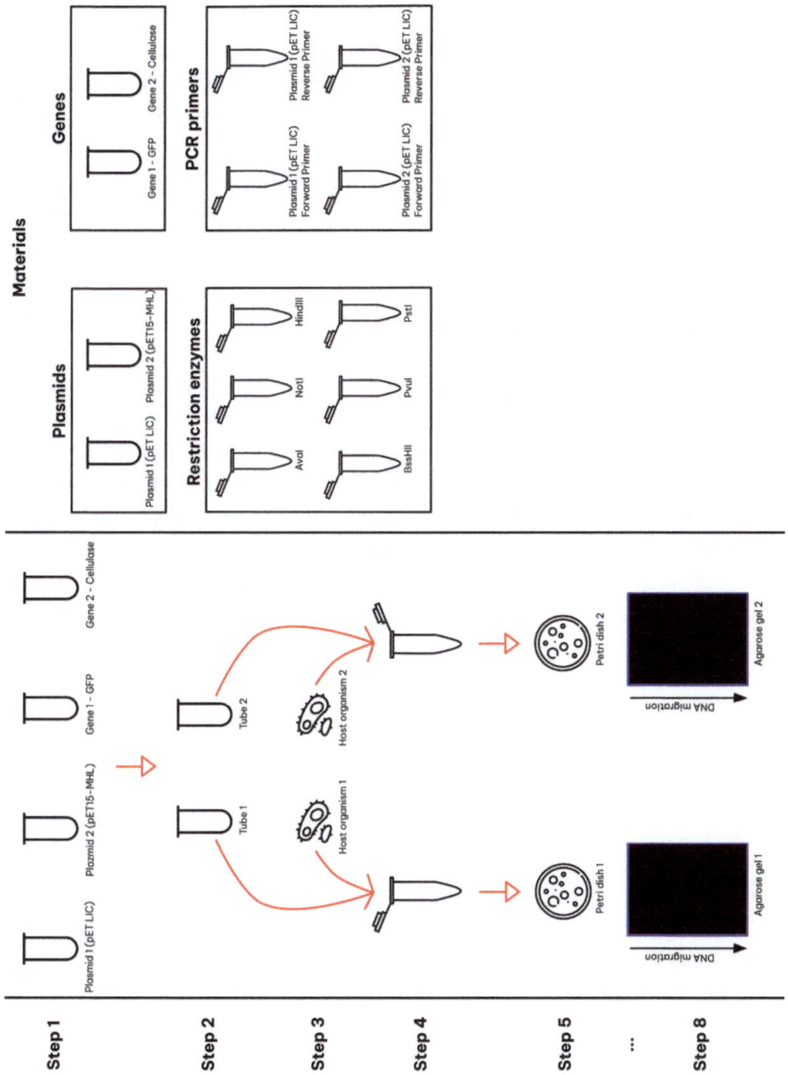

Fig. 18.9 Laboratory setup

for individuals that took up the recombinant DNA? Explain your answer.

6. Hopefully, you have some colonies that grow on your Petri dish. If you did the previous step correctly, the colonies consist only of individuals that have taken the plasmid. But we still need to verify whether the plasmid has the gene we intended to insert. What would you do to prove your organism has the plasmid with the gene in it?

Your colleagues provided you with sequences[3] for the PCR primers:

Primers for plasmid 1 amplify regions between the following sequences (including the sequences):

Forward primer	5′ AGATATCCCAACTC-CATAAGG 3′
Reverse primer	5′ AGGAAGCTGAGTTGGCTGCT-GCCACCG 3′

Primers for plasmid 2 amplify regions between the following sequences (including the sequences):

Forward primer	5′ ACCATGAGCAGCGGCAGA-CATATGA 3′
Reverse primer	5′ AGGAAGCTGAGTTGGCTGCT-GCCACCG 3′

7. How long should the PCR products be if they contain your inserted genes? How long will the PCR products be if they do not contain the genes?
8. Denote the lengths of the fragments on agarose gel (Fig. 18.9, Step 8).

Congratulations, you have successfully completed your first recombinant DNA experiment and inserted it into an organism, creating a genetically modified organism! Hopefully, you saw some cheerful results, and all the lab work worked as you designed it. But do not despair if it does not work in the first trial. Many things can go wrong that we did not cover here. Scientists face multiple challenges, and most of the time, experiments do not work as expected on the first attempt. Fixing issues on the go is a common practice; this is known as troubleshooting, and it's a skill that you'll learn over time.

This project was developed by Jarmila Šimková.

[3] To make it simple, sequences for all primers are shown on the same strand—the coding strand.

19 The Exciting World of Genes and Their Origin

Katarína Juríková and Kristína Hudáková

▶ The genomes of a tomato plant and a wild cat have a lot in common, but they are obviously not the same. While they share some of their genes, some exist only in one species and not in the other. How did this happen? How are new genes born? Can a gene die? Are new genes important for the emergence of new traits? In this chapter, we will dive into the amazing world of genes and investigate their origins.

19.1 What are Genes?

The question is not as simple as it might seem. Defining a gene is not easy, even for geneticists. When scientists started to recognise DNA as the carrier of genetic information, they thought that the relationship between DNA and genes was much simpler than it later turned out to be.

Today, we generally say that a gene is a piece of DNA whose RNA *or* protein product serves a particular function in the organism. Interestingly, a huge portion of the organism's DNA does not contain any genes, especially in some organisms. Often, DNA sequences other than protein-coding genes are called noncoding DNA. To the best of our knowledge, such noncoding sequences form a staggering 98% of human DNA. The other 2% of our DNA (which we roughly estimate as 120 million base pairs in every cell) contains approximately 20,000 protein-coding genes. Some of these genes are responsible for the fundamental functions of the organism-like cells' ability to divide or communicate. Some genes determine the essential structural characteristics of our bodies–such as the appearance of a human muscle or the structure of a liver cell. Some genes determine the traits we perceive as different in different people, for example, eye colour or hair type. But if a gene is a functional unit of DNA, how is that function expressed?

19.2 From a Gene to Its Function

Let us consider a gene that determines the structure of amylase, an enzyme responsible for breaking down starch. The information for the amylase structure is encoded in DNA. How do we get from information in the DNA's A-C-T-G alphabet to a functioning enzyme in our saliva?

All organisms on Earth, from bacteria to sequoia trees, can perform this essential

K. Juríková (✉)
Department of Genetics, Faculty of Natural Sciences, Comenius University in Bratislava, Mlynská dolina, Bratislava, Slovakia
e-mail: katarina.jurikova@uniba.sk

K. Hudáková
eduB, Bratislava, Slovakia
e-mail: kristina.hudakova@edub.sk

B. Trubenová et al. (eds.), *Everyday Evolution: Unveiling Life's Hidden Patterns*,
https://doi.org/10.1007/978-3-032-07050-0_19

information transfer trick. In a process known as **transcription**, the information in DNA is copied into RNA, and then, in the process of **translation**, the information from RNA is transferred to proteins when the latter are synthesised. Proteins are usually the actual functional machines of cells, as they often work as building blocks or as enzymes, catalysing the conversion of one chemical molecule into another.

As we have already learned, transcription is performed by RNA polymerase, an enzyme capable of producing RNA according to the DNA template. In eukaryotes, this process takes place in the nucleus. Then, the produced mRNA associates with ribosomes, and with the help of tRNAs, the protein is produced. However, this is not the whole story. We did not yet mention that an especially important step, typical in eukaryotic cells, is **splicing**. Many eukaryotic genes have two types of sequences: **exons**, the parts that code for proteins, and **introns**, the parts that are removed before the mRNA gets into the cytoplasm, where it associates with ribosomes. The RNA that is generated by RNA polymerase during transcription is usually referred to as pre-mRNA, and only after it goes through splicing and has its introns removed do we call it mRNA. Note that in prokaryotes, splicing does not usually take place. In addition, both transcription and translation take place in the same cellular compartment, as there is no membrane-separated nucleus.

Although the flow of information is usually in the direction of DNA →RNA →protein, it is also possible to transfer information from RNA to DNA. This process is called **reverse transcription**, and it is carried out by an enzyme called (unsurprisingly) **reverse transcriptase**. It is a process that is very important in the life cycle of some RNA viruses (such as HIV) that normally carry their genetic information as RNA, but in some stage of the life cycle, also have their RNA reverse transcribed into DNA. However, this process also occurs in many organisms' normal, healthy cells. A good example of an important reverse transcriptase is telomerase–the enzyme that elongates telomeres (see Chap. 20).

The most common use of reverse transcriptase, however, probably has nothing to do with cell survival. In the human genome, there is a suspiciously high number of genes coding for reverse transcriptase. This is because the human genome and many other eukaryotic genomes have been invaded by genetic elements called **retrotransposons**. Retrotransposons are often considered to be "selfish" genetic elements because (at least at first glance) they do not contribute to the host's biology: They bring neither advantage nor protein (or RNA) product important for the survival of their host organism. Their only "goal" is to amplify themselves and expand as much as possible in the genome they inhabit.

Retrotransposons are pieces of DNA that are transcribed into RNA and then reverse-transcribed back into DNA, which then integrates itself into the genomic DNA at another site (Fig. 19.1). One can imagine retrotransposons as if they were amplifying in the genomes of eukaryotes in a copy-paste manner. Sometimes, retrotransposons also carry their own gene for reverse transcriptase, but otherwise, they just contain sequences recognised by a reverse transcriptase encoded by another retrotransposon. Whether or not they encode reverse transcriptase, these elements have been rather successful in spreading across genomes–for example, they form around 40% of mammalian DNA and even more in some plants such as maize. What is the function of these elements? There is a possibility that these elements are, in fact, selfish, and there is nothing in retrotransposons that their carriers would benefit from. However, they may also play a role in shaping the evolution of the genomes they inhabit–for example, two sites in one genome containing copies of the same retrotransposon may make the genome prone to recombine at these very loci, possibly driving genome rearrangements.

Despite the fact that it is mostly proteins that carry out the functions encoded in DNA, there are a number of RNA molecules produced in the process of transcription that are not translated into proteins. Such RNAs are often called noncoding

From One Sequence to Many Proteins

Introns are pieces of sequence which are removed when a functional RNA or a protein is being produced–what could their function possibly be?

One of the explanations may lie in the existence of **alternative splicing**. This means that under some circumstances, sequences that are usually introns are kept in the mature mRNA as exons, and vice versa, or only a subset of exons is kept. As a result, functional parts of proteins are removed or added, often completely changing their localisation in the cell and their function, sometimes leading to a completely different physiology or behaviour of the cells or individuals carrying the alternative variant. For example, in fruit flies, a gene called *fruitless* is spliced differently in male and female flies. This tiny difference is responsible for a number of behavioural differences, from male courtship behaviour to sex differences in aggression patterns during conflicts. Many of the alternatively spliced genes that have been studied seem to be involved in brain physiology, ensuring correct memory function or brain development.

The existence of alternative splicing enables cells to encode a huge variety of RNAs and proteins in a relatively small number of genes. Make no mistake, alternative splicing is not an obscure eventuality that cells rarely use. In fact, while there are around 20,000 genes in the human genome, the number of known proteins they code for goes up to 90,000, and alternative splicing has been found to occur in more than 95% of human genes that contain several exons. Also, the exon-intron structure of a gene is typical for the vast majority of genes in higher eukaryotes.

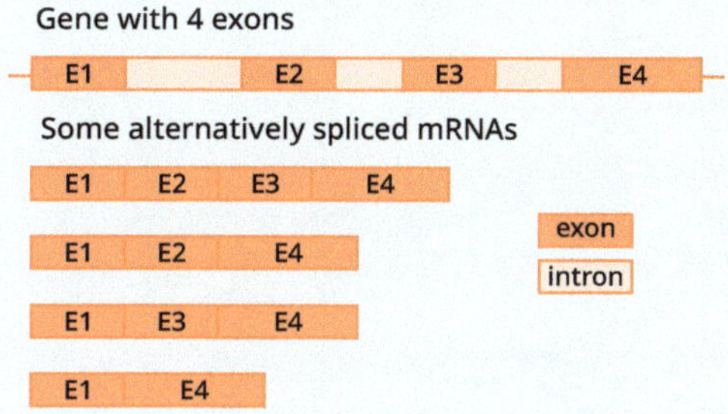

Alternative splicing. Multiple, different transcript variants can be built from a gene's exons. The first one with the start codon and the last one with the stop codon usually (but not always) need to be present. Note that alternative splicing sometimes also means that an intron sequence is kept in the final mRNA

Interestingly, alternative splicing contributes to differences between species. Many genes between mice and humans are very similar (we will learn more about this below), but their products differ because of alternative splicing; in different species, different pieces of a gene contribute to the final protein product. Not only that, but alternative splicing within any species can also contribute to differences between individuals.

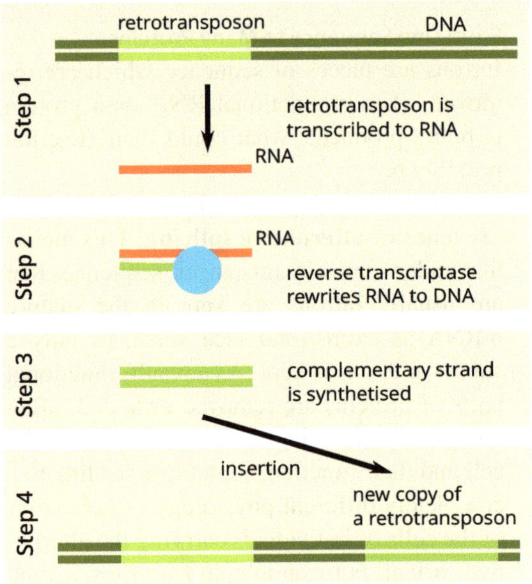

Fig. 19.1 Retrotransposon spread. Retrotransposon DNA is transcribed into RNA and then reverse-transcribed to DNA and integrated into chromosomal DNA. (Figure based on Campbell, N. A., and Reece, J. B. (2008). Biology. 8th ed. Pearson Education, Inc)

Human:	TCATGACCACAGTCCATGCCATCACTGCCACCCAGAAGACTGTGGATGGCCCCTCCGGGAAACTG
Chimpanzee:	TCATGACCACGGTCCATGCCATCACTGCCACCCAGAAGACTGTGGATGGCCCCTCCGGGAAACTG
Yeast:	TGATGACCACTGTTCACTCCATGACCGCCACTCAAAAGACTGTTGATGGTCCATCCCACAAGGAC

Fig. 19.2 Sequence alignment of part of the glucose-6-phosphate gene from a human, chimpanzee and yeast. Note that the gene sequence is very similar between the three organisms, despite the evolutionary distance between yeast and primates. This is because this gene is an essential metabolic gene in all eukaryotes

since they do not encode any information for protein production. However, these RNAs often have their own important roles and may carry out various functions. Well-known classes of such RNAs are transfer RNAs, which participate in the translation process, or ribosomal RNAs, which are important building blocks of ribosomes. However, there are many more such RNAs–the scientists in the ENCODE project, whose aim is to describe all functional parts in human DNA, estimate that there are around 25,500 genes that encode functional RNAs in the human genome. That means there are more functional RNA-encoding genes than there are protein-coding genes. Such RNAs can have very diverse functions–they can influence the expression of other genes, help to maintain the structure of the genome or help cells successfully divide.

19.3 The Things We Share

When we look closely at the genome of an organism and compare it with the DNA of other known organisms on Earth, we notice that there are some similarities and many differences. For example, if we take the genomes of humans and chimpanzees, we will see that more than 90% of the DNA is the same. However, considering species that are much less similar to humans–for example, baker's yeast–we still find surprising similarities. Take the glucose-6-phosphate dehydrogenase (also shortened GAPDH), an enzyme that is crucial in the metabolism of sugar. In Fig. 19.2, we can see that some sequences of this gene are identical in humans, chimpanzees and baker's yeast.

Types of RNA

The commonly known types of RNA are the coding messenger RNA (mRNA), which is translated into protein, and two types of noncoding RNAs, ribosomal RNA (rRNA) and transfer RNA (tRNA), that participate in the translation.

However, there are other types of RNA with other important functions in the cell:

Small nuclear RNAs (snRNAs) are found in the nuclei of eukaryotic cells. Five snRNA molecules, together with proteins, form a complex called a spliceosome. The RNAs within this complex bind to short nucleotide sequences at the ends of an intron and also within an intron of a pre-mRNA. The intron is then cut out and degraded, while the spliceosome joins together the two exons. The snRNAs recognise the splicing site and catalyse the splicing.

MicroRNAs (miRNAs) are small single-stranded RNA molecules, 19 to 24 nucleotides in length. These small RNAs can bind to complementary sequences in mRNA molecules. The miRNAs form complexes with proteins, and such binding of an miRNA-protein complex on an mRNA molecule blocks the translation of this mRNA, or the mRNA is degraded by the protein part of the miRNA-protein complex. In this way, miRNA plays an important role in gene expression regulation.

Long noncoding RNAs (lncRNAs) are RNA molecules longer than 200 nucleotides (up to several hundreds of nucleotides) that are not translated into proteins. It has been estimated that in the human body, more than 100,000 different lncRNAs may be transcribed, with functions ranging from the remodelling of chromatin structure through regulating gene expression to regulating immune cell activation.

Catalytic RNAs (ribozymes) are RNA molecules that can catalyse biochemical reactions in a similar way as protein enzymes do. Ribozymes are involved in various processes within a cell, including the cleavage of RNA and the formation of peptide bonds. Ribozymes are present within ribosomes, where they play a role in linking amino acids into a polypeptide chain during translation. When ribozymes were discovered, scientists realised that RNA has a unique capability to work both as a genetic material and as an enzyme. Ribozymes may thus be remnants of the evolutionary past, from the times before DNA took the role of the main genetic material and proteins took the role of enzymes.

This is because this gene is so important to the cell's survival that nearly one billion years of evolution in the lineages leading to humans and yeast were not enough for the DNA to change. The genes that share a common ancestor sequence are called **homologs**. Quite often, when we talk about homologs, we mean specifically **orthologs**, i.e. homologs in two different species. The above-mentioned GAPDH is a good example–we would say that the human GAPDH is homologous to the yeast GAPDH, and vice versa, or that these two genes are homologs/orthologs. There are also examples of two homologs in the same genome, though–such homologs are called **paralogs**.

Sometimes, in two species, there are genes that have the same function but have very different sequences–although the DNA sequence could have been identical in the common ancestor, during the years in which the two descendant species evolved separately, the DNA sequences diverged while preserving the function. This can happen in species that are evolutionarily very distant (i.e. a lot of time has passed since they shared a common ancestor). However, the conservation of the function may be very robust. To test the extent to which function is conserved, scientists tried to replace genes in yeast with their human homologs. It turned out that despite great differences in their sequences and despite the fact that their last common ancestor lived one billion years ago, many human genes were able to functionally replace yeast genes. To illustrate how far apart the sequences can be, we may look at the gene called EFB1 in yeast. This gene encodes a translation elongation factor, a protein that helps with the smooth progression of translation at ribosomes. If we replace this gene with its human counterpart, EEF1B2, the yeast cells survive just fine, even though they die if this gene is removed without replacement. However, if we compared the DNA sequences of these two genes, we would not identify them as homologs.

Interesting Fact

Sometimes, even closely related species may share homologous genes that are very different in sequence. This can happen as a result of strong selective pressure or when the conservation of the gene sequence is not so important for its function. The latter can be the case for many genes encoding noncoding RNAs–in these molecules, their secondary structure is often more important for their function than their particular sequence.

Genes that share a common ancestor, whether they are orthologs or paralogs, are often grouped into so-called **gene families** (Fig. 19.3). This grouping is a useful tool for geneticists trying to uncover the function of a studied gene. Even if we cannot find a gene's function by observing individuals carrying a mutated version of the gene, a useful approach can be to look at the gene's sequence and try to find its gene family because homologous genes often share similar functions. This is a very common approach in the scientific field of bioinformatics. Bioinformaticians, among many other things, create and use databases containing millions of DNA sequences and invent and apply algorithms that can efficiently find homologs of a given gene in such databases.

In every genome, there are a number of genes for which we *cannot* find a homolog in any other organism, even if we look in the most closely related species. Such genes are called **orphan genes**, as they do not belong to any gene family and do not have any "genetic relatives". In some cases, these orphan genes are actually homologs, but their DNA sequences have changed too much to identify the relationship. Sometimes, however, there is really no existing homolog, and the gene simply has no related genes. How can this happen?

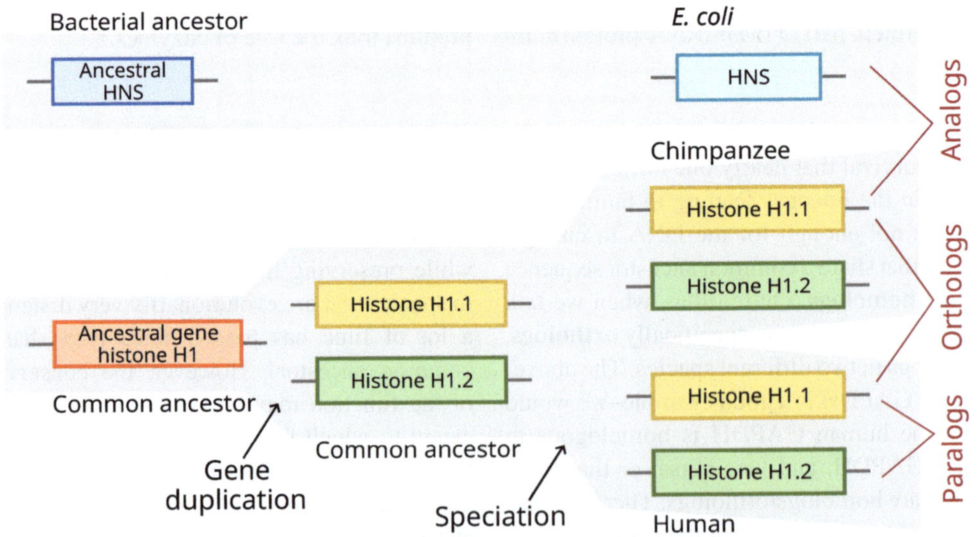

Fig. 19.3 **A gene family.** All homologous genes (paralogs and orthologs) share a common origin. In comparison, analogous genes are similar in function but don't share a common ancestor. (Figure based on Thomas Shafee, CC BY-SA 4.00 https://creativecommons.org/licenses/by-sa/4.0, via Wikimedia Commons)

X Chromosome Inactivation

One example of a long noncoding RNA in mammals is XIST, which is transcribed from chromosome X. Its function is to bind back to its DNA and start a process that ends up with one randomly chosen X chromosome being inactivated in (almost) every cell–gene expression in this chromosome is suppressed, and unlike other chromosomes, it is physically compacted into a little ball that is placed close to the nuclear membrane. Such inactivated X chromosomes are normally present in the cells of female individuals of a species, where females have two X chromosomes and males have X and Y (as in humans!). The goal of having one X chromosome inactivated is to provide males and females with the same amount of gene products (RNA and proteins) from the X chromosome–as many very important genes are encoded here, the shortage (in males) or overdose (in females) of these genes' products would negatively impact normal physiological processes.

Calico cat. The gene for fur pigment is located on the X chromosome. There are two alleles of this gene: one for black colour and another for orange colour. The black and orange patches result from the inactivation of different X chromosomes in different parts of the skin during development. The white parts are produced by another gene, which controls whether the pigment is distributed in the hair. Photo courtesy of Enikő Szép, reproduced with permission

19.4 How New Genes Are Born

We have learned a lot about genes, but we still don't know how they come to exist or how some genes can arise without belonging to a gene family. The number of genes in the species now living on Earth is not fixed; in fact, new genes are born quite often. The most common mechanism by which new genes emerge is the process of gene duplication.

Gene duplication can take place as a molecular event when the cell machinery that replicates and recombines DNA makes a mistake. The result is that a sequence that originally occurred only once is now present in the genome twice, meaning the organism now has two copies of the given gene instead of one (Fig. 19.4). What happens after such an event? One copy of the gene remains functional and carries out the function it had prior to duplication. The other copy usually ends up

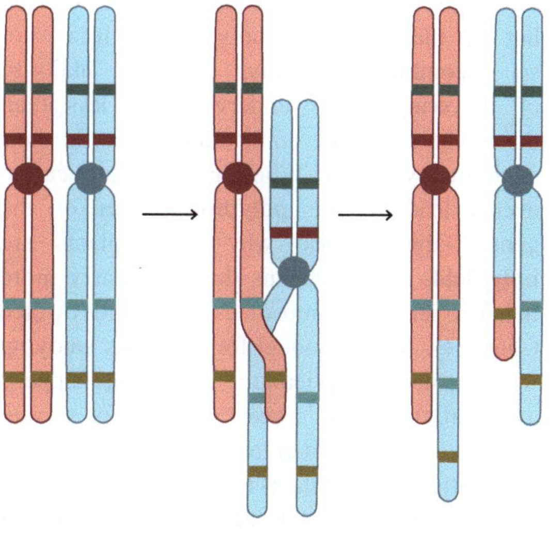

Fig. 19.4 Gene duplication by unequal crossing over. The blue gene is doubled on one of the recombined chromosomes

becoming useless, as an extra copy of this gene does not bring any advantage to its bearer. As a result, there is no selection pressure to keep its sequence intact. Over time, such a sequence accumulates random mutations that eventually destroy the previous functionality of the gene. The sequence may accumulate so many mutations that it becomes a noncoding sequence. In this case, we call it a **pseudogene**–a sequence that used to be a gene but lost its function. Alternatively, the extra copy of the gene may (purely by chance) accumulate mutations that do not disrupt its function but change it–such a gene copy would thus gain a new function, and if this function is useful to the organism, the duplicated gene, with its changed sequence, may be retained in the genome.

The masters of gene duplication are species that, during their evolution, went through a **whole-genome duplication** event. This often happens as a result of a mistake during gamete production. For instance, when a cell that should undergo meiosis–and generate daughter cells with half the number of its chromosomes–does not complete this process successfully, it can result in a gamete that carries all the chromosomes of the mother cell. This process rarely gives rise to viable offspring in animals, but in plants or microorganisms, it is not so uncommon. Actually, the vast majority of flowering plants underwent no less than three whole-genome duplication events in their evolutionary history. Whole-genome duplication events lead to the **polyploidisation** of the genome–this is also why many plants are **polyploids**, which means that they carry more than two sets of chromosomes, as diploids do (see Chap. 11). A great model for scientists studying whole-genome duplication (and what happens afterwards) is baker's yeast. This is for several reasons: It underwent at least one whole-genome duplication, its genome is relatively small (unlike many plant genomes), and the functions of many genes are well studied. Baker's yeast is also a good example of what happens a long time after the whole-genome duplication takes place. Even though this yeast was a **tetraploid** after the genome duplication, in the 100 million years that followed this event, it went through a **diploidisation** process. During this time, the chromosomes diversified in their sequence and their overall structure. Similarity was conserved only in that today, there are pairs of homologous chromosomes that are similar (not **tetrads**, as one would expect in a **tetraploid**). However, there have been other whole-genome duplication events in some yeasts, and so, for instance, the strains of yeasts used to produce Tokaj wine are again tetraploid! Sometimes, tracking genome evolution can be very complicated.

However, gene duplication is not the only way a new gene can arise from the sequence of another previously existing gene. Another path to a new gene is **retroduplication**. We already know that by reverse transcription (=retrotranscription), RNA can be copied into DNA. However, if an mRNA is caught in this process by mistake and is retrotranscribed and then inserted back into the genome as a DNA, a special case of gene duplication occurs. Importantly, in this case, the new, retroduplicated copy of a gene is not its perfect copy–it lacks intronic sequences. For example, the gene encoding fibroblast growth factor in dogs has been subjected to this kind of event (Fig. 19.5). The result was the creation of the FGF4 retrogene: Its product binds to the same receptor as the original fibroblast growth factor. However, these receptors are "overwhelmed" by the amount of bound growth factors in dogs possessing both genes, leading to the impairment of leg growth. The result is dog breeds with extremely short legs, such as dachshunds and basset hounds.

Another way new genes are born is through exon shuffling (Fig. 19.6). In this process, exons from different genes are duplicated and stitched together in another place in the genome–either via gene duplication or retroduplication. The result is a chimeric gene–a gene carrying parts of other genes. One example is a gene for the tissue-type plasminogen activator (tPA; a protein involved in blood clot dissociation), which carries four exons from three other functional genes.

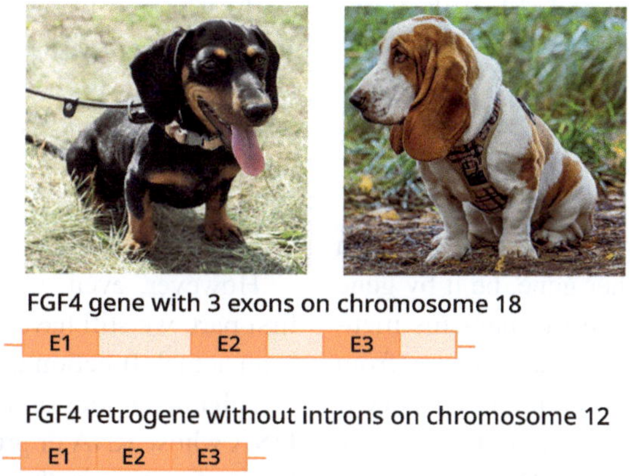

Fig. 19.5 New genes that originated by retroduplication can have a large impact on the phenotype. Both the dachshund (top left) and the basset hound (top right) have short legs due to the existence of a fibroblast growth factor (FGF) retrogene (bottom). (Figure based on Dickinson PJ, Bannasch DL. Current Understanding of the Genetics of Intervertebral Disc Degeneration. Front Vet Sci. 2020 Jul 24;7:431. https://doi.org/10.3389/fvets.2020.00431)

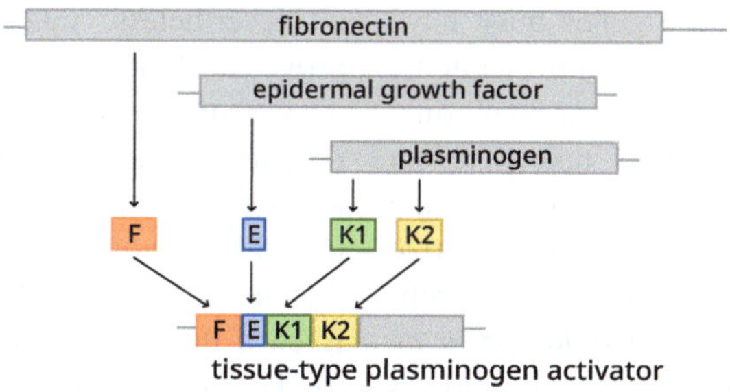

Fig. 19.6 Exon shuffling gave rise to the tissue-type plasminogen activator gene. The tissue-type plasminogen activator gene originated from the exons of three other genes with different functions: The F domain is responsible for the binding of fibrin (a blood clotting protein), the E domain may stimulate cell division, and the K1 and K2 domains enable binding to fibrin clots. (Figure based on Juríková K., 2013, Diploma Thesis: Non conventional yeasts as a model for the study of telomeres and orphan genes)

An unexpected way to acquire a new gene is to steal it from someone else. As improbable as this sounds, there are a number of documented events in which a gene from one organism is incorporated into the genome of a completely distinct species and remains functional. This process is called **horizontal gene transfer (HGT)** and usually occurs between species that exist in close proximity, for instance, different bacterial species in the same environment or a host-parasite tandem. Prokaryotes are especially prone to horizontal gene acquisition–for example, antibiotic resistance genes may be spread among bacteria of different species (see Chap. 6). Bacterial genes can also travel into eukaryotic cells, which is how a certain roundworm species, *Prystionchus pacificus*, gained the gene for cellulase, an enzyme that enables cellulose digestion, which expands the dietary options of this worm. Although even rarer, horizontal gene transfer can also occur between eukaryotes. The most notable example of this is bdelloid rotifers, tiny water invertebrates that reproduce only asexually. The genome of one of these animals may contain up to 10% of genes acquired horizontally from bacteria as well as unicellular eukaryotes and fungi. Scientists speculate that in the absence of sex, horizontal gene transfer may be an important source of genetic variability for these animals.

The tightest of relationships–endosymbiosis–is, of course, also accompanied by horizontal gene transfer. Mitochondria and chloroplasts are prime examples of this (see Chap. 15).

19.5 Something from Nothing

It may appear that new genes are usually born from the sequence of another gene, be it by gene duplication or another of the mechanisms mentioned above. Can new genes also come from DNA sequences that do not contain any genes? After all, at the beginning of life on Earth, there were obviously no genes and no DNA sequences encoding meaningful information. All the genes of the first living things must have originated through the shuffling of random DNA sequences without any older genes that could have been duplicated. We say that these genes must have originated *de novo*, or simply "from scratch". However, geneticists believed for a long time that this process does not really take place in the genomes of currently living organisms, to the extent that a famous biologist and a Nobel Prize winner, François Jacob, once said that the "probability that a functional protein would appear *de novo* by random association of amino acids is practically zero". Recently, however, scientists started to uncover evidence that *de novo* genes do, in fact, still arise in genomes and that these genes can even play an important role in the survival of organisms.

Scientists studying *de novo* genes often have to think about what makes a gene a gene. When they find a particular gene that they think may be a *de novo* gene, they first have to show that this gene does not have any existing homologs. Why? Well, if two genes are homologs, then they have a common evolutionary history–they originated from a common ancestral gene, for instance, by gene duplication. Thus, the existence of a homolog means that a gene did not arise from a noncoding sequence. Now, it becomes clear that *de novo* genes are always orphan genes. However, the reverse is not true! Try to think of an example of how a gene can be an orphan gene but not a *de novo* gene. Sometimes, even the inability to find homologs is not enough, and what the scientists do is they look at exactly the same site (or locus) in the genome of a closely related species and check whether there is a noncoding sequence that is very similar to the sequence of the putative *de novo* gene. Only if we find no homologs and no noncoding sequences on the same locus in several closely related species can we be fairly sure that this candidate *de novo* gene has no gene family.

However, even when we are sure about this first part, we still have to show that this sequence is an actual, functional gene! Now, recall the crucial characteristics of a gene: A gene is a piece of DNA whose RNA or protein product serves a particular function in the organism. To demonstrate that a given sequence is a gene, scientists look for the presence of an RNA transcribed from the gene and/or the existence of a protein translated from its sequence.

Nevertheless, this is often not enough to prove the functionality of a gene because cells of various organisms transcribe and translate random sequences surprisingly often, which leads to the accumulation of transcription and translation "noise"–RNA and protein molecules that are not functional and are quickly degraded. Even if these gene products bring no advantage to their owner and their presence in the cell is a result of "molecular luck", we could still find them. Interestingly, various organisms increase the levels of randomly transcribed and translated sequences when they are under stress–for example, budding yeast cells do this when they lack nitrogen in the medium. Scientists speculate that this may help cells to increase the chance they will find something random in the genome that will help them survive hostile conditions.

In the following paragraphs, we will focus more on protein-coding *de novo* genes. The reason is that it is much more difficult to identify an RNA-coding *de novo* gene, and scientists so far have been focusing mostly on the protein-coding examples of these evolutionarily young genes.

So, how does one show that a gene is truly functional if the presence of an RNA or protein product is not proof enough? To show this, one could disable the gene, either by introducing mutations in its sequence or by removing its product from the cell. If these changes have an impact on the survival, development or reproduction of the studied organism, it means that the gene in question plays an important role in certain processes and may thus be considered functional.

The first *de novo* genes were discovered only in 2006 in the genome of the fruit fly *Drosophila melanogaster*. This is because establishing if a gene is *de novo* is not so simple: Not only do we need to know the DNA sequence of the species we are studying (in this case, *D. melanogaster*), but also the sequences of the genomes of its closely related species. Otherwise, we would not be able to check whether the gene in question has any homologs in other species. It might be difficult to imagine how a *de novo* gene could be born from a noncoding DNA sequence. After all, there are a number of features that a gene must have, among other things, a functional promoter, the presence of an open reading frame (that is, a start codon and a stop codon) and, quite importantly, a sequence that encodes a protein that is able to perform some meaningful function in the cell. In fact, it is this complexity that at first made scientists question the very possibility of *de novo* gene emergence: How can all these features be produced at once and by chance?

The trick is that these conditions can be fulfilled sequentially, not necessarily simultaneously, over tens of millions of years. The order of the steps may vary; for example, a sequence may first acquire a promoter that would allow its transcription just by the accumulation of random mutations. Once the DNA sequence is transcribed, the RNA may associate with ribosomes. For example, in baker's yeast, many RNAs bind to ribosomes, even without coding for meaningful proteins, although the reason for this is still unclear. Further mutations may introduce a start and a stop codon, even though these are often present in noncoding sequences, too, appearing purely by chance.

What happens after a previously noncoding DNA sequence has been transcribed and translated? Now, the cell (or organism) contains a protein that probably does not have any meaningful role in cellular processes. At this point, natural selection is very important. The proteins without any role in the cell are a burden to the cell's or organism's resources, and so their translation may represent a disadvantage in comparison with the individuals that do not "waste" their resources translating meaningless proteins. This disadvantage may become even more pronounced if the protein forms aggregates in the cytoplasm, as such protein "clumps" may interfere with normal cellular processes. On the other hand, if this protein, by chance, interacts with other cellular proteins or has some other, even marginally useful role in the cell, this may represent an advantage for the individual that carries it.

It is not yet completely clear in what order noncoding DNA most commonly becomes a *de novo* gene. However, it seems that at least in some cases, *de novo* genes originate from sequences that have been transcribed in the common ancestor of the species with the *de novo* gene and its close relatives. This has been shown, for example, by one group of scientists who found 13 *de novo* genes in the human genome. For most of these genes, they found that in the genome of the chimpanzee or rhesus macaque, the homologous sequences are transcribed into long noncoding RNAs. This suggests that before becoming translated genes, these sequences had also been long noncoding RNAs in the human genome. This means that transcription may be one of the first steps on the road to *de novo* gene emergence.

Considering how many complicated and random steps lead to a *de novo* gene, we may ask how often these genes even arise. Do they play any significant role in the survival of the organisms and new traits that emerge in evolution? Unfortunately, these are not easy questions to answer. As for the first question, we have at least some estimates. Different scientific teams estimate there to be from three to 60 *de novo* genes in the human genome, depending on the exact methods they use to find them. As these *de novo* genes are not present in other primates, including the two chimpanzee species most related to humans, it must mean that they originated in the human genome in the seven million years since the evolutionary lineages leading to humans and to the two chimpanzee species split. However, these numbers may be very different for different groups of organisms. This is the result of many

How to Disable a Gene

Molecular geneticists often study the function of a gene by disabling it, and there are a number of ways to do this. In many cases, the method we pick depends on how permanent we want the malfunction to be.

Say we want to disable a gene only temporarily–if we work with yeast or human cancer cells that grow on a dish, that might mean that we only add something (like a chemical) to their growth medium. Then, when we want the gene to regain its functionality, we simply use a growth medium without the additive. When it comes to these temporary solutions, scientists interested in mammalian cells, such as cancer cells, have a significant advantage–they can use so-called RNA interference. RNA interference (or RNAi) uses a kind of RNA known as miRNA and similar short RNAs. These RNAs, if delivered to the cell, can bind the mRNA whose sequence they match and cause its degradation. So, if we need to disable a human gene, we just provide the cells we are growing with the appropriate short RNA, and the mRNA of our gene of interest is no longer present in the cell.

Unfortunately for all scientists studying budding yeast, *Saccharomyces cerevisiae*, this little eukaryote is a huge exception: It does not have a functional RNA interference pathway. Yeast researchers have, therefore, had to come up with other strategies. One popular method involves **conditional mutants**–strains of yeast that have a particular gene functional at room temperature but not at higher temperatures ("conditional" thus refers to the fact that the gene only works under certain conditions).

But what if we want to get rid of a gene permanently? In this case, we need to perform a so-called gene deletion, or **gene knock-out**. Currently, the most popular way to do this is with the CRISPR/Cas9 technique, which enables scientists to delete or interrupt the sequence of a particular gene quite efficiently (see Chap. 17).

factors, including the rate at which mutations appear in the genome and the very different selection pressures the individuals of different species face.

As for the second question, the problem is that after a long enough time, it becomes very difficult to tell a *de novo* gene from a very distantly related homolog. For example, the genes that are responsible for the emergence of the cnidocytes, or stinging cells, in cnidarians (jellyfish and hydras) are certainly orphan genes–we do not know, however, whether they started out as *de novo* genes or as duplicated genes. We just know that today, we can no longer find any homology with other genes. However, there are examples of actual *de novo* genes that have important roles. In the fruit fly, six *de novo* genes were found to be essential for their survival. Specifically, if these genes are disabled, the flies are unable to complete their life cycle and die at the pupal stage. In general, *de novo* genes may be often involved in organisms' response to rough conditions, be it a pathogen invasion or a low amount of nutrients in the environment.

The emergence of new traits in evolution is tightly connected to the origination of new genes. A new gene can arise in many ways, and gene duplication and its variants are the most common ones. However, *de novo* genes, or genes "from scratch", have been found in every organism surveyed. Although it is still too soon to estimate their importance, and it is still hard to find them, we now have enough examples to see that *de novo* genes play a role in the adaptation of organisms to their environment.

DEFINITIONS

Gene: A segment of DNA, encoding either functional RNA or a protein.

Homologs: Genes that derive from the same ancestral sequence.

Orthologs: Homologous genes that are related by descent from a common ancestor. They encode proteins with the same function in different species.

Paralogs: Homologous genes that have evolved by gene duplication. They encode proteins with similar, but NOT identical, functions.

Pseudogene: A nonfunctional segment of DNA that resembles a functional gene.

***De novo* gene:** A novel gene that originates from a previously non-genic DNA sequence

Gene family: A set of similar genes created by the duplication of a single original gene. They have similar biochemical functions.

Open reading frame (ORF): A continuous sequence of codons that begins with a start codon (AUG) and ends with a stop codon (UAA, UAG or UGA).

SUGGESTED READING

- Kaessmann H. Origins, evolution, and phenotypic impact of new genes. Genome Res. 2010 Oct;20(10):1313–1326.
- Van Oss SB, Carvunis AR. De novo gene birth. PLoS Genet. 2019 May 23;15(5):e1008160.
- Levine MT et al. Novel genes derived from noncoding DNA in Drosophila melanogaster are frequently X-linked and exhibit testis-biased expression. Proc Natl Acad Sci U S A. 2006 Jun 27;103(26):9935–9939.

19.6 Questions

1. **Which of the following statements is/are true?**
   (A) Ortholog genes have different DNA but code exactly the same protein.
   (B) Paralog genes have different DNA but code exactly the same protein.
   (C) Ortholog genes originate via gene duplication.
   (D) Paralog genes originate via gene duplication.
   (E) Homolog genes are identical in sequence and in function.
2. **It has been estimated that the human body is capable of producing about 80,000 to 400,000 proteins. Not all of them are present at the same time in the human body, and not all of them are produced by all cells. How can such a huge number of different proteins be produced if, according to the current estimates, the number of human genes is about 20,000?**
   (A) A given gene may be spliced differently into several different mRNAs, which may then be translated into different proteins.
   (B) The same gene can be read and translated from both sides and on both strands; therefore, it can code up to four different proteins.
   (C) Some parts of noncoding DNA can also be used to produce proteins.
   (D) Some proteins are made from RNA genes.
   (E) Some additional proteins are encoded by plasmids.
3. **What are the possible consequences of horizontal gene transfer (HGT)?**
   (A) HGT may transfer genes from one chromosome to another within a single cell.
   (B) HGT can lead to antibiotic resistance in normally sensitive bacteria.
   (C) HGT can lead to the acquisition of a new gene by an unrelated species.
   (D) A species may lose a gene it previously possessed as the gene is transferred to a related species.
   (E) A copy of a gene may be transferred to a different chromosome within a cell, and these two copies may then evolve independently.
4. **Which of the following could be the result of an unequal crossing over between these two chromosomes, assuming only one point of crossover?**
   Chromosome 1: A B C D E F
   Chromosome 2: a b c d e f
   (A) A B C c d e f
   (B) A B C d e F
   (C) A B C D F f
   (D) a b c d e E F
   (E) a b F
5. **Which of the following statements about *de novo* genes is/are correct?**
   (A) They have an open reading frame.
   (B) They can have paralogs but not homologs.
   (C) They originate in gene duplication followed by extensive mutations.
   (D) They code for functional protein or RNA.
   (E) *De novo* genes are never critically important for the survival of the organism.
6. **Some experiments have shown that you can replace a gene in the baker's yeast genome with its human counterpart, and the yeast will survive–even though the yeast dies if this gene is removed without replacement. This is possible because**
   (A) All the mutations accumulated during the years of evolution are present only in the introns of the genes. Thus, the mature mRNA sequences are identical, and in turn, the resulting protein is identical.
   (B) Most of these mutations are likely silent mutations–even if the DNA sequences differ significantly, these changes do not significantly change the amino acid sequence in the resulting polypeptide chain.
   (C) Of the many mutations that occurred over time, only those that did not affect the function of the resulting protein persisted in the populations. This can be because the mutations are in the introns and thus are silent or because they change the amino acid within the polypeptide chain, but only to one with similar properties, so the functionality of the resulting protein is not affected.

(D) The proteins corresponding to the genes that may be interchanged in this way are not very specific in the reactions they catalyse. Thus, even if such a protein is replaced by a similar protein, the reaction or cellular process can still be performed.

(E) The proteins coded for by these genes do not catalyse key processes for the survival of cells and may not be crucial for the survival of the organism, so replacing them with different genes is not likely to affect the survival either.

19.7 Project: Exploring the Evolutionary Relationships Between Genes in the Globin Gene Families

An interesting example of a gene family is the globin gene family–actually two families–that we find within our genome as well as in the genomes of other vertebrates. Scientists can reconstruct the evolutionary history of genes within a multigene family by comparing the gene sequences. This can be done either by comparing the DNA nucleotide sequences or by comparing the sequences of amino acids in polypeptides coded for by these genes. In this project, you will become acquainted with some of the online tools used by scientists and analyse the results of your comparisons.

Below is a link to the DNA sequences of eight genes, some of which belong to one of the globin gene families and the rest to the other family. The sequences show the DNA coding strand–the one complementary to the template strand (which is used for transcription), so it shows the same nucleotide sequence as the mRNA, just with T instead of U.[1] The introns have also been removed from the sequence, so the entire length is translated into a polypeptide chain.

The gene sequences are available at this link: https://en.edub.sk/everyday-evolution

Part 1: Comparing DNA Sequences

1. Estimate the percent identity value for each pair of globin genes using the pairwise sequence alignment tool Needle: https://www.ebi.ac.uk/jdispatcher/psa/emboss_needle. This tool enables you to align either protein or DNA sequences, so choose the DNA option. Otherwise, use the default settings. Wait for a while to get the results, then scroll down the result page to the "Identity", and write down the percentage. The percent identity reflects the number of positions with identical nucleotides relative to the total number of nucleotides in the sequence.
2. Write the results of the pairwise alignment of the globin genes in Table 19.1. Some pairs have already been completed for you.
3. Look at the percent identity values in the table and identify "sister genes"–pairs of genes with the highest percent identity values. These pairs of genes are the most closely related. You can represent their relatedness as a simple branching.
4. Try to identify a "cousin gene" for each of these "sister-pairs" (this will not be possible for all of them, as there are eight genes). Adding these means that branching gets more complex.
5. Continue with the branching pattern, tracing the relatedness of genes until you get to the common ancestor of all of them. Draw a branching diagram showing the relationships between the genes.
6. How confident are you about the proposed evolutionary relationships of the globin genes? Are there any alternatives you have doubts about? Explain.
7. Translate the DNA sequences into polypeptide chains. Use the DNA to protein translation here: https://web.expasy.org/translate/. Use the default tool settings, select to translate the forward DNA strand, and use the results, which show all the sequences within a single open reading frame. Write the sequences in Table 19.2. The gene for the Gamma A protein has been translated for you.

Part 2: Comparing Polypeptide Chains

1. Estimate the percent identity value for each pair of globin proteins using the same pairwise sequence alignment tool Needle: (https://www.ebi.ac.uk/jdispatcher/psa/emboss_needle). This time, choose the protein option. Now, the percent identity reflects the number of positions with identical amino acids relative to the total number of amino acids in the sequence.

[1] Strange as this may seem, this is just the way scientists do it.

Table 19.1 DNA sequence identity table (percent)

	Alpha 2	Beta	Gamma A	Gamma G	Delta	Epsilon	Zeta
Alpha 1							67.9
Alpha 2		61					
Beta					93		
Gamma A				99.8			
Gamma G							
Delta						76.7	
Epsilon							54.6

Table 19.2 Polypeptide chains. Each chain should be between 140 and 150 amino acids long

Alpha 1:	MVLSPADKTNVK....	
Alpha 2:		
Beta:		
Gamma A:	MGHFTEEDKATITSLWGKVNVEDAGGETLGRLLVVYPWTQRFFDSFGNLS	50
	SASAIMGNPKVKAHGKKVLTSLGDAIKHLDDLKGTFAQLSELHCDKLHVD	100
	PENFKLLGNVLVTVLAIHFGKEFTPEVQASWQKMVTAVASALSSRYH*	148
Gamma G:		
Delta:		
Epsilon:		
Zeta::		

Table 19.3 Amino acid identity table

	Alpha 2	Beta	Gamma A	Gamma G	Delta	Epsilon	Zeta
Alpha 1					43.6		
Alpha 2		43.6		40.3		38.3	
Beta							34.9
Gamma A							
Gamma G							
Delta							
Epsilon							

2. Write the results of the pairwise alignment of the globin genes in Table 19.3. Some pairs have already been completed for you.
3. The proteins Gamma A and Gamma G show 99.3% identity. This means that these proteins differ just by .... amino acids (fill this number in). What is the position of this/these amino acid/s in the protein chain? What mutation/s in the DNA sequence could have led to this/these difference/s?
4. Look at the percent identity values in the table and use a similar approach as before to estimate the relatedness of the eight proteins. Draw a branching diagram showing the relationship between them based on the amino acid sequences.
5. How confident are you about the proposed evolutionary relationships of the globin proteins? Are there any alternatives you have doubts about? Explain.

Analysis

1. Compare the branching diagrams you made in Parts I and II. Do the diagrams show the same evolutionary relationships?
2. Look at the identity values for the pairs of genes in Table 19.1 and for the pairs of their products in Table 19.3. Why are the values different? Hint: Think also about how the percentages are counted in each case.

3. As said above, these eight genes belong to two globin families. Each of these families is located on a different chromosome; that is, members of one family are located on one chromosome and members of the other family are located on another chromosome. Assign each of the eight genes to one of the two families. Explain your reasoning.
4. Suggest some advantages and disadvantages of using DNA and protein sequences for studying evolutionary questions. What problems or questions would be better addressed using DNA sequences? When would it be more useful to use protein sequences?

This project was developed by Kristína Hudáková and Katarína Juríková.

Why Can't We Live Forever?

20

Katarína Juríková

▶ Eukaryotes, from unicellular fungi and algae to sycamore trees and humans, are composed of cells with membrane-bound organelles and have more DNA than prokaryotes such as bacteria or archaea. The eukaryotic nuclear genome–all the DNA in the nucleus–is typically divided into several linear DNA molecules called chromosomes. But does it have to be so? Why are eukaryotic genomes linear? Was this an inevitable evolutionary step? Are there any challenges connected with having linear chromosomes instead of circular ones? Let's try to answer some of these questions and learn more about a very special feature of linear chromosomes: telomeres.

20.1 Problems at the Ends of Eukaryotic Chromosomes

Eukaryotic chromosomes are usually composed of tens to hundreds of millions of DNA nucleotides bound by various proteins. Most of the time, chromosomes look like long filaments, and only for cell division, they are doubled and compacted into the well-known X shape. One crucial part of the chromosome is the centromere, which ensures proper attachment of microtubules during cell division. It looks like a constriction, often around the middle of the chromosome. Another special part of a chromosome is its end–the telomere. Telomeres help the cell cope with the problems that arise at chromosome ends.

There are two big problems that are common to all linear chromosomes. The first problem is caused by the imperfection of cells' own enzymatic machinery, and the second is a result of cells being a little too efficient.

At the heart of the first problem are DNA and RNA polymerases. These are enzymes that copy DNA during cellular replication and produce RNA in the process of transcription. Their function is essential for all processes in living organisms. They are also incredibly efficient: DNA polymerase can synthesise DNA at rates approaching 800 bp per second, and it makes only one mistake for every 10,000 or even 100,000 bases. However, all known DNA and RNA polymerases have one key limitation–they only synthesise new nucleic acids in the $5' \rightarrow 3'$ direction.

To make a new strand of nucleic acid, both DNA and RNA polymerase require a template strand, to which the new nucleotides will be paired according to the complementarity of bases. The RNA polymerase only needs this template

K. Juríková (✉)
Department of Genetics, Faculty of Natural Sciences, Comenius University in Bratislava, Mlynská dolina, Bratislava, Slovakia
e-mail: katarina.jurikova@uniba.sk

B. Trubenová et al. (eds.), *Everyday Evolution: Unveiling Life's Hidden Patterns*,
https://doi.org/10.1007/978-3-032-07050-0_20

The $5' \rightarrow 3'$ Directionality

Why can nucleic acids be synthesised only in the $5'$ to $3'$ direction? Each free nucleotide that is added to the growing DNA strand is a triphosphate–it contains not just one but three phosphate groups. If the word *triphosphate* sounds familiar, you are right–the incoming nucleotides are actually very similar to ATP–the basic unit of cellular energy (just with the difference that the DNA nucleotides contain deoxyribose instead of ribose and any of the nitrogenous bases may be present). Like in the case of ATP, the triphosphate tails provide the energy necessary to add the nucleotide to the growing strand. The incoming nucleotides thus bring the energy needed for synthesis, and this process is catalysed by DNA polymerase. The necessity of attaching its $5'$ end (with the phosphates) to an exposed $3'$ end of the previous nucleotide is the reason why each new DNA and RNA strand can grow only in the $5'$ to $3'$ direction.

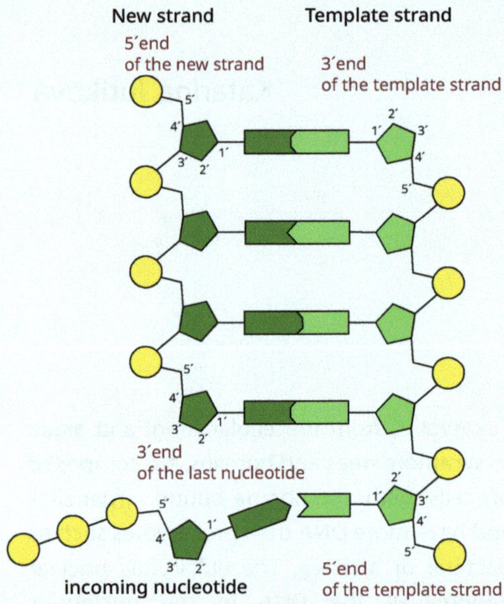

DNA synthesis. When adding a new nucleotide to a growing strand, the two phosphates break off and a covalent bond is formed between the phosphate group and the $3'$ carbon of the sugar of the previous nucleotide

DNA; it can start producing the RNA without any previously existing RNA strand to which it would be adding nucleotides. In other words, it can start the new strand from scratch. Conversely, the DNA polymerase must also add the incoming nucleotide to a nucleotide already present in the new strand. Because of this limitation, the two polymerases collaborate even during DNA replication. First, a short sequence of RNA (a primer) is made by RNA polymerase, using only the template DNA strand. Then, DNA polymerase can add DNA nucleotides, starting at the primer and proceeding according to the base pairing rules. (The DNA polymerase does not mind that the primer is RNA and not DNA.)

The two strands of a DNA molecule are antiparallel, meaning they run in opposite directions (see Box: The $5' \rightarrow 3'$ Directionality). At each end of the double-stranded DNA molecule, one strand terminates with a $3'$ end, while the other strand terminates with a $5'$ end.

When a linear eukaryotic chromosome begins to replicate, the two strands of the DNA molecule separate at a specific location known as the **origin of replication** (not at the ends), creating a replication bubble.[1] RNA polymerase synthesises an RNA primer in the middle of this replication bubble, and then, DNA polymerase adds DNA nucleotides to the growing strand. For the so-called leading strand that is synthesised in the $5' \rightarrow 3'$ direction, this process is straightforward. As the replication bubble expands along the DNA molecule, new nucleotides are continuously added in the $5'$ to $3'$ direction, following the exposure of the template strand.

By contrast, the other (lagging) strand is synthesised in short stretches called Okazaki fragments. As the replication bubble opens up, new primers are synthesised at short intervals, to which DNA polymerase then adds nucleotides

[1] There are many such origins of replication and thus also replication bubbles along each linear chromosome.

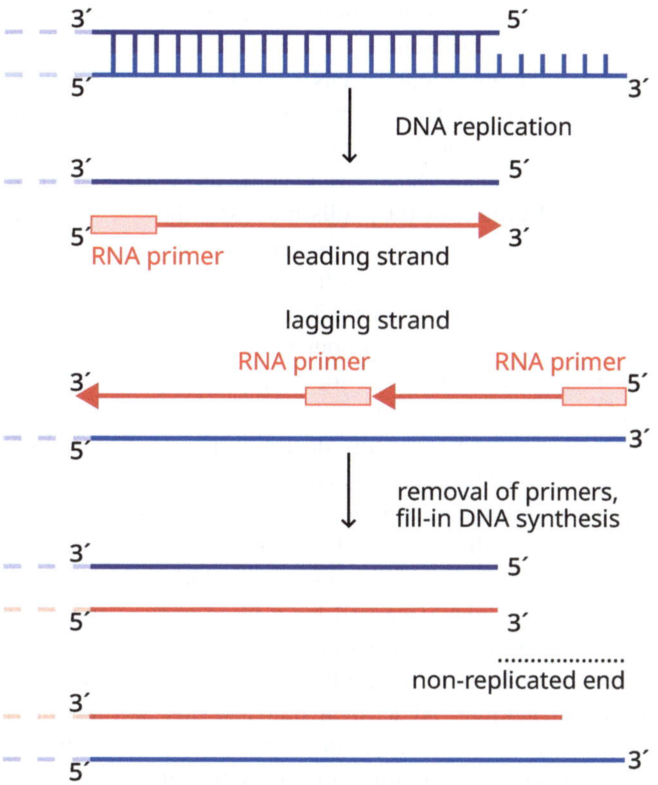

Fig. 20.1 The end replication problem. One of the newly synthesised DNA molecules is shorter than the parental one in every cell division. This is due to the presence of the 3' overhang in the original DNA, the DNA polymerase's need for a primer and the 5′ → 3′ directionality of DNA synthesis. (Figure based on Lingner J, Cooper JP, Cech TR. Telomerase and DNA end replication: no longer a lagging strand problem? Science. 1995 Sep 15;269(5230):1533-4. https://doi.org/10.1126/science.7545310)

in the usual 5′ to 3′ direction. To understand the problem with linear chromosome replication fully, we need to know another fact–the ends of eukaryotic chromosomes are not blunt but have a single-stranded 3′ overhang at their very end.

A major issue arises during leading-strand replication. As the template strand is shorter because of the 3′ overhang, the DNA polymerase copies it perfectly, but the resulting chromosome misses the final part that was single-stranded in the old chromosome. Thus, every time a cell replicates its DNA, which occurs every time the cell divides, telomeres are shortened-a phenomenon called the **end replication problem** (Fig. 20.1).

On the strand replicated by lagging-strand synthesis, the RNA primer occupies its very end. The removal of the RNA primer at the very end leaves behind an unfilled gap, and the DNA polymerase, in need of a primer, is unable to fill it. Therefore, as the primer is removed, the newly made strand will again have the 3′ single-stranded overhang and will cause the end replication problem in the next round of replication.

The end replication problem causes the chromosomes to shorten slightly in each cell division. After around 60 cell divisions, the telomeres become critically short–the cell recognises them as a problem. It then stops dividing and enters a state known as senescence, in which the cell's functions gradually fail. Senescence is often followed by apoptosis–programmed cell death.

The second problem at the ends of chromosomes stems from cells being too efficient in identifying damaged DNA. On one hand, it helps them to repair actual DNA damage quickly and avoid mutations. The most troublesome DNA damage that can occur in a cell is the so-called double-stranded DNA break. This happens when both strands of the DNA double helix are broken simultaneously in the same spot. Double-stranded breaks are dangerous because cells cannot always determine which broken ends belong together, if there are more of them. As a result, the cell's repair machinery may end up connecting two broken DNA ends that were not joined before. This may lead, in the worst case, to genome aberrations such as translocations, insertions or deletions of large parts of chromosomes.

Because double-stranded DNA breaks are so dangerous, cells try to quickly join any two free DNA ends. The trouble is that when a cell has linear chromosomes, each chromosome has two free ends. Because of this, cells must somehow distinguish the natural chromosome ends from broken DNA and shield them from being joined by the DNA repair machinery. This is called the *end protection problem*–without structures that prevent chromosomes from joining, chromosomal aberrations may occur.

So, how do cells normally deal with the two challenges of *end replication* and *end protection*? The answer to the end protection problem lies in the proteins that bind to the telomeric DNA and prevent it from being recognised as a double-stranded DNA break (we will learn more about them soon-see Fig. 20.3). Surprisingly, the solution to the end replication problem is most often just to do nothing. At least in humans, this is true–most cells in our body do nothing special to fight telomere shortening. Our somatic cells, the majority of cells in the human body, divide until their telomeres get too short and then die or never reach the critical number of cell divisions. However, there is one cellular type that cannot exist and function if it ignores telomere shortening: stem cells.

Stem cells are present in both the adult body and the developing embryo and give rise to all other cells. If embryonic stem cells were unable to extend their telomeres, they would not be able to divide a sufficient number of times to populate the whole body. Additionally, many organs in the adult body would not function without continuous renewal, which also requires the division of stem cells. (Just think of organs such as the skin or the lining of the gut, where millions of cells die and are replaced every day.)

There is also one pathology that requires telomere maintenance for its existence: cancer. Cancer is a disease of unrestricted cell division in which healthy cells lose their normal regulation and divide over and over again. Normally, if a cell accumulates many mutations in its DNA, cellular mechanisms recognise the problem and guide such a damaged cell to senescence and apoptosis. However, sometimes, these protective mechanisms fail. Some of the mutated genes can then drive the cell to divide uncontrollably and become cancerous. There are several physiological barriers that a human cell must overcome to become a cancer cell. One of them is the inability to renew telomeres.

Telomerase

All stem cells and a majority of cancer cells use the same mechanism to elongate their telomeres. They use an enzyme called **telomerase**. Telomerase has a peculiar structure, composed of two components or subunits: One is the catalytic subunit, a protein that acts as a reverse transcriptase, synthesising DNA according to an RNA template. The other subunit is the RNA template itself, a small piece of its sequence being retrotranscribed into telomeric DNA. In humans and many other species, the RNA subunit of telomerase carries the template sequence CCCUAA that is copied into the TTAGGG telomeric repeat. The repeated addition of the TTAGGG sequence copied from the RNA template by the reverse transcriptase gives a telomere its typical repetitive sequence.

Around 15% of human cancers have found another way to elongate their telomeres, without using telomerase. They use recombination (remotely similar to meiotic crossing-over) to extend their telomeres. This recombination-based telomere extension is called **alternative lengthening of telomeres (ALTs)**. Scientists are still trying to understand how this process is regulated and why it develops in some cancers more often. One factor that seems to be involved is TERRA, a long noncoding RNA that is transcribed from telomeres and never translated. Its precise role in human cells is still not fully understood, but it appears to stimulate telomeric recombination.

Is there a reason why telomerase is not present in all our cells? Possible explanations come from looking beyond human telomere biology. It turns out that some other organisms have active telomerase in their cells all the time. Let's consider budding yeast *Saccharomyces cerevisiae* and fission yeast *Schizosaccharomyces pombe*. In these two yeasts, telomerase is practically always present and active, extending telomeres in each cell di-

Apoptosis

Apoptosis–programmed cell death–is a very important process for both normal embryonic development and tissue homeostasis in later life stages. Disrupted regulation of apoptosis has been linked to a variety of human diseases, including cancer and neurodegenerative diseases such as Parkinson's or Alzheimer's. In embryos, apoptosis removes cells from where they are no longer needed, such as between the fingers of a developing hand. In adults, apoptosis removes dangerous cells (such as precancerous or infected cells) and also cells that are normal but not needed anymore (e.g. cells producing antibodies after an infection is over).

Apoptosis may be triggered by a signal from outside the cell. These can be physical or chemical challenges in the environment–such as toxins, heat or radiation–or a signalling molecule binding to a death receptor on the cell membrane (we call this the extracellular pathway of apoptotic signalling). Alternatively, it may be triggered from within the cell, e.g. by a stress signal (intracellular pathway).

Unlike necrosis, where the cell just bursts and releases its contents, apoptosis is a complex process characterised by specific changes in cell morphology, energy use and protein synthesis that regulate specific genes and biochemical pathways within the dying apoptotic cell. The most important apoptotic changes occur in the mitochondria, nucleus and plasma membrane. DNA is degraded into short fragments, and the chromatin condenses within the nucleus. The cell shrinks, and the cytoskeleton and plasma membrane form outward protuberances called blebs. As apoptosis continues, these membrane blebs enclosing cellular debris detach, forming **apoptotic bodies**, which are then engulfed by phagocytic cells.

As the cell enters apoptosis, it releases immunostimulatory molecules. These molecules attract macrophages that recognise the apoptotic bodies and phagocytise them. Thus, apoptosis not only prevents the leakage of cell contents into the tissue but also ensures that the tidily wrapped contents of the dead cells are recognised and taken care of by immune cells.

Apoptosis occurs quite rapidly–it is typically completed within a few hours–and the changes described above may be easily observed through a light microscope.

vision (as in human stem cells). If we force yeast to abandon telomerase, for example, by deleting the gene that encodes it, then the yeast can still survive and maintain its telomeres by a process very similar to the ALT observed in human cancer. This may suggest that active telomerase is a default option for many organisms and that ALT-like maintenance of telomeres is a backup mechanism that the ALT cancer cells possibly hijack for their own benefit.

But why would humans lose the active telomerase in their somatic cells? One hypothesis is inspired by comparisons with short-lived, small animals such as mice. Mice have active telomerase in all their cells throughout their life. Scientists think that the inactivation of telomerase in the somatic cells of large, long-lived organisms, such as humans, may be a way of preventing cancer. The reasoning is that the more cells divide and the longer they live, the more mutations they accumulate in their DNA. Telomere shortening leads to the death of cells that are too old or have divided too many times–cells that could otherwise potentially become cancerous. However, this is not a big issue for small, short-lived animals, such as mice, which typically die in the wild after a few months–their cells simply do not have enough time to develop cancer.

20.2 Telomere Structure

Now, we know what the function of telomeres is. But what do telomeres actually look like, beyond the TTAGGG DNA repeats introduced by telomerase?

Chromosome Aberrations

If the normal chromosome structure is heavily reorganised, we call the resulting mutation a chromosomal aberration. These are often the result of faulty rejoining of DNA after a DNA double-stranded break occurs in two different sites at the same time. An example of such an event is the so-called Philadelphia chromosome–a reciprocal translocation, which involves a mutual exchange of fragments between chromosomes 9 and 22. This particular chromosome aberration causes the fusion of two genes, which normally reside on different chromosomes, and creates a chimeric gene that drives the development of several types of leukaemia in the affected individuals.

Interestingly, chromosome aberrations may also result from extreme telomere shortening. In some cases, cells with critically short telomeres fail to undergo senescence and apoptosis. Such eroded telomeres are then recognised by the cell's repair mechanisms as chromosome breaks and are treated as such, causing the individual chromosomes to be joined into one. This can either lead to dangerous gene fusions (as in the Philadelphia chromosome) or create chromosomes with double centromeres. Such chromosomes cannot normally go through cell division and are broken at random places during anaphase, leading to repeated breaking and inappropriate fusion of chromosomes.

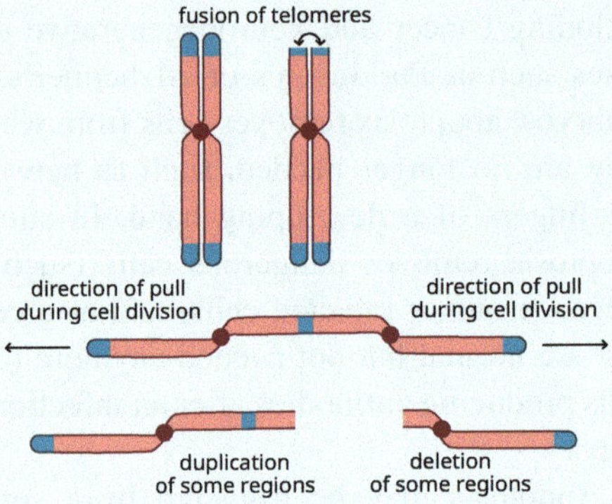

Fusion of chromosome ends. The fusing of chromosome ends creates problems during cell division when two chromatids are pulled apart by a mitotic spindle. This can lead to chromosome breakage, creating chromosomes with extra segments (insertions) or missing segments (deletions), while also creating chromosomes with sticky ends, which are likely to be rejoined by the DNA repair machinery again. (Figure based on Maciejowski J, de Lange T. Telomeres in cancer: tumour suppression and genome instability. Nat Rev Mol Cell Biol. 2017 Mar;18(3):175–186. https://doi.org/10.1038/nrm.2016.171. Epub 2017 Jan 18. Correction in: Nat Rev Mol Cell Biol. 2019 Apr;20(4):259. https://doi.org/10.1038/s41580-019-0113-7).

Scientists noticed early that the ends of chromosomes usually have a particular structure, different from other parts of the genome. Telomeres were named upon their discovery in 1938 by Hermann Muller, from the Greek *telos* for "end" and *meros* for "part". Typically, no genes are found in telomeres. In humans, as we already know, the telomeric repeat sequence is TTAGGG (CCCTAA on the complementary strand), and it covers up to 10,000 bp at the end of each chromosome. As we have already learned in Figs. 20.1 and 20.2, telomeres end in a 3′ overhang. In humans, the overhang is formed by the DNA strand containing "TTAGGG" and is also called the G-overhang. This single-stranded DNA then bends back, enters the double-stranded portion of telomeric DNA and forms what is known as a telomeric loop or t-loop (Fig. 20.3).

In other parts of the genome, DNA is tightly packed around proteins called histones to form bead-shaped units called nucleosomes. In contrast, telomeric DNA is bound by a different set of proteins, unique to telomeres. Some of these proteins bind telomeric DNA directly, either binding to the double-stranded portion of telomeres or to the single-stranded overhang. Additional proteins are associated with telomeres through protein-protein interactions, in other words, through binding to proteins that are already directly attached to telomeric DNA.

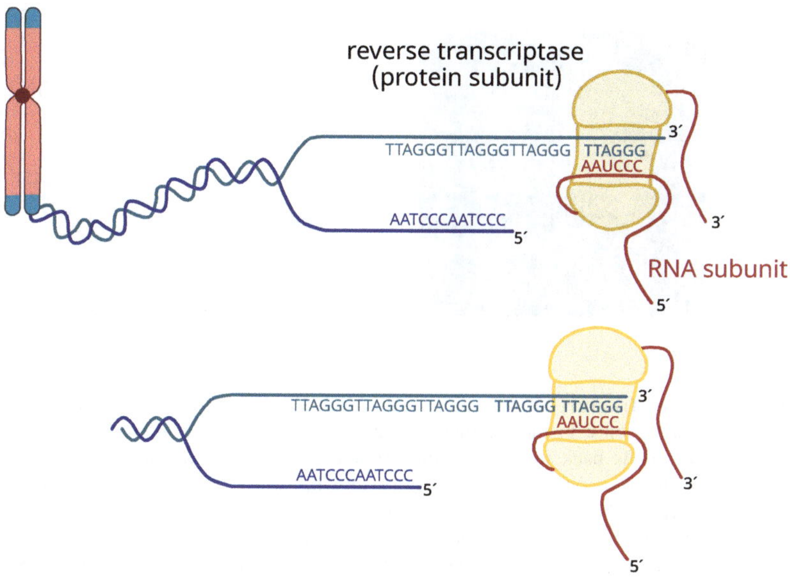

Fig. 20.2 Telomerase extends telomeres one repeat at a time. The protein subunit retrotranscribes the RNA template onto the end of a chromosome over and over and over. (Figure redrawn in Inkscape, based on a design from an unpublished figure created with BioRender.com)

Interesting Fact

In humans, there are six core telomeric proteins, collectively known as shelterin (Fig. 20.3b). TRF1 and TRF2 bind directly to the double-stranded telomeric DNA and recognise the DNA sequence TTAGGG/AATCCC. POT1 specifically binds to the single-stranded telomeric DNA. The remaining three proteins are part of the shelterin complex and bind to the above-mentioned DNA-binding proteins: RAP1 associates with telomeres by binding to TRF2, and TPP1 and TIN2 bind to POT1 and TRF1+TRF2, respectively, bridging the double-stranded and single-stranded parts of the telomere. It is these proteins that, together with the t-loop structure, help the cell to distinguish a normal telomere end from a harmful double-stranded DNA break. Apart from these main telomeric proteins, many other proteins bind to telomeres under different conditions and in different phases of the cell's life. One such crucial interaction partner of telomeres is the enzyme telomerase, which we have already encountered.

20.3 Extending Telomeres: A Road to Immortality?

We have already discussed how cells lacking telomerase eventually stop dividing once their telomeres are critically shortened. This has a consequence at the level of the entire organism. As we age and our telomeres shorten, cells in many tissues lose their regenerative potential, causing the functionality of various tissues and organs to deteriorate. Could treatment with telomerase then restore telomere length in old people and "cure" ageing? There are indications that this appealing idea could actually work. In an experiment in mice that normally have active telomerase, scientists artificially inactivated their telomerase gene, causing their telomeres to shorten with ageing (as is normal in humans). After several generations, as the telomeres of these mice became shorter, they started showing signs of early brain degeneration and organ failure. In fact, the reactivation of telomerase reversed these artificial ageing symptoms in mice. However, there are two main problems with applying these findings to cure

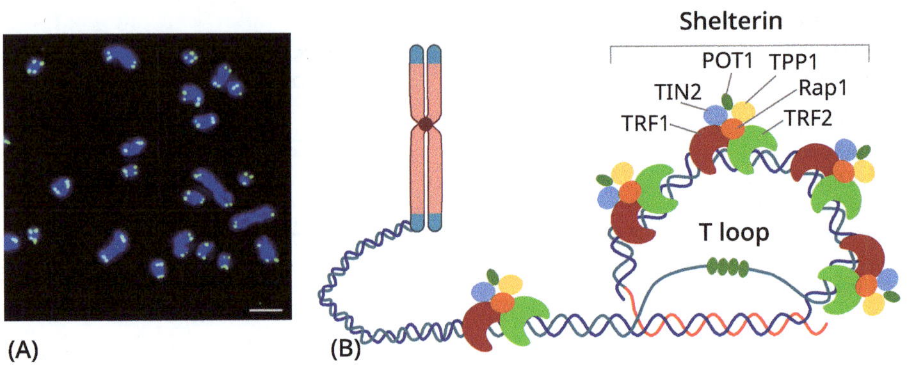

Fig. 20.3 **The structure of telomeres.** (**a**) Microscopy image of telomeres, visualised by light green fluorescent dye binding only to the telomeric sequence. Scale bar = 5 μm. (**b**) The very end of a telomere is single-stranded and bends back into the double-stranded part of the telomere to form a lariat-like structure called a t-loop. The proteins specific for telomeres are shown, too — they are collectively called shelterin. (Figure based on Calado RT, Young NS. Telomere maintenance and human bone marrow failure. Blood. 2008 May 1;111(9):4446-55. https://doi.org/10.1182/blood-2007-08-019729. Epub 2008 Jan 31. Photo adapted from Adam, N., Degelman, E., Briggs, S. et al. Telomere analysis using 3D fluorescence microscopy suggests mammalian telomere clustering in hTERT-immortalised Hs68 fibroblasts. Commun Biol 2, 451 (2019). https://doi.org/10.1038/s42003-019-0692-z. CC-BY-4.0 https://creativecommons.org/licenses/by/4.0/)

human ageing. First, these rapidly ageing mice are an artificial system. In contrast to mice, telomerase is normally *inactive* in humans. Thus, introducing telomerase does not restore the normal state but brings about something new, possibly leading to the development of cancer, as discussed above. Second, unfortunately, ageing seems to be a more complex process than just telomere shortening. Even though telomere erosion certainly accompanies ageing in humans, many other factors are also involved. These include the accumulation of mutations, mitochondrial damage, stem cell exhaustion and improper functioning of the immune system. However, who knows–one day, the activation of telomerase may become a part of rejuvenation therapy.

20.4 Diversity of Telomeres

The structure of telomeres is surprisingly similar across all eukaryotes. In almost every species studied, telomeres are composed of repetitive sequences associated with dedicated telomeric proteins, and even the presence of TERRA–the telomeric RNA–seems to be ubiquitous. However, there are a few exceptions of species where the chromosome ends look different.

One of the most striking examples of unique telomere biology is found in insects from the order Diptera, such as flies and mosquitoes. The majority of species in this group do not use telomerase but employ a recombination-based strategy, similar to ALT, as their main telomere maintenance mechanism. A small number of dipterans have an even stranger way of renewing chromosome ends: Fruit flies (*Drosophila melanogaster*) do not have the typical telomeric sequence composed of short repeats. Instead, their telomeric DNA contains three retrotransposons called Het-A, TART and TAHRE, and their telomeres are maintained by the retroposition of these elements. This is notable from two perspectives: First, retrotransposons are typically considered "selfish" genetic elements–they often do not serve any useful role for the organism. Second, even in this unusual system, the process of reverse transcription is important for maintaining telomeres, just as in the case of telomerase. This leads us to the question: Do these two mechanisms have a common evolutionary origin?

20.5 Origin of Telomeres

Before examining the evolutionary origin of telomeres and telomerase, another mystery remains to be solved. Apart from rare exceptions, all eukaryotes have linear chromosomes. Why did chromosomes become linear in the first place?

One hypothesis is that circular DNA is not compatible with the process of sexual reproduction, which is typical in eukaryotes. In particular, meiotic crossing over on circular chromosomes would, in theory, lead to the emergence of chromosomes with two centromeres or entangled chromosomes. Linearisation of DNA may help to prevent this. However, we now know that many bacteria have enzymes that help cut such entangled circular DNA. That probably means that this hypothesis for the origin of linear chromosomes does not adequately explain why cells would not be able to cope with circular chromosomes.

There is another possible explanation for the existence of linear chromosomes. At the dawn of eukaryotic life, alpha-proteobacteria were engulfed by an archaeal host and evolved into mitochondria–one of the key organelles in eukaryotic cells (see Chap. 15). This endosymbiont brought with it a special class of selfish genetic elements, known as mobile group II introns. These are sequences of DNA that encode both an RNA and a reverse transcriptase enzyme. Using the process of reverse transcription, these selfish elements started migrating from the genome of the alpha-proteobacterial symbiont to the host genome. In other words, they became part of the nuclear genome. As more and more copies of these mobile elements accumulated in the nuclear DNA, a random double-stranded DNA break might have occurred just between two such retrotranscribed elements. As a result, both fragments of the broken molecule might have had this selfish sequence at their very end.

Some of these fragments may have been simply rejoined together to restore the circular chromosome. However, in some cases, the retrotransposon sequence in the very end might have looped back and paired with another retrotransposon sequence inside the chromosome. This would create a loop at the end of the chromosome, resembling the t-loop that is part of the telomere protection structure in present-day telomeres (Fig. 20.3). This would have solved the end protection problem of this primordial telomere.

The end replication problem might have been solved by the reverse transcription of the retrotransposon. The sequence of the selfish DNA might have been amplified by the reverse transcriptase, cementing the ends of chromosomes with this sequence.

If this way of telomere maintenance sounds somewhat familiar, it is because we already know one example of telomere maintenance via retrotransposons in present-day organisms–the *Drosophila* telomeres. Some scientists believe that the telomeric retrotransposons in fruit flies may resemble what occurred in early eukaryotes. However, telomeres in fruit flies have likely evolved secondarily after the loss of telomerase in this evolutionary branch and probably do not represent a conserved example of this ancient method of maintaining telomeres.

How did the switch to telomerase occur once the chromosomes were linear? Comparing the telomerase of different species across the eukaryotic tree of life, it turns out that the protein subunit, and in particular, the part that performs reverse transcription, is very similar between yeasts, protists and humans. Mobile group II introns seem to be a very close relative of telomerase–it seems that telomerase indeed originated from these selfish genetic elements during evolution. We can then speculate that the linearisation of chromosomes and the appearance of telomerase are a result of one event that took place at the beginning of eukaryotic life: the invasion of mobile group II introns in the nuclear genome that gave rise to both the problem–the linear chromosome–and the solution–the reverse transcriptase that later evolved into telomerase.

There are many aspects of the function and evolution of telomeres that we still do not understand. For example, how did the role of the telomeric RNA, TERRA, change during evolution, and why did it originate in the first place? Or why do the majority of species have a telomeric sequence that is so rich in guanines? We know that guanine-rich DNA can fold into special secondary structures, called G-quadruplexes, that are different from the double helix of DNA–do these play a role at telomeres? And will we ever be able to manipulate telomere biology to cure cancer or slow down ageing? These are some of the many open questions in the field of telomere biology–and in the quest to answer them, we are surely in for some surprises along the way.

DEFINITIONS

Primer: A short RNA sequence that serves as a starting point for DNA polymerase during DNA replication

Telomere: The end of a linear chromosome, typically composed of repetitive DNA and specialised telomeric proteins

Telomerase: Enzyme capable of telomere elongation

End replication problem: The shortening of linear DNA ends during every cell division

End protection problem: The need to protect the natural chromosome ends (telomeres) from being recognised and repaired as DNA breaks

SUGGESTED READING

- Jones-Weinert C, Mainz L, Karlseder J. Telomere function and regulation from mouse models to human ageing and disease. Nat Rev Mol Cell Biol. 2025 Apr;26(4):297–313.
- Dilley RL, Greenberg RA. ALTernative Telomere Maintenance and Cancer. Trends Cancer. 2015 Oct 1;1(2):145–156.
- de Lange T. A loopy view of telomere evolution. Front Genet. 2015 Oct 20;6:321.
- Nosek J, Kosa P, Tomaska L. On the origin of telomeres: a glimpse at the pre-telomerase world. Bioessays. 2006 Feb;28(2): 182–190.

20.6 Questions

1. **Which of the following statements correctly describes the link between telomeres and cancer?**
   (A) Cells that lose their telomeres completely are those most likely to become cancer cells.
   (B) The telomeres of cancer cells do not shorten with successive cell divisions as telomeres in normal cells do.
   (C) In precancer and cancer cells, we can usually observe increased telomerase activity.
   (D) Cells with the shortest telomeres are those most likely to become cancer cells.
   (E) The telomeres of precancer and cancer cells shorten about twice as rapidly as telomeres in gametes.
2. **If the rate of prokaryotic DNA replication is 800 base pairs per second, how long would it take to replicate the *Escherichia coli* genome, which is 4.6 million base pairs long?**
   (A) 1 hour, 35 minutes and 50 seconds
   (B) 95 minutes and 30 seconds
   (C) 5705 seconds
   (D) 90 minutes and 50 seconds
   (E) 1 hour, 35 minutes and 5 seconds
3. **Which of the following statements about telomerase is/are true?**
   (A) Telomerase enables the cell to synthesise the lagging strand in one go, without the Okazaki fragments.
   (B) Telomerase provides a template for lengthening shortened telomeres.
   (C) Telomerase catalyses the retrotranscription of an RNA template into a DNA strand.
   (D) Telomerase consists of two subunits: an RNA subunit and a DNA subunit.
   (E) Some eukaryotes have telomerase active in all cells during their entire lifetime.
4. **Telomere is...**
   (A) Structurally similar to centromeres, just located at the end of the chromosome.
   (B) The site where the replication of a DNA molecule starts.
   (C) The end of a linear eukaryotic chromosome.
   (D) The end of a linear prokaryotic chromosome.
   (E) Composed of a repeated sequence TTAGGG in humans.
5. **Which of the following cells/organisms would be affected the most by adding an active telomerase? How would these cells/organisms be affected?**
   (A) Prokaryotic cells. Their circular chromosomes would become linear.
   (B) Single-celled eukaryotes. Their cells would age and lose the ability to reproduce.
   (C) Somatic cells of eukaryotes. They would lose the ability to regulate the number of cell divisions.
   (D) Human stem cells. They would lose the ability to differentiate.
   (E) Asexually reproducing eukaryotes. They would lose the ability to go through mitosis repeatedly.
6. **Which of the following statements best describes the general differences between eukaryotic and prokaryotic chromosomes?**
   (A) Eukaryotic cells usually have one circular chromosome, whereas prokaryotic cells have several linear ones.
   (B) Ends of chromosomes in prokaryotes shorten with each replication, but those in eukaryotes are maintained at a constant length.
   (C) Prokaryotic DNA is organised around histone proteins, while eukaryotic chromosomes rely on shelterin complexes for overall DNA packaging.
   (D) Prokaryotic DNA consists entirely of coding regions, while eukaryotic DNA has both coding and noncoding segments.
   (E) Prokaryotic chromosomes have neither the end replication nor the end protection problem, while eukaryotic chromosomes have both these problems.

20.7 Project: Evaluating New Drug Candidates

Restoring telomerase function could combat ageing, but it risks cancer development. A potential solution is the transient activation of telomerase for worn-down telomeres in specific tissues.

Developing new drugs is a complex process, involving initial discovery and rigorous preclinical testing in labs and animals for safety and efficacy, followed by multiphase clinical trials in humans.

Scientists at a hypothetical biotech company identified a promising drug candidate, **TelomereRestore**, designed for transient telomerase activation. In this project, you will examine and evaluate a series of hypothetical studies on TelomereRestore, analysing their methodology and interpreting their results within the context of drug development challenges.

Part 1: Preclinical Evaluation

Step A: Evaluating TelomereRestore in Cell Cultures

To begin the preclinical evaluation of TelomereRestore (TR), scientists investigated its potential benefits as an anti-ageing agent. While its exact chemical structure is proprietary, TelomereRestore was hypothesised to transiently activate telomerase, thus offering a potential strategy to combat cellular ageing. Scientists applied TelomereRestore to cultured mammalian cells and analysed the telomere lengths of these cells over time. Figure 20.4 below illustrates the effect of TelomereRestore on telomere length in cultured mammalian cells.[2]

Question 1: Explain what cell cultures are and list some advantages of their use in experiments evaluating potential drugs.

Question 2: Summarise the observed effect of TelomereRestore on telomere length. Based on data shown in Fig. 20.4, what initial conclusions can be drawn about TelomereRestore's effectiveness?

Step B: Evaluating TelomereRestore in Mice

Following promising in vitro results, scientists next used TelomereRestore to study its effects in mouse models. Beyond directly measuring telomere length, they also analysed changes in TERT mRNA levels in cells from various tissues, as TERT is the catalytic subunit of telomerase and its mRNA levels indicate telomerase gene expression. Figure 20.5 below shows the change in TERT mRNA levels in different tissues of TelomereRestore-treated one-year-old mice compared to age-matched controls.

Question 3: Suggest possible advantages of using model organisms, such as mice, instead of cell cultures. Give some other examples of model organisms commonly used by scientists. What features do these organisms have in common?

Question 4: What changes in TERT mRNA levels can be observed in different mouse tissues in Fig. 20.5? Which organs seem to be the most affected by the treatment? Which seems to be the least affected?

Scientists were also interested in assessing any potential long-term adverse effects of TelomereRestore on the mice. They monitored the lifespan of the treated mice (which was found to be unaffected) and meticulously tracked the development of cancer. Figure 20.6 below presents the percentage of mice developing specific tumour types at their time of death in both TelomereRestore-treated and control groups.

Question 5: After successful preclinical trials on animals, a new drug candidate must undergo rigorous clinical trials in humans. This progression occurs only when the drug is deemed genuinely promising and, crucially, safe, based on comprehensive animal study results. Based on the results presented in Figs. 20.5 and 20.6, would you recommend TelomereRestore to proceed to human clinical trials?

[2]The p represents the number of times a cell culture has been subcultured or passaged. The passage number is an indicator of the cells' "in vitro age".

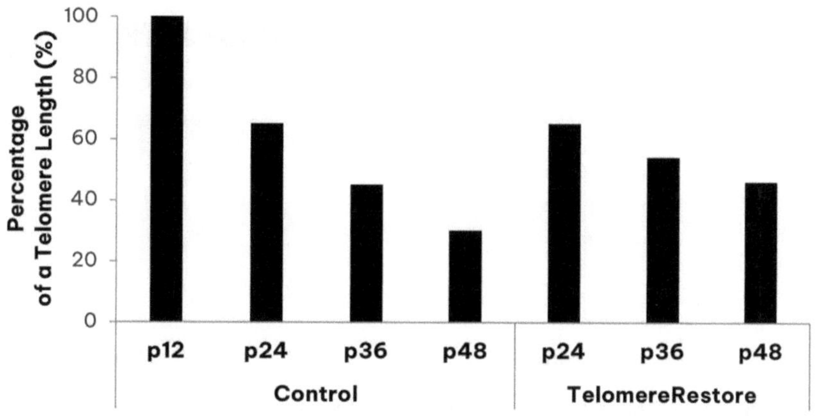

Fig. 20.4 Post-treatment telomere length data from cell passages 12, 24, 36 and 48 for each treatment group are shown

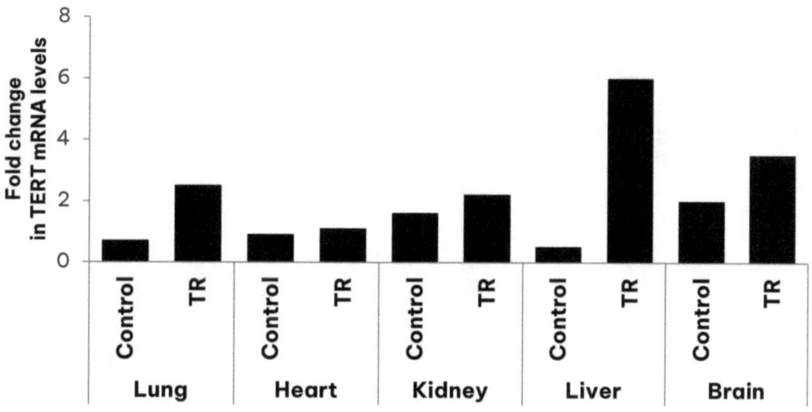

Fig. 20.5 Post-treatment change in TERT mRNA levels in different tissues

Part 2: Clinical Evaluation in Humans

Several studies on tissue cultures and model organisms supported TelomereRestore's ability to increase telomere length and improve various biological markers associated with ageing. Consequently, human clinical studies were initiated, focusing on TelomereRestore's capacity to both increase telomere length and improve health conditions.

Step A: Evaluating Effect on Human Telomeres

Figure 20.7 illustrates the initial distribution of participant ages and their median telomere length at the beginning of the study (baseline). This study involved 100 volunteers, 80 of whom successfully completed the one-year trial. Figure 20.8 shows the change in the median telomere length measured at different time points throughout and after the study.

Participants were divided into three treatment groups:

- **Low-dose group:** 26 individuals received one TelomereRestore capsule and three placebo capsules.
- **High-dose group:** 31 individuals received four TelomereRestore capsules.
- **Placebo group:** 23 individuals received four placebo capsules.

Telomere length measurements were taken at day 0 (baseline) and at 3, 6 and 12 months after the study began.

Question 6 Analyse Fig. 20.7 to answer the following:

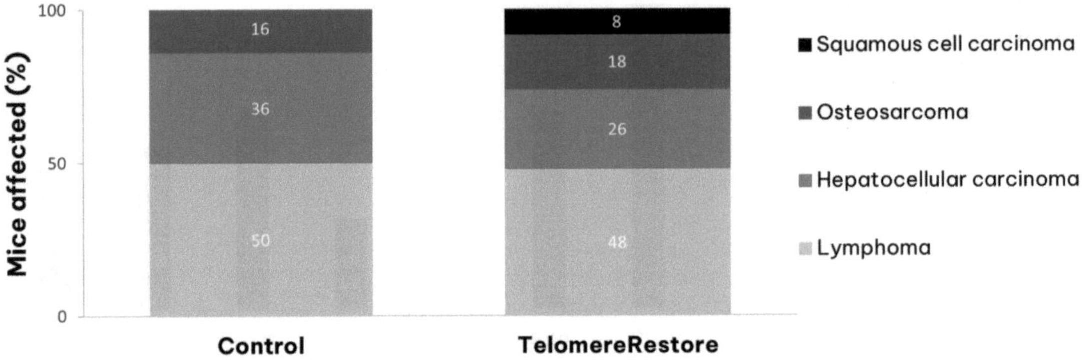

Fig. 20.6 Percentage of mice with the indicated tumour at their time of death. The observed differences between the treated group and control were not statistically significant

Fig. 20.7 Initial distribution of participant ages and their median telomere length at the beginning of the study (the baseline)

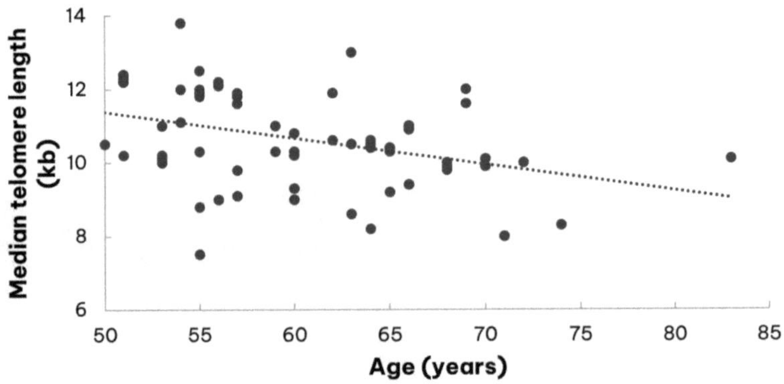

(a) What was the approximate age range of the participants included in this study?
(b) Describe the general relationship between a participant's age and their baseline median telomere length, as suggested by this graph.
(c) Based on the distribution of dots, which age group appears to be the most represented in this study?

Question 7: One of the key findings reported from this study was that the low dose of TelomereRestore significantly increased telomere lengths, suggesting potential clinical significance.

(a) Based on the results shown in Fig. 20.8, evaluate whether this finding is supported by the data.
(b) What can you conclude about the high-dose TelomereRestore treatment based on Fig. 20.8?

Step B: Evaluating Effect on Human Health

Having observed the promising capacities of TelomereRestore to influence telomere length in humans, scientists decided to investigate its possible effects on human health conditions. This is a critical step, as a drug's ultimate value lies in its impact on patient well-being, not just molecular markers.

It is well established that adults aged 65 and older are more susceptible to cardiovascular diseases, and telomere shortening has been found to play a crucial role in the development of these ageing-associated conditions. Therefore, this study focused on an older demographic with relevant health concerns. Table 20.1 presents data on inflammatory biomarkers in patients with high-risk factors for cardiovascular disease after taking TelomereRestore or a placebo for 4 months.

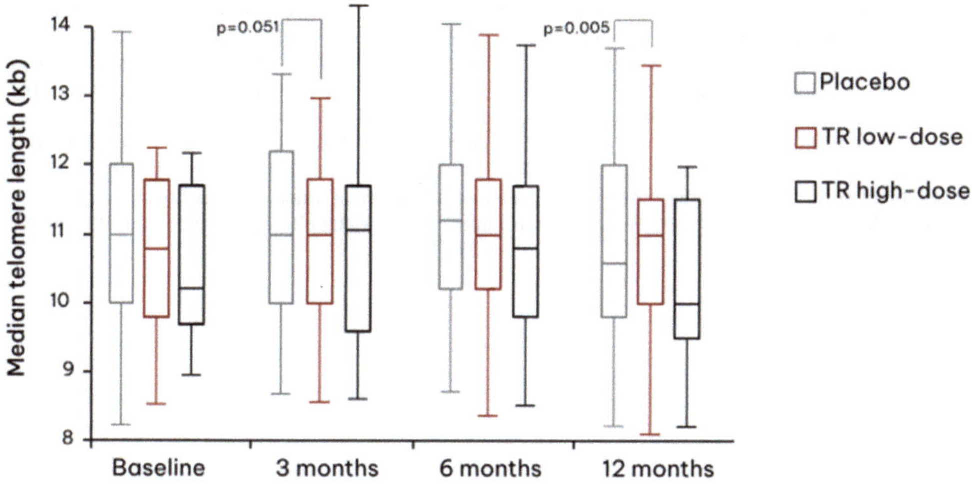

Fig. 20.8 Change in the median telomere length in the placebo group. Low-dose and high-dose groups measured at different time points throughout and after the study

Table 20.1 Inflammatory biomarkers in high-risk cardiovascular patients after the treatment: C-reactive protein (CRP), interleukin-6 (IL-6), tumour necrosis factor-alpha (TNF-alpha) and interleukin-1 beta (IL-1beta)

Parameter	TelomereRestore	Placebo	p-value
CRP (mg/L)	4.1 ± 2.3	4.2 ± 2.7	NS
IL-6 (pg/mL)	4.3 ± 2.8	4.9 ± 2.5	NS
TNF-alpha (pg/mL)	5.4 ± 1.9	6.3 ± 3.1	0.04
IL-1beta (pg/mL)	4.9 ± 3.1	5.6 ± 2.1	NS

Question 8: Based on Table 20.1, what, if any, observable effects did TelomereRestore have on the inflammatory biomarkers in patients with high-risk factors for cardiovascular disease? Do these findings suggest a potential health benefit?

Analysis

This project walked you through the complex, multistage process of evaluating TelomereRestore, a hypothetical drug. You saw how initial lab and animal studies led to human clinical trials, each phase providing vital data on a drug's efficacy and safety.

Question 9: Considering all the data presented across this project (cell culture, mouse models and human trials for both telomere length and health markers), summarise your findings. What is your overall assessment of TelomereRestore as a potential therapeutic agent for ageing-related conditions?

Question 10: Beyond the data presented, what additional long-term studies or safety monitoring would be crucial before TelomereRestore could be considered for widespread clinical use?

Understanding experimental results requires critical thinking, careful interpretation and an awareness of the ethical considerations in research. Science is an iterative process–each study builds on those before it to gradually reveal a treatment's full potential and safety. This process is particularly important today, as anti-ageing products and increased life expectancy are widely discussed topics. Beyond cosmetics, age-related health challenges are becoming more prevalent, especially in developed countries. In this project, you have learned how to critically evaluate such claims and to appreciate the rigorous scientific process involved in developing safe and effective treatments for age-related conditions.

This project was developed by Kristína Hudáková and Katarína Juríková.

The project was inspired by the following studies: Karsono AH, Tandrasasmita OM, Berlian G, Tjandrawinata RR. Potential Antiaging Effects of DLBS1649, a Centella asiatica Bioactive Extract. J Exp Pharmacol. 2021 Aug 11;13:781-795. ; Bernardes de Jesus B, Schneeberger K, Vera E, Tejera A, Harley CB, Blasco MA. The telomerase activator TA-65 elongates short telomeres and increases health span of adult/old mice without increasing cancer incidence. Aging Cell. 2011 Aug;10(4):604-21. ; Salvador L, Singaravelu G, Harley CB, Flom P, Suram A, Raffaele JM. A Natural Product Telomerase Activator Lengthens Telomeres in Humans: A Randomized, Double Blind, and Placebo Controlled Study. Rejuvenation Res. 2016 Dec;19(6):478-484. ; Fernandez ML, Thomas MS, Lemos BS, DiMarco DM, Missimer A, Melough M, Chun OK, Murillo AG, Alyousef HM, Medina-Vera I. TA-65, A Telomerase Activator improves Cardiovascular Markers in Patients with Metabolic Syndrome. Curr Pharm Des. 2018;24(17):1905-1911.

MIX
Papier aus verantwortungsvollen Quellen
Paper from responsible sources
FSC® C105338

If you have any concerns about our products,
you can contact us on
ProductSafety@springernature.com

In case Publisher is established outside the EU,
the EU authorized representative is:
Springer Nature Customer Service Center GmbH
Europaplatz 3, 69115 Heidelberg, Germany

Printed by Libri Plureos GmbH
in Hamburg, Germany